Genetic Engineering

Principles and Methods

Volume 18

GENETIC ENGINEERING
Principles and Methods

Genetic Engineering

Principles and Methods

Volume 18

Edited by

Jane K. Setlow

Brookhaven National Laboratory

Upton, New York

Plenum Press • New York and London

The Library of Congress cataloged the first volume of this title as follows:

Genetic engineering: principles and methods, v. 1–
New York, Plenum Press, [1979–
v. ill. 26 cm.
Editors: 1979– J. K. Setlow and A. Hollaender
Key title: Genetic engineering, ISSN 0196-3716

1. Genetic engineering—Collected works. I. Setlow, Jane K. II. Hollaender, Alexander, [DATE]

QH442.G454 575.1 79-644807
MARC-S

ISBN 0-306-45371-1

A Division of Plenum Publishing Corporation
233 Spring Street, New York, N.Y. 10013

10 9 8 7 6 5 4 3 2 1

Printed in the United States of America

CONTENTS OF EARLIER VOLUMES

VOLUME 1 (1979)

Introduction and Historical Background • *Maxine F. Singer*

Cloning of Double-Stranded cDNA • *Argiris Efstratiadis and Lydia Villa-Komaroff*

Gene Enrichment • *M. H. Edgell, S. Weaver, Nancy Haigwood and C. A. Hutchison III*

Transformation of Mammalian Cells • *M. Wigler, A. Pellicer, R. Axel and S. Silverstein*

Constructed Mutants of Simian Virus 40 • *D. Shortle, J. Pipas, Sondra Lazarowitz, D. DiMaio and D. Nathans*

Structure of Cloned Genes from Xenopus: A Review • *R. H. Reeder*

Transformation of Yeast • *Christine Ilgen, P. J. Farabaugh, A. Hinnen, Jean M. Walsh and G. R. Fink*

The Use of Site-Directed Mutagenesis in Reversed Genetics • *C. Weissmann, S. Nagata, T. Taniguchi, H. Weber and F. Meyer*

Agrobacterium Tumor Inducing Plasmids: Potential Vectors for the Genetic Engineering of Plants • *P. J. J. Hooykaas, R. A. Schilperoot and A. Rörsch*

The Chloroplast, Its Genome and Possibilities for Genetically Manipulating Plants • *L. Bogorad*

Mitochondrial DNA of Higher Plants and Genetic Engineering • *C. S. Levings III and D. R. Pring*

Host-Vector Systems for Genetic Engineering of Higher Plant Cells • *C. I. Kado*

Soybean Urease—Potential Genetic Manipulation of Agronomic Importance • *J. C. Polacco, R. B. Sparks, Jr. and E. A. Havir*

VOLUME 2 (1980)

Cloning of Repeated Sequence DNA from Cereal Plants • *J. R. Bedbrook and W. L. Gerlach*

The Use of Recombinant DNA Methodology in Approaches to Crop Improvement: The Case of Zein • *Benjamin Burr*

Production of Monoclonal Antibodies • *Sau-Ping Kwan, Dale E. Yelton and Matthew D. Scharff*

Measurement of Messenger RNA Concentration • *S. J. Flint*

DNA Cloning in Mammalian Cells with SV40 Vectors • *D. H. Hamer*

Adenovirus–SV40 Hybrids: A Model System for Expression of Foreign Sequences in an Animal Virus Vector • *Joseph Sambrook and Terri Grodzicker*

Molecular Cloning in *Bacillus subtilis* • *D. Dubnau, T. Gryczan, S. Contente and A. G. Shivakumar*

Bacterial Plasmid Cloning Vehicles • *H. U. Bernard and D. R. Helinski*

Cloning with Cosmids in *E. coli* and Yeast • *Barbara Hohn and A. Hinnen*

DNA Cloning with Single-Stranded Phage Vectors • *W. M. Barnes*

Bacteriophage Lambda Vectors for DNA Cloning • *Bill G. Williams and Frederick R. Blattner*

VOLUME 3 (1981)

Constructed Mutants Using Synthetic Oligodeoxyribonucleotides as Site-Specific Mutagens • *M. Smith and S. Gillam*

Evolution of the Insertion Element IS1 That Causes Genetic Engineering of Bacterial Genomes *In Vivo* • *E. Ohtsubo, K. Nyman, K. Nakamura and H. Ohtsubo*
Applications of Molecular Cloning to Saccharomyces • *M. V. Olson*
Cloning Retroviruses: Retrovirus Cloning? • *W. L. McClements and G. F. Vande Woude*
Repeated DNA Sequences in *Drosophila* • *M. W. Young*
Microbial Surface Elements: The Case of Variant Surface Glycoprotein (VSG) Genes of African Trypanosomes • *K. B. Marcu and R. O. Williams*
Mouse Immunoglobulin Genes • *P. Early and L. Hood*
The Use of Cloned DNA Fragments to Study Human Disease • *S. H. Orkin*
Physical Mapping of Plant Chromosomes by *In Situ* Hybridization • *J. Hutchinson, R. B. Flavell and J. Jones*
Mutants and Variants of the Alcohol Dehydrogenase-1 Gene in Maize • *M. Freeling and J. A. Birchler*
Developmentally Regulated Multigene Families in *Dictyostelium discoideum* • *R. A. Firtel, M. McKeown, S. Poole, A. R. Kimmel, J. Brandis and W. Rowekamp*
Computer Assisted Methods for Nucleic Acid Sequencing • *T. R. Gingeras and R. J. Roberts*

VOLUME 4 (1982)
New Methods for Synthesizing Deoxyoligonucleotides • *M. H. Caruthers, S. L. Beaucage, C. Becker, W. Efcavitch, E. F. Fisher, G. Galluppi, R. Goldman, P. deHaseth, F. Martin, M. Matteucci and Y. Stabinsky*
An Integrative Strategy of DNA Sequencing and Experiments Beyond • *J. Messing*
Transcription of Mammalian Genes *In Vitro* • *J. L. Manley*
Transcription of Eukaryotic Genes in Soluble Cell-Free Systems • *N. Heintz and R. G. Roeder*
Attachment of Nucleic Acids to Nitrocellulose and Diazonium-Substituted Supports • *B. Seed*
Determination of the Organization and Identity of Eukaryotic Genes Utilizing Cell-Free Translation Systems • *J. S. Miller, B. E. Roberts and B. M. Paterson*
Cloning in Streptomyces: Systems and Strategies • *D. A. Hopwood and K. F. Chater*
Partial Sequence Determination of Metabolically Labeled Radioactive Proteins and Peptides • *C. W. Anderson*
Molecular Cloning of Nitrogen Fixation Genes from *Klebsiella pneumoniae* and *Rhizobium meliloti* • *F. M. Ausubel, S. E. Brown, F. J. deBruijn, D. W. Ow, G. E. Riedel, G. B. Ruvkun and V. Sandaresan*
The Cloning and Expression of Human Interferon Genes • *R. M. Lawn*
Cloning by Complementation in Yeast: The Mating Type Genes • *J. B. Hicks, J. N. Strathern, A. J. S. Klar and S. L. Dellaporta*
Construction and Screening of Recombinant DNA Libraries with Charon Vector Phages • *B. A. Zehnbauer and F. R. Blattner*

VOLUME 5 (1983)
Microcloning of Microdissected Chromosome Fragments • *V. Pirrotta, H. Jackle and J. E. Edstrom*
Transient Expression of Cloned Genes in Mammalian Cells • *J. Banerji and W. Schaffner*
Transposable Elements in Archaebacteria • *W. F. Doolittle, C. Sapienza, J. D. Hofman, R. M. Mackay, A. Cohen and W. -L. Xu*
The Application of Restriction Fragment Length Polymorphism to Plant Breeding • *B. Burr, S. V. Evola, F. A. Burr and J. S. Beckmann*
Antibodies against Synthetic Peptides • *G. Walter and R. F. Doolittle*
Wheat α-Amylase Genes: Cloning of a Developmentally Regulated Gene Family • *D. Baulcombe*
Yeast DNA Replication • *J. L. Campbell*
Chromosome Engineering in Wheat Breeding and Its Implications for Molecular Genetic Engineering • *C. N. Law*

Bovine Papillomavirus Shuttle Vectors • *N. Sarver, S. Miltrani-Rosenbaum, M. -F. Law, W. T. McAllister, J. C. Byrne and P. M. Howley*
Chemical Synthesis of Oligodeoxyribonucleotides: A Simplified Procedure • *R. L. Letsinger*

VOLUME 6 (1984)
Cloning of the Adeno-Associated Virus • *K. I. Berns*
Transformation in the Green Alga *Chlamydomonas reinhardii* • *J. -D. Rochaix*
Vectors for Expressing Open Reading Frame DNA in *Escherichia coli* Using *lacZ* Gene Fusions • *G. M. Weinstock*
An Enigma of the Leghemoglobin Genes • *J. S. Lee and D. P. S. Verma*
Yeast Transposons • *G. S. Roeder*
Rearrangement and Activation of C-MYC Oncogene by Chromosome Translocation in B Cell Neoplasias • *K. B. Marcu, L. W. Stanton, L. J. Harris, R. Watt, J. Yang, L. Eckhardt, B. Birshtein, E. Remmers, R. Greenberg and P. Fahrlander*
Screening for and Characterizing Restriction Endonucleases • *I. Schildkraut*
Molecular Studies of Mouse Chromosome 17 and the T Complex • *L. M. Silver, J. I. Garrels and H. Lehrach*
Use of Synthetic Oligonucleotide Hybridization Probes for the Characterization and Isolation of Cloned DNAs • *A. A. Reyes and R. B. Wallace*
Hybridization of Somatic Plant Cells: Genetic Analysis • *Yu. Yu. Gleba and D. A. Evans*
Genetic Analysis of Cytoskeletal Protein Function in Yeast • *P. Novick, J. H. Thomas and D. Botstein*
Use of Gene Fusions to Study Biological Problems • *L. Guarente*
The Use of the Ti Plasmid of *Agrobacterium* to Study the Transfer and Expression of Foreign DNA in Plant Cells: New Vectors and Methods • *P. Zambryski, L. Herrera-Estrella, M. De Block, M. Van Montagu and J. Schell*
Analysis of Eukaryotic Control Proteins at Their Recognition Sequences by Scanning Transmission Electron Microscopy • *P. V. C. Hough, M. N. Simon and I. A. Mastrangelo*
The Mass Culture of a Thermophilic Spirulina in the Desert • *K. Qian, G. H. Sato, V. Zhao and K. Shinohara*
DNA-Mediated Gene Transfer in Mammalian Gene Cloning • *F. H. Ruddle, M. E. Kamarck, A. McClelland and L. C. Kühn*

VOLUME 7 (1985)
Biochemical and Genetic Analysis of Adenovirus DNA Replication *In Vitro* • *B. W. Stillman*
Immunoscreening λGT11 Recombinant DNA Expression Libraries • *R. A. Young and R. W. Davis*
In Situ Hybridization to Cellular RNAs • *R. C. Angerer, K. H. Cox and L. M. Angerer*
Computer Methods to Locate Genes and Signals in Nucleic Acid Sequences • *R. Staden*
Biochemical and Molecular Techniques in Maize Research • *N. Fedoroff*
Analysis of Chromosome Replication with Eggs of *Xenopus laevis* • *R. A. Laskey, S. E. Kearsey and M. Mechali*
Molecular Genetic Approaches to Bacterial Pathogenicity to Plants • *M. J. Daniels and P. C. Turner*
Synthesis of Hybridization Probes and RNA Substrates with SP6 RNA Polymerase • *P. A. Krieg, M. R. Rebagliati, M. R. Green and D. A. Melton*
Identification and Isolation of Clones by Immunological Screening of cDNA Expression Libraries • *D. M. Helfman, J. R. Feramisco, J. C. Fiddes, G. P. Thomas and S. H. Hughes*
Molecular Studies on the Cytomegaloviruses of Mice and Men • *D. H. Spector*
Gene Transfer with Retrovirus Vectors • *A. Bernstein, S. Berger, D. Huszar and J. Dick*
HPRT Gene Transfer as a Model for Gene Therapy • *T. Friedmann*
Catabolic Plasmids: Their Analysis and Utilization in the Manipulation of Bacteria Metabolic Activities • *S. Harayama and R. H. Don*

Transcription of Cloned Eukaryotic Ribosomal RNA Genes • *B. Sollner-Webb, J. Tower, V. Culotta and J. Windle*
DNA Markers in Huntington's Disease • *J. F. Gusella*

VOLUME 8 (1986)
Regulation of Gene Activity During Conidiophore Development in *Aspergillus nidulans* • *W. E. Timberlake and J. E. Hamer*
Regulation of Expression of Bacterial Genes for Bioluminescence • *J. Engebrecht and M. Silverman*
Analysis of Genome Organization and Rearragements by Pulse Field Gradient Gel Electrophoresis • *C. L. Smith, P. E. Warburton, A. Gaal and C. R. Cantor*
Structural Instability of *Bacillus subtilis* Plasmids • *S. D. Ehrlich, Ph. Noirot, M. A. Petit, L. Jannière, B. Michel and H. te Riele*
Geminiviruses, The Plant Viruses with Single-Stranded DNA Genomes • *A. J. Howarth*
The Use of Bacterial Plasmids in the Investigation of Genetic Recombination • *A. Cohen*
Shuttle Mutagenesis: A Method of Introducing Transposons into Transformable Organisms • *H. S. Seifert, M. So and F. Heffron*
Genetic Advances in the Study of *Rhizobium* Nodulation • *S. R. Long*
Galactokinase Gene Fusion in the Study of Gene Regulation in *E. coli Streptomyces*, Yeast and Higher Cell Systems • *M. Rosenberg, M. Brawner, J. Gorman and M. Reff*
Structure and Function of the Signal Recognition Particle • *V. Siegel and P. Walter*
Alteration of the Structure and Catalytic Properties of Rubisco by Genetic Manipulation • *S. Gutteridge*
Electrophoresis of DNA in Denaturing Gradient Gels • *L. S. Lerman*
Caulimoviruses as Potential Gene Vectors for Higher Plants • *R. J. Shepherd*
An Insect Baculovirus Host–Vector System for High-Level Expression of Foreign Genes • *D. W. Miller, P. Safer and L. K. Miller*
Preparation of cDNA Libraries and the Detection of Specific Gene Sequences • *J. Brandis, D. Larocca and J. Monahan*
Construction of Human Chromosome Specific DNA Libraries: The National Laboratory Gene Library Project • *L. L. Deaven, C. E. Hildebrand, J. C. Fuscoe and M. A. Van Dilla*
New Approaches to the Expression and Isolation of a Regulatory Protein • *D. Bastia, J. Germino, S. Mukherjee and T. Vanaman*

VOLUME 9 (1987)
Gene Transfer in the Sea Urchin • *B. R. Hough-Evans and E. H. Davidson*
Properties and Uses of Heat Shock Promoters • *H. Pelham*
The Expression of Introduced Genes in Regenerated Plants • *D. Dunsmuir J. Bedbrook, D. Bond-Nutter, C. Dean, D. Gidoni and J. Jones*
Control of Maize Zein Gene Expression • *R. S. Boston and B. A. Larkins*
Dnase I Footprinting as an Assay for Mammalian Gene Regulatory Proteins • *W. S. Dynan*
Use of Gene Transfer in the Isolation of Cell Surface Receptor Genes • *D. R. Littman and M. V. Chao*
A New Method for Synthesizing RNA on Silica Supports • *D. J. Dellinger and M. H. Caruthers*
Activity Gels: Reformation of Functional Proteins from SDS-Polyacrylamide Gels • *R. P. Dottin, B. Haribabu, C. W. Schweinfest and R. E. Manrow*
Plasmid Vectors Carrying the Replication Origin of Filamentous Single-Stranded Phages • *G. Cesareni and J. A. H. Murray*
High Level Production of Proteins in Mammalian Cells • *R. J. Kaufman*
Plant Microinjection Techniques • *R. J. Mathias*
Genetic Transformation to Confer Resistance to Plant Virus Disease • *R. N. Beachy, S. G. Rogers and R. T. Fraley*
Alternative Splicing: Mechanistic and Biological Implications of Generating Multiple Proteins from a Single Gene • *B. Nadal-Ginard, M. E. Gallego and A. Andreadis*

VOLUME 10 (1988)
Genomic Footprinting • *P. B. Becker and G. Schütz*
Theoretical and Computer Analysis of Protein Primary Sequences: Structure Comparison and Prediction • *P. Argos and P. McCaldon*
Affinity Chromatography of Sequence-Specific DNA-Binding Proteins • *C. Wu, C. Tsai and S. Wilson*
Applications of the Firefly Luciferase as a Reporter Gene • *S. Subramani and M. DeLuca*
Fluorescence-Based Automated DNA Sequence Analysis • *L. M. Smith*
Phosphorothioate-Based Oligonucleotide-Directed Mutagenesis • *J. R. Sayers and F. Eckstein*
Design and Use of *Agrobacterium* Transformation Vectors • *M. Bevan and A. Goldsbrough*
Cell Commitment and Determination in Plants • *F. Meins, Jr.*
Plasmids Derived from Epstein–Barr Virus: Mechanisms of Plasmid Maintenance and Applications in Molecular Biology • *J. L. Yates*
Chromosome Jumping: A Long Range Cloning Technique • *A Poustka and H. Lehrach*
Isolation of Intact MRNA and Construction of Full-Length cDNA Libraries: Use of a New Vector, λgt22, and Primer-Adapters for Directional cDNA Cloning • *J. H. Han and W. J. Rutter*
The Use of Transgenic Animal Techniques for Livestock Improvement • *R. M. Strojek and T. E. Wagner*
Plant Reporter Genes: The GUS Gene Fusion System • *R. A. Jefferson*
Structure of the Genes Encoding Proteins Involved in Blood Clotting • *R. T. A. MacGillivray, D. E. Cool, M. R. Fung, E. R. Guinto, M. L. Koschinsky and B. A. Van Oost*

VOLUME 11 (1989)
DNA Methylases • *A. Razin*
Advances in Direct Gene Transfer Into Cereals • *T. M. Klein, B. A. Roth and M. E. Fromm*
The Copy Number Control System of the 2μm Circle Plasmid of *Saccharomyces cerevisiae* • *B. Futcher*
The Application of Antisense RNA Technology to Plants • *W. R. Hiatt, M. Kramer and R. E Sheehy*
The Pathogenesis-Related Proteins of Plants • *J. P. Carr and D. F. Klessig*
The Molecular Genetics of Plasmid Partition: Special Vector Systems for the Analysis of Plasmid Partition • *A. L. Abeles and S. J. Austin*
DNA-Mediated Transformation of Phytophathogenetic Fungi • *J. Wang and S. A. Leong*
Fate of Foreign DNA Introduced to Plant Cells • *J. Paszkowski*
\Generation of cDNA Probes by Reverse Translation of Amino Acid Sequence • *C. C. Lee and C. T. Caskey*
Molecular Genetics of Self-Incompatibility in Flowering Plants • *P. R. Ebert, M. Altschuler and A. E. Clarke*
Pulsed-Field Gel Electrophoresis • *M. V. Olson*

VOLUME 12 (1990)
Folding of Eukaryotic Proteins Produced in *Escherichia coli* • *R. F. Kelley and M. E. Winkler*
Human Retinoblastoma Susceptibility Gene • *C. -C. Lai and W. -H. Lee*
α-Oligodeoxynucleotides (α-DNA): A New Chimeric Nucleic Acid Analog • *F. Morvan, B. Rayner and J. -L. Imbach*
The Utility of Streptomycetes and Hosts for Gene Cloning • *P. K. Tomich and Y. Yagi*
From Footprint to Function: An Approach to Study Gene Expression and Regulatory Factors in Transgenic Plants • *E. Lam*
Purification of Recombinant Proteins with Metal Chelate Adsorbent • *E. Hochuli*
Determinants of Translation Efficiency of Specific mRNAs in Mammalian Cells • *D. S. Peabody*
The Polymerase Chain Reaction • *N. Arnheim*
Regulation of Alternative Splicing • *M. McKeown*
Structure and Function of the Nuclear Receptor Superfamily for Steroid, Thyroid Hormone and Retinoic Acid • *V. Giguère*

Identification and Functional Analysis of Mammalian Splicing Factors • *A. Bindereif and M. R. Green*
The Genes Encoding Wheat Storage Proteins: Towards a Molecular Understanding of Bread-Making Quality and Its Genetic Manipulation • *V. Colot*
Control of Translation Initiation in Mammalian Cells • *R. J. Kaufman*
Electroporation of Bacteria: A General Approach to Genetic Transformation • *W. J. Dower*
The Isolation and Identification of cDNA Genes by Their Heterologous Expression and Function • *G. G. Wong*
Molecular Cloning of Genes Encoding Transcription Factors with the Use of Recognition Site Probes • *H. Singh*

VOLUME 13 (1991)
The *Mutator* Transposable Element Family of Maize • *V. Walbot*
Protein Phosphorylation and the Regulation of Cellular Processes by the Homologous Two-Component Systems of Bacteria • *A. J. Ninfa*
The Peculiar Nature of Codon Usage in Primates • *S. Zhang and G. Zubay*
The Role of Nodulation Genes in Bacterium–Plant Communication • *A. Kondorosi, E. Kondorosi, M. John, J. Schmidt, and J. Schell*
Regulation of Gene Expression by Epidermal Growth Factor • *L. G. Hudson and G. N. Gill*
Machinery of Protein Import into Chloroplasts and Mitochondria • *D. Pain, D. J. Schnell, H. Murakami, and G. Blobel*
High-Level Expression of Foreign Genes in Mammalian Cells • *S. E. Kane*
Aromatic Hydrocarbon Degradation: A Molecular Approach • *G. J. Zylstra and D. T. Gibson*
Employment of Fibroblasts for Gene Transfer Applications for Grafting into the Central Nervous System • *M. D. Kawaja, J. Ray, and F. H. Gage*
The Molecular Biology of Amino Acid Biosynthesis in Plants • *T. Brears and G. M. Coruzzi*
Genetic Manipulation of *Bacillus thuringiensis* Insecticidal Crystal Protein Genes in Bacteria • *C. Gawron-Burke and J. A. Baum*
Progress Towards Gene Targeting in Plants • *J. I. Yoder and E. Kmiec*
Molecular Biology of Mating-Type Determination in *Schizophyllum commune* • *R. C. Ullrich, C. A. Specht, M. M. Stankis, H. Yang, L. Giasson, and C. P. Novotny*
Functions of Intracellular Protein Degradation in Yeast • *M. Hochstrasser*
Transgenic Fish for Aquaculture • *G. L. Fletcher and P. L. Davies*

VOLUME 14 (1992)
Cleavage-Site Motifs in Protein Targeting Sequences • *G. von Heijne*
Complications of RNA Heterogeneity for the Engineering of Virus Vaccines and Antiviral Agents • *E. Domingo and J. J. Holland*
The Quaternary Structures of SV40 Large T Antigen and Tumor Suppressor p53: Analysis by Gel Electrophoresis • *J. E. Stenger, G. A. Mayr, K. Mann, S. Ray, M. E. Anderson, and P. Tegtmeyer*
Assembly of Antibodies and Mutagenized Variants in Transgenic Plants and Plant Cell Cultures • *A. Hiatt, Y. Tang, W. Weiser, and M. B. Hein*
Maize Endosperm Tissue as an Endoreduplication System • *R. V. Kowles, G. L. Yerk, F. Srienc, and R. L. Phillips*
Study of Chlorate-Resistant Mutants of Arabidopsis: Insights into Nitrate Assimilation and Ion Metabolism of Plants • *N. M. Crawford*
Approaches and Progress in the Molecular Cloning of Plant Disease Resistance Genes • *J. L. Bennetzen and J. D. G. Jones*
Is GRP78 a Sensor of Cellular Secretory Activity? • *T. Leustek*
The Molecular Biology of Pathogenesis in *Ustilago maydis* • *B. J. Saville and S. A. Leong*
Molecular Design of Oligomeric Channel Proteins • *A. Grove, J. M. Tomich, and M. Montal*
Regulation of Gene Expression by Thyroid Hormones and Retinoic Acid • *S. M. Lipkin, M. G. Rosenfeld, and C. K. Glass*

RNA *Trans*-Splicing • *X. -Y. Huang and D. Hirsh*
Structural Constraints on Residue Substitution • *J. Overington*
Molecular and Functional Analysis of the *A* Mating Type Genes of *Coprinus cinereus* • *U. Kües and L. A. Casselton*
Physical Mapping of Human Chromosomes • *G. A. Evans and D. L. McElligott*

VOLUME 15 (1993)
Application of Computational Neural Networks to the Prediction of Protein Structural Features • *S. R. Holbrook*
Human Cellular Protein Patterns and Their Link to Genome Data Mapping and Sequencing Data: Towards an Integrated Approach to the Study of Gene Expression • *J. E. Celis, H. H. Rasmussen, H. Leffers, P. Madsen, B. Honore, K. Dejgaard, P. Gromov, E. Olsen, H. J. Hoffman, M. Nielsen, B. Gesser, M. Puype, J. Van Damme, and J. Vandekerckhove*
Regulation of Translation in Plants • *A. Danon, C. B. Yohn, and S. P. Mayfield*
On the Origins, Structures and Functions of Restriction-Modification Enzymes • *J. Heitman*
Manipulation of Amino Acid Balance in Maize Seeds • *T. Ueda and J. Messing*
Investigational Approaches for Studying the Structures and Biological Functions of Myeloid Antimicrobial Peptides • *M. E. Selsted*
Progress in the Cloning of Genes for Plant Storage Lipid Biosynthesis • *V. C. Knauf*
Genes for Crop Improvement • *J. Bennett*
Molecular Biology and Genetics of Protective Fungal Endophytes of Grasses • *C. L. Schardl and Z. An*
Prospects for Human Gene Therapy • *A. B. Moseley and C. T. Caskey*
The Use of Microparticle Injection to Introduce Genes into Animal Cells *In Vitro* and *In Vivo* • *S. A. Johnston and D-C. Tang*

VOLUME 16 (1994)
RNA Polymerase III Transcription in the Yeast *Saccharomyces cerevisiae* • *Stephen Buratowski*
Lens Oncogenesis and Differentiation • *Heiner Westphal*
Genetic Engineering of Cardiac Muscle Cells: *In vitro* and *In vivo* • *Stephen J. Fuller and Kenneth R. Chien*
Genetic Control of Plant Ureases • *Joseph C. Polacco and Mark A. Holland*
Gene Discovery in *Dictyostelium* • *William F. Loomis, Adam Kuspa, and Gad Shaulsky*
Transfer of YACs to Mammalian Cells and Transgenic Mice • *Clare Huxley*
Plant Genetic Engineering and Future Agriculture • *S. Riazuddin*
Internal Initiation of mRNA Translation in Eukaryotes • *Ann Kaminski, Sarah L. Hunt, Catherine L. Gibbs, and Richard J. Jackson*
Genetic Recombination Analysis Using Sperm Typing • *Karin Schmitt and Norman Arnheim*
Genetic Regulation in Plant Pathogenic Pseudomonads • *David K. Willis, Jessica J. Rich, Thomas G. Kinscherf and Todd Kitten*
Defense-Related Gene Induction in Plants • *Danny Alexander, Kay Lawton, Scott Uknes, Eric Ward and John Ryals*
The P1 Vector System for the Preparation and Screening of Genomic Libraries • *Nancy S. Shepherd and David Smoller*
The Unmasking of Maternal mRNA During Oocyte Maturation and Fertilization • *James L. Grainger*
Recognizing Exons in Genomic Sequences Using Grail II • *Ying Xu, Richard Mural, Manesh Shah and Edward Uberbacher*
Gene Expression of Plant Extracellular Proteins • *Beat Keller*

VOLUME 17 (1995)
The Molecular Biology of Nucleotide Excision Repair and Double-Strand Break Repair in Eukaryotes • *Alan R. Lehman*
Manipulating and Mapping DNA with RecA-Assisted Restriction Endonuclease (RARE) Cleavage • *Lance J. Ferrin*
Molecular Studies on the Virulence of *Listeria monocytogenes* • *Michael Kuhn and Werner Goebel*
Indirect Use of Immobilized Metal Affinity Chromatography for Isolation and Characterization of Protein Partners • *Michèle Sawadogo and Michael W. Van Dyke*
Structure and Function of RNA Pseudoknots • *C. W. A. Pleij*
Role of Molecular Chaperones in the Initiation of Plasmid DNA Replication • *Dhruba K. Chattoraj*
Structure, Function and Engineering of *Bacillus thuringienis* Toxins • *Mark A. Thompson, H. Ernest Schnepf, and Jerald S. Feitelson*
Uses for GAL4 Expression in Mammalian Cells • *Ivan Sadowski*
Protein Thiol Modification of Glyceraldehyde-3-Phosphate Dehydrogenease • *Bernhard Brüne and Eduardo G. Lapetina*
The Genetics of Nuclear Migration in Fungi • *Susan M. Beckwith, Christian H. Roghi, and N. Ronald Morris*
Structure and Function of the Platelet-Derived Growth Factor Family and Their Receptors • *Kristen C. Hart, Brendan D. Galvin, and Daniel J. Donoghue*
Recombination Between Prokaryotic and Eukaryotic DNA: Integration of *Agrobacterium tumefaciens* T-DNA into the Plant Genome • *Bruno Tinland and Barbara Hohn*
Metal Precipitation by Marine Bacteria: Potential for Biotechnological Applications • *Bradley M. Tebo*

ACKNOWLEDGMENT

The final processing of the manuscripts has been efficiently accomplished by Carol A. Lamb of Janice Lamb, Office Services, Stony Brook, NY. The Editor is grateful for her labors.

CONTENTS

CLONING AND CHARACTERIZATION OF DNAs WITH PALINDROMIC SEQUENCES 1
David R.F. Leach

DNA ISOLATION, MANIPULATION AND CHARACTERIZATION FROM OLD TISSUES 13
Rob DeSalle and Elizabeth Bonwich

GROWTH FACTORS AND NEURAL CONNECTIVITY 33
Sarah McFarlane and Christine E. Holt

GENE IDENTIFICATION BY 3′ TERMINAL EXON TRAPPING 49
David B. Krizman

ENGINEERING TRANSGENES FOR USE IN THE MAMMARY GLAND 57
Sinai Yarus, Darryl Hadsell and Jeffrey M. Rosen

PROBLEMS THAT CAN LIMIT THE EXPRESSION OF FOREIGN GENES IN PLANTS: LESSONS TO BE LEARNED FROM *B.t.* TOXIN GENES 83
Scott H. Diehn, E. Jay De Rocher and Pamela J. Green

RENATURATION AND RECONSTITUTION OF FUNCTIONAL HOLOENZYME FROM RECOMBINANT SUBUNITS OF CASEIN KINASE II EXPRESSED AS INCLUSION BODIES IN *E. coli* 101
Wey-Jinq Lin, Rolf Jakobi and Jolinda A. Traugh

PLANT ACYL-ACP THIOESTERASES: CHAIN-LENGTH DETERMINING ENZYMES IN PLANT FATTY ACID BIOSYNTHESIS 111
Toni Voelker

GENETIC ENGINEERING OF AN INSECT PARASITE 135
Randy Gaugler and Sarwar Hashmi

THE STOP SIGNAL CONTROLS THE EFFICIENCY OF RELEASE FACTOR-MEDIATED TRANSLATIONAL TERMINATION 157
Warren P. Tate, Mark E. Dalphin, Herman J. Pel and Sally A. Manning

MECHANISM OF REPLICATION AND COPY NUMBER CONTROL OF PLASMIDS IN GRAM-POSITIVE BACTERIA 183
Saleem A. Khan

PATHWAYS OF PROTEIN REMODELING BY *Escherichia coli* MOLECULAR CHAPERONES 203
Marie Pak and Sue H. Wickner

PHEROMONES AND PHEROMONE RECEPTORS AS MATING-TYPE DETERMINANTS IN BASIDIOMYCETES 219
Lisa J. Vaillancourt and Carlene A. Raper

SYNTHESIS AND APPLICATIONS OF PHOSPHOPEPTIDES 249
Kazuyasu Sakaguchi, Peter P. Roller and Ettore Appella

INDEX 279

CLONING AND CHARACTERIZATION OF DNAs WITH PALINDROMIC SEQUENCES

David R.F. Leach

Institute of Cell and Molecular Biology
University of Edinburgh
King's Buildings
Edinburgh EH9 3JR

INTRODUCTION

A sentence contains an ordered set of letters that is interpreted by reading in one direction, but a palindromic sentence such as "Madam I'm Adam" can be read in either direction. Palindromic sentences are relatively rare because languages use a large number of letters and we normally write sense. The example given above, contains only three letters (apart from the central I) and conveys a limited amount of information. By contrast, DNA is composed of only four bases, does not always make sense (as defined by an encoded polypeptide) and can be read in either direction (though always 3′ - 5′ on a given strand to give an mRNA with 5′ - 3′ polarity). A DNA palindrome is defined with respect to the double strand as a sequence with two-fold rotational symmetry as shown in Figure 1. This rotational symmetry has the consequence that a palindromic DNA sequence read in a given polarity from left to right is identical to that read with the same polarity from right to left. Another way of thinking about a DNA palindrome is that it is composed of two halves that are identical but repeated in inverted orientation with respect to each other. It is a DNA inverted repeat with no unique sequence separating the repeated sequences. Many short palindromes exist in genomes and are best known as recognition targets for the type II restriction-modification systems that are widely used for genetic engineering. Some imperfect palindromes are the recognition targets for other proteins such as transcriptional regulators or recombinases. Still other palindromes are found at origins of DNA replication, sites of transcription termination or RNA processing and roles for palindromes in chromosome organization have been proposed. In bacteria, perfect palindromes tend to be short (< 40 bp in length) and imperfect palindromes fairly short (< 130 bp in length). In eukaryotic genomes, long palindromes are more common and can cause problems when attempts are made to clone them in bacteria. It appears that many if not all of these long palindromes may in fact be inverted repeats that are separated by short central spacers. However many of the problems associated with long palindromes are shared by long inverted repeats with short spacer regions. These sequences are therefore included in this discussion and referred to as interrupted palindromes. Long

Genetic Engineering, Vol. 18, Edited by J.K. Setlow
Plenum Press, New York, 1996

palindromic sequences not only cause problems for cloning but also for certain *in vitro* reactions such as hybridization, amplification by the polymerase chain reaction (PCR) and DNA sequencing. Some discussion of these problems is included here.

PALINDROME-MEDIATED INVIABILITY AND INSTABILITY IN *E. coli*

Approximately 20 years ago, several groups working in the newly-emerging field of genetic engineering observed that occasionally by chance two copies of one DNA fragment were ligated into a vector adjacent to one another and recovered in a recombinant *E. coli* clone. This was expected as the consequence of a trimolecular ligation reaction. However it was not expected that the two copies should always be oriented as direct repeats of each other. Nevertheless that is what was observed (see for example ref. 1). Inverted repeats were not recovered and this suggested that a palindromic structural organization caused a viability problem in bacteria. When long palindromes were deliberately constructed and introduced into *E. coli* it was found that they were associated with two fates. Either clones containing the constructs were not recovered as though the palindrome had interfered with the propagation of the entire replicon or a low frequency of clones was recovered but showed evidence of deletion of the palindromic sequence (2-7). These two fates were denoted respectively as inviability and instability and are believed to be caused by intrastrand base pairing to form hairpin or cruciform DNA structures as shown in Figure 1.

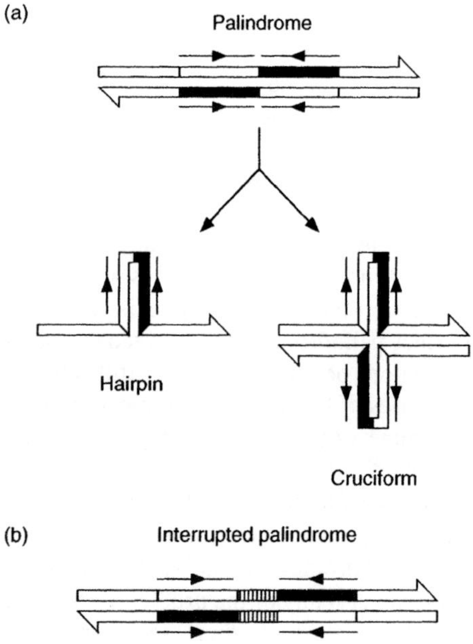

Figure 1. Palindromic DNA sequences. (a) DNA palindromes, hairpins and cruciforms. A DNA palindrome is a sequence with two-fold rotational symmetry. The complementary arms of the palindrome are shown in black and white and an arbitrary direction to the sequence is indicated by a line with an embedded arrow. Intra-strand base pairing is possible to generate a hairpin structure or a cruciform if both strands pair. (b) An interrupted palindrome. If some unique DNA separates the arms of the palindromic sequence, an interrupted palindrome can be said to exist. This is really an inverted repeat sequence with only a small DNA spacer separating the repeats. As in (a), complementary sequences within the palindrome are shown in black and white and an arbitrary direction to the repeat is given by the line with an embedded arrow. The central spacer is shown as a box with vertical shading.

In bacteria, there is an upper limit of approximately 150-200 base pairs total length for a palindromic DNA sequence before it confers inviability (8,9). On the other hand, instability has been detected for palindromes as short as 22 base pairs and increases with the length of palindrome (10). Instability is also a function of position, since identical DNA palindromes are deleted at very different frequencies from different positions in plasmids. In fact, two identical palindromes located only one base pair apart are deleted at frequencies that differ by a factor of more than 10^3 (10). This last observation suggests that differences in deletion frequencies cannot be satisfactorily explained by long-range position effects but must involve precise effects of local DNA sequences. Both inviability and instability are significantly reduced by the presence of a central spacer between the inverted repeats (8). A spacer of 57 base pairs was found to partially overcome the inviability associated with palindromes inserted in plasmids (8) and the viability in λ is restored by a central insert of two directly repeated copies of a 75 bp sequence but not by one copy of this DNA (11). Long perfect palindromes and inverted repeats with small spacer regions can be propagated in *E. coli sbcC* mutant hosts as described below and these hosts show an increase in stability which may be accounted for by the removal of a replication disadvantage of the palindrome-containing DNA.

Palindromes are deleted by recombination between short directly repeated DNA sequences close to their ends (see 12). However, palindromes flanked symmetrically by short direct repeats such as restriction target sequences, or transposon-mediated target-site duplications, are seldom deleted *via* recombination between these sequences. Instead, other direct repeats are used, with a preference for one repeat to be located just within the palindrome and the other to be located just outside the palindrome. This is consistent with a model in which replication enters the palindrome and only stalls after progressing some distance. The nascent strand then dissociates and can hybridize to any complementary sequence that is nearby. Such a sequence is typically found downstream of the palindrome. This asymmetry permits us to define a donor and a target repeat in the deletion reaction. The donor repeat is that located within the palindrome and the target repeat is that located outside. In a long DNA palindrome, the target repeat may be located within the palindrome and deletion can generate a shorter palindrome composed of the outer parts of the original sequence bounding a central spacer that was originally part of the larger palindrome (2) (see Figure 2).

PALINDROME INSTABILITY IN OTHER ORGANISMS

Palindrome mediated instability is not restricted to *E. coli*. It has also been detected in *Bacillus* (13), *Streptococcus* (14), *Streptomyces* (15) and *Saccharomyces cerevisiae* (16-18). Whether long palindromes cause similar problems in other genomes remains unknown. It is certainly true that fast re-annealing DNA is relatively abundant in many eukaryotic genomes, indicating the existence of long inverted repeats (see 19,20). However the techniques used to study these sequences have not been sufficiently accurate to determine whether any of the long inverted repeats are in fact palindromes, or if they all contain spacer sequences at their centers. The DNA of certain organisms (e.g., *Physarum polycephalum*) contains a large proportion of long inverted repeats and is hard to clone in wild-type *E. coli*. That this DNA is easier to clone in *recBC sbcB* mutants suggests that wild-type *E. coli* cannot tolerate palindromic DNA sequences that are tolerated in their natural host (21). Similar conclusions have been reached for human DNA (22-24). Perhaps there are different levels of palindrome-mediated instability and inviability in different organisms. Perhaps also, palindromes are tolerated at some chromosomal locations and not at others.

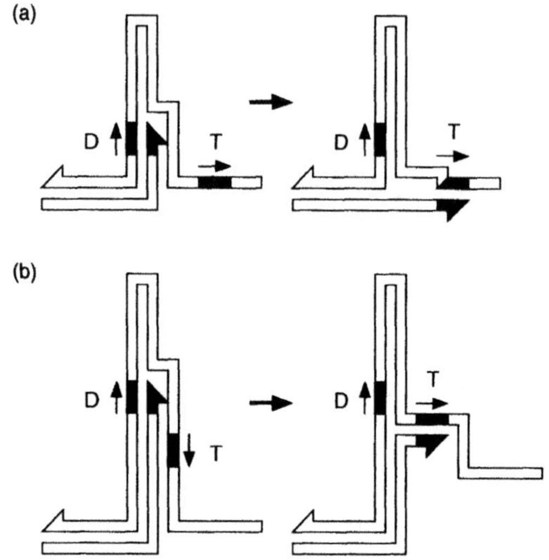

Figure 2. Preferred deletion sites in palindromic DNA. (a) Palindrome deletion by strand-slippage. Replication enters a hairpin DNA structure and stalls. If a short sequence of bases that has just been copied is also found on the downstream side of the palindrome, the newly-replicated strand can dissociate from its template and re-associate with the second copy. This mechanism identifies a donor repeat and a target repeat denoted D and T in the figure. (b) Palindrome shrinkage by strand-slippage. Palindrome shrinkage occurs by a similar mechanism to palindrome deletion. However, in this reaction the target repeat for strand-slippage is located within the palindromic sequence. The new shorter palindrome is composed of the outer regions of the original palindrome and contains a short central spacer that comes from one arm of the parental sequence.

THE REPLICATION OF LONG PALINDROMES IN WILD-TYPE AND *sbcCD* MUTANTS OF *E. coli*

DNA molecules containing long palindromes have been introduced into wild-type *E. coli* in order to investigate the mechanism of inviability and instability. Actively replicating molecules were poorly recovered relative to palindrome-free controls, whereas unreplicated molecules were preserved (25,26). Molecules that had lost their palindromes were present at an elevated frequency in the replicated material (26). Furthermore, when DNA strands were labelled by Dam-methylation and injected into a *dam* mutant host, hemi-methylated molecules were efficiently recovered while unmethylated molecules were under-represented (27). At first sight, this result is unexpected, since hemi-methylated DNA may arise from multiple rounds of replication (each round of replication generating a hemi-methylated product from a hemi-methylated parent). These experiments indicate that parental DNA strands remain present within replicated material despite the poor recovery of material with two non-parental strands. This is consistent with slow replication of DNA containing long palindromes and is not consistent with a destructive mechanism that results in significant loss of parental strands.

SbcC and *sbcD* mutants of *E. coli* allow the replication of DNA containing long palindromes (28-30). This alleviation of inviability is apparently complete for inverted repeats with a small central spacer but perfect palindromes retain some viability problems in this genetic background. This is reflected in a reduced plaque size for bacteriophage λ derivatives containing long palindromes. Furthermore, it has been shown that plaque size is determined by the central sequence of the palindrome (31,32). The data argue that the residual inviability found in *sbcC* mutants is determined by intra-strand base pairing.

Viability is reduced by G/C base pairs at positions except for the central two, consistent with the fact that hydrogen bonds can never form between these central bases in a hairpin or cruciform structure. The data also suggest that the formation *in vivo* of tight loops containing only two unpaired bases is possible in DNA sequences which can form two-base loops *in vitro*. The center-dependence of this residual inviability suggests that there is a center-dependent pathway of cruciform formation that leads to SbcCD-independent inviability. This structure may also be sensitive to SbcCD in wild-type cells (see Figure 3).

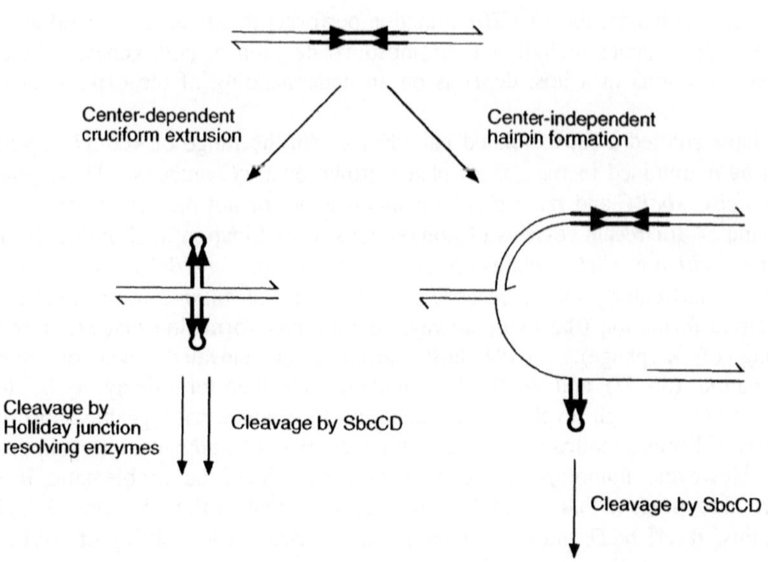

Figure 3. Model pathways of cruciform and hairpin formation inside cells. Data are consistent with the existence of at least two pathways of secondary structure formation by long palindromic sequences within cells. A center-dependent pathway that may result in the formation of cruciform DNA can lead to some inhibition of DNA recovery even in *sbcC* mutant cells. It is hypothesized that this inhibition of recovery may be caused by the operation of Holliday junction-resolving enzymes. Alternatively another SbcCD-independent mechanism that prevents recovery could operate. A center-independent pathway generates hairpin structures that are sensitive to the nucleolytic activity of SbcCD protein. The processing of the hairpins formed by this second pathway is primarily determined by the action of the SbcCD protein. This model is consistent with *sbcCD* mutants being permissive for the replication of interrupted palindromes and partially permissive for perfect palindromes.

The centers of long palindromes propagated in *sbcC* mutants are poor substrates for methylation by the Dam methylase of *E. coli* (11). This methylase normally methylates adenine residues in the recognition sequence GATC but is unable to modify the DNA when it is in an unusual conformation. If this target is inserted at the center of a long palindrome its methylation *in vivo* is reduced. This observation is consistent with the formation of a structure that can inhibit methylation but this structure may not be the same as that which causes residual inviability in *sbcC* mutants since under-methylation is seen for both inverted repeats with small central asymmetries and perfect palindromes. The under-methylation patterns observed suggest that a center-independent pathway for the formation of DNA hairpins exists and that structures with loops below approximately 100 base pairs in length are sensitive to SbcCD action *in vivo* (see Figure 3). This and other potential explanations of the effects of central asymmetry on methylation-inhibition and plaque size in *sbcC* mutants are discussed in detail elsewhere (11).

The SbcC and SbcD polypeptides have been purified and shown to have double-strand exonuclease and hairpin cleavage activities (J. Connelly and D. Leach, unpublished data). These activities can explain the behavior of *sbcC* and *sbcD* mutants on palindromic DNA *in vivo* as shown in Figure 3 (see also 12).

STRATEGIES FOR CLONING LONG PALINDROMES IN *E. coli*

The only *E. coli* genes that have been found to have a direct effect on the propagation of palindromic sequences are *sbcC* and *sbcD*. Clones of eukaryotic DNA that may contain palindromic sequences should therefore be recovered in either *sbcC, sbcD* or *sbcCD* mutants. In our hands, the *sbcC201* mutation performs this function as well as any other mutation in these genes including a chromosomal deletion of both genes. The choice of additional mutations in a host depends on an understanding of the experimental system used.

We have studied a set of related palindromes (in the range of 460-571 bp in length) that can be maintained in *red gam* λ phage grown on *sbcC* mutants. These phage grow well on *recBC sbcBC* and *recD sbcC* mutants whether or not the vector carries a *chi* site (see 33 and 34 for recent reviews of homologous recombination in *E. coli*). If the vector does carry a *chi* site, *sbcC* mutants are also potential hosts. *recBC sbcBC* and *recD sbcC* mutants are particularly useful because they are both permissive for recombination and rolling-circle formation (the two pathways of multimer formation that are necessary for packaging of λ phage). *recD* hosts promote an elevated level of homologous recombination (35-37) and *recD sbcC* mutants are therefore likely to be the more permissive hosts. High levels of recombination to produce packageable substrate under conditions of limited replication will give the least selective advantage to palindrome-free DNA. However, homologous recombination may itself be problematic if repeated sequences are present that could lead to rearrangement within the cloned DNA. To prevent this, *recA*, *recD* and *sbcC* mutants can be used. The viability of *recA* and *recD* strains is poor and they are difficult to use for the preparation of high titer lysates. However, the introduction of the *sbcC* mutation improves the viability of these hosts allowing them to be used with little difficulty. It must be remembered that the primary effect of *sbcC* is to alleviate inviability, but instability is also improved in hosts of this genotype. This is because of the reduction in competitive disadvantage of the palindrome-containing molecule relative to its palindrome-free deletion derivatives. Nevertheless, long palindromes can be unstable in *sbcC* hosts. The 3.2 kb perfect palindrome that we initially constructed *in vitro* to initiate our work on palindromic sequences (28), is deleted in a *recBC sbcBC* host to generate a series of interrupted palindromes of between 500 bp and 700 bp in length. The deletions remove the central sequences and occur between small, directly repeated sequences (J. Lindsey and D. Leach, unpublished data; see Figure 2).

λ phage containing palindromes can form plaques in *sbcC*$^+$ hosts if the phage are *gam*$^+$ (38,39). This effect of *gam* was shown to be independent of the *recBCD* genotype of the host, suggesting that the Gam protein could interact directly with the SbcCD protein to interfere with its action on palindromic DNA (39). This raises the possibility that *gam*$^+$ vectors may be useful for the cloning of DNA palindromes. However, it was observed that palindrome instability remained higher under *gam*$^+$ *sbcC*$^+$ conditions than *gam*$^-$ *sbcC*$^-$ conditions (38). This suggests that the Gam protein only partially inactivates the SbcCD protein and that the use of *sbcC* mutants is advised when DNA palindromes are cloned, whatever the *gam* genotype of the phage.

We have attempted to subclone palindromes constructed in phage λ into a plasmid vector. The palindromes were ligated into the polylinker of the plasmid pUC18 and introduced into an *sbcC* mutant host. We succeeded in obtaining clones for a long inverted repeat (571 bp total length) with a 15 bp short central sequence but several perfect palindromes of 460 bp were not recovered (C. Blake, A. Chalker, D. Pinder and D. Leach, unpublished data). The 571 bp interrupted palindrome is deleted at a frequency of approximately 10^{-4} in λ *red gam* but is deleted at a frequency of 50% in the plasmid. The difficulty in cloning long palindromes in a plasmid and the extreme instability of the construct obtained may suggest that multicopy plasmids are poor vectors for these sequences. The central 109 bp of this interrupted palindrome was stably subcloned in the plasmid pUC18 but could not be propagated in the M13 mp18 vector (38). This suggested a more severe problem with the replication of palindromic sequences in M13 vectors than in plasmids. Our results are somewhat anecdotal but fit with anecdotal observations that other workers have reported to me over the years that palindromes are most stably replicated in phage λ followed by plasmids (with a possible advantage to low copy number plasmids) and that M13 is the least suitable vector system for these sequences. The *sbcC* mutation is required for replication of the 571 bp interrupted palindrome in pUC18 but this plasmid can subsequently be introduced into a *recA sbcC* host to prevent homologous recombination with little adverse effect on palindrome maintenance (C. Blake and D. Leach, unpublished data). Some increase in instability is seen and this may be dependent on the size, preferred multimeric form and copy number of the vector used. It is therefore suggested that either an *sbcC* host or a *recA sbcC* host should be used for the recovery of palindromes in plasmids and cosmids. *recBC*, *sbcBC* and *recD sbcC* strains are not useful for plasmids as they promote linear and circular multimer formation respectively which result in plasmid instability (37,40,41). The poor maintenance of even relatively small palindromes in M13 suggests that M13-plasmid hybrids (phagemids) may be the best vectors to use if single-strand DNA isolation is necessary. A table of host strains constructed and characterized in collaboration with N. Murray is shown in Table 1 (42). Further information on these strains is available from NBL Gene Sciences (UK).

HYBRIDIZATION, PCR AND SEQUENCING OF LONG PALINDROMES

The characterization of long palindromes by hybridization, PCR and sequencing is complicated by their self-complementarity (C. Blake, A. Davison, E. Okely and D. Leach, unpublished data). We have not found it possible to probe for long palindromic sequences using internal oligonucleotides in Southern hybridization or plaque hybridization experiments. It is likely that this difficulty is caused by self-complementarity of the palindromic sequence excluding hybridization with the probe. Similarly, we have not succeeded in amplifying palindrome centers by PCR using one internal and one external primer. We were able to amplify a whole palindrome of 526 bp in length by PCR using two flanking primers. However, the yield of product was low and consisted of double strands, hairpin snapbacks and other unidentified products. We were unable to determine the sequence of this product by di-deoxy chain termination (43). The sequencing of long palindromes by di-deoxy chain-termination is problematic whether the source of the DNA is a PCR fragment, M13 single strand, phagemid single strand, plasmid double strand or λ double strand. If there is sufficient asymmetry at the center of an interrupted palindrome, it is possible to amplify the two arms separately by PCR with the use of two pairs of primers, one to the center and the other to one or other unique flanking sequence (R. Durbin, personal communication). Similarly, if a unique restriction site exists in the central spacer region, an interrupted palindrome can be decomposed into two non-identical halves that can be subcloned and sequenced. However, if the palindrome is too perfect for either of these strategies, we have successfully used subcloning of central fragments of 32

Table 1. Host strains constructed and characterized

Strain	Background	Genotype	Phenotype	Use
DL709	K12SH-28 (prototrophic)*	Δ(*mrr*, *hsdRMS*,*mcrBC*)2, *mcrA* (*e14*°) *sbcC201*, *supE*	Restriction deficient Palindrome permissive Amber suppressor	Plasmid and cosmid cloning (genomic libraries and subcloning)
DL795	K12SH-28 (prototrophic)*	Δ(*mrr*, *hsdRMS*,*mcrBC*)2, *mcrA* (*e14*°) *sbcC201*, *recA*::*Cm*^{R} *supE*	Restriction deficient Palindrome permissive Recombination deficient Chloramphenicol resistant Amber suppressor	Plasmid and cosmid cloning (genomic libraries and subcloning)
NM772	K12SH-28 (prototrophic)*	Δ(*mrr*, *hsdRMS*,*mcrBC*)2, *mcrA* (*e14*°) *sbcC201*, *recD1009*, *supE*	Restriction deficient Palindrome permissive Hyper-rec Amber suppressor	Phage λ cloning (genomic libraries)
DL847	K12SH-28 (prototrophic)*	Δ(*mrr*, *hsdRMS*,*mcrBC*)2, *mcrA* (*e14*°) *sbcC201*, *recD1009*, *recA*::*Cm*^{R} *supE*	Restriction deficient Palindrome permissive Recombination deficient Chloramphenicol resistant Amber suppressor	Phage λ cloning (genomic libraries)

* These strains carry a mutation in the *deoA* gene (also known as *tpp*) encoding thymidine phosphorylase activity (46).

and 109 base pairs in plasmids followed by "double-strand" di-deoxy chain termination sequencing (31,38). A similar strategy with the use of phagemids is also likely to be successful. In the final resort, chemical sequencing by the method of Maxam and Gilbert (44) is possible. However, if the palindrome is long, its symmetry is likely to interfere with strategies for the labelling of unique ends.

LONG PALINDROMES IN THE GENOMIC DNA OF *C. elegans*

At present the nematode, *Caenorhabditis elegans*, is the multicellular eukaryotic organism whose genome is best understood at the DNA sequence level (45). It is therefore of interest to ask whether palindromic sequences have, in fact, caused cloning and sequencing problems and how these problems have been overcome. The information given here has kindly been provided by A. Coulson, M. Berks, R. Durbin and J. Sulston of the Sanger Centre. Computational analysis of 16,141,317 bp of *C. elegans* genomic DNA has revealed 473,910 bp located in 2645 inverted repeat structures. This information was obtained by aligning sequences and scoring +1 for matches, -1 for mismatches and -3 for insertion/deletions. The minimum threshold was set at 20 and maximum length of the inverted repeat (end to end) was set at 1000 bp. This translates on average as one inverted repeat every 6.1 kb of genomic DNA sequence.

Among inverted repeats that can be recovered in cosmid vectors, many have been found that cannot be efficiently subcloned in M13 derivatives. Phagemid vectors have been used to clone and sequence these DNAs. Two examples are the inverted repeats found in cosmids C53B4 and C52G5. The former was found to contain an inverted repeat of 80 bp separated by a spacer of 7 adenine residues (total length, 167 bp) 55 bp of which was missing from M13 libraries. The latter contained an inverted repeat of approximately 160 bp with a spacer of approximately 50 bp (total length, approximately 370 bp) 170 bp of which was missing from M13 libraries. In the former case the sequence was verified by PCR amplification of the region of the cosmid with primers well removed from the repeated sequence and sequencing of the PCR product.

Many inverted repeat sequences may exist that cannot be cloned in cosmids but are present in genomic DNA. Some 20% of the genome is unclonable in cosmids but can be recovered in YACs. An interesting example of a palindrome that is deleted in bacteria is found in the overlap between cosmids B0393 and F17C8. B0393 contains the outside 420 bp of an inverted repeat found elsewhere in the genome but the structure of F17C8 does not correspond to that found in the genome. Southern hybridization suggests that F17C8 carries a deletion of 3 kb of DNA that includes a putative 2 kb genomic palindrome. Flanking sequences can be used to amplify this region from the cosmid by PCR but this amplification is not successful from genomic DNA consistent with the existence of a long palindromic sequence.

CONCLUSIONS

Long DNA palindromes cause problems both *in vitro* and *in vivo* as a consequence of their ability to make intra-strand hydrogen bonds to form hairpin and cruciform structures. These structures interfere with biochemical steps used in their characterization and in their replication and stability inside cells. These problems can be overcome in a number of ways. If the palindrome is interrupted by a central spacer region, this may be used to separate the two halves so that the sequences to be analyzed are no longer palindromic. If this is not possible, a reduction in the overall length of the palindrome may be possible. *In vivo*, instability can be minimized by reducing all unnecessary DNA replication between the formation of a construct and its analysis and the use of host-vector systems that minimize the replicational advantage of palindrome-free deletion derivatives. In *red gam* λ vectors, the inviability associated with interrupted palindromes is entirely overcome in *sbcC* mutant hosts and the viability of phage containing perfect palindromes is enhanced to the extent that DNA recovery is possible. The effect of *sbcC* on viability has also been reproduced in the plasmid pUC18 but instability is greater than in λ. M13 provides the least stable vector system for palindromic sequences but this instability can be reduced by using phagemids.

ACKNOWLEDGMENTS

I would like to thank Richard Durbin, John Sulston and their colleagues from the Sanger Centre for providing unpublished information, and Thorsten Allers, Catherine Blake, John Connelly and John Darlow for their comments on the manuscript. Further information about the strains listed in Table 1 can be obtained from NBL Gene Sciences Ltd., South Nelson Rd, Cramlington, Northumberland, NE23 9WF, UK (FAX +44 (0) 1670 730 454). The work in my laboratory has been supported by the MRC and the BBSRC.

REFERENCES

1 Collins, J. and Hohn, B. (1978) Proc. Nat. Acad. Sci. U.S.A. 75, 4242-4246.
2 Collins, J. (1981) Cold Spring Harbor Symp. Quant. Biol. 45, 409-416.
3 Lilley, D.M.J. (1981) Nature 292, 380-382.
4 Collins, J., Volkaert, G. and Nevers, P. (1982) Gene 19, 139-146.
5 Mizuuchi, K., Mizuuchi, M. and Gellert, M. (1982) J. Mol. Biol. 156, 229-243.
6 Hagan, C.E. and Warren, G.J. (1982) Gene 19, 147-151.
7 Hagan, C.E. and Warren, G.J. (1983) Gene 24, 317-326.
8 Warren, G.J. and Green, R.L. (1985) J. Bacteriol. 161, 1103-1111.
9 Yoshimura, H., Yoshino, T., Hirose, T., Nakamura, Y., Higashi, M., Hase, T., Yamaguchi, K., Hirokawa, H. and Masamune, Y. (1986) J. Gen. Appl. Microbiol. 32, 393-404.
10 Das Gupta, U., Weston-Hafer, K. and Berg, D.E. (1987) Genetics 115, 41-49.
11 Allers, T.A. and Leach, D.R.F. (1995) J. Mol. Biol. 252, 70-85.
12 Leach, D.R.F. (1994) BioEssays 16, 893-900.
13 Peeters, B.P.H., De Boer, J.H., Bron, S. and Venema, G. (1988) Mol. Gen. Genet. 212, 450-458.
14 Behnke, D., Malke, H., Hartmann, M. and Walter, F. (1979) Plasmid 2, 605-616.
15 Kieser, T. and Melton, R. (1988) Gene 65, 83-91.
16 Henderson, S.T. and Petes, T.D. (1993) Genetics 113, 57-62.
17 Gordenin, D.A., Lobachev, K.S., Degtyareva, N.P., Malkova, A.L., Perkins, E. and Resnick, M.A. (1993) Mol. Cell. Biol. 13, 5315-5322.
18 Ruskin, B. and Fink, G.R. (1993) Genetics 133, 43-56.
19 Bernardi, G. (1962) Biochem. J. 83, 32-33.
20 Wilson, D.A. and Thomas, C.A. (1974) J. Mol. Biol. 84, 115-144.
21 Nader, W.F., Edlind, T.D., Huehermann, A. and Waver, H.W. (1985) Proc. Nat. Acad. Sci. U.S.A. 82, 2698-2702.
22 Wyman, A.R., Wolfe, L.B. and Botstein, D. (1985) Proc. Nat. Acad. Sci. U.S.A. 82, 2880-2884.
23 Wyman, A.R., Wertman, K.F., Baruer, D., Helms, C. and Petri, W.H. (1986)

Gene 49, 263-271.
24 Wertman, K.F., Wyman, A.R. and Botstein, D. (1986) Gene 49, 253-262.
25 Leach, D.R.F. and Lindsay, J.C. (1986) Mol. Gen. Genet. 204, 322-327.
26 Shurvinton, C.E., Stahl, M.M. and Stahl, F.W. (1987) Proc. Nat. Acad. Sci. U.S.A. 84, 1624-1628.
27 Lindsey, J.C. and Leach, D.R.F. (1989) J. Mol. Biol. 206, 7024-7027.
28 Leach, D.R.F. and Stahl, F.W. (1983) Nature 305, 448-451.
29 Chalker, A.F., Leach, D.R.F. and Lloyd, R.G. (1988) Gene 71, 201-205.
30 Gibson, F.P., Leach, D.R.F. and Lloyd, R.G. (1992) J. Bacteriol. 174, 1222-1228.
31 Davison, A. and Leach, D.R.F. (1994) Genetics 137, 1-8.
32 Davison, A. and Leach, D.R.F. (1994) Nucl. Acids Res. 22, 4361-4363.
33 Kowalczykowski, S.C., Dixon, D.D., Eggleston, A.K., Lauder, S.D. and Rehrauer, W.M. (1994) Microbiol. Rev. 58, 401-465.
34 Myers, R.S. and Stahl, F.W. (1994) Annu. Rev. Genet. 28, 49-70.
35 Chaudhury, A.M. and Smith, G.R. (1984) Proc. Nat. Acad. Sci. U.S.A. 81, 7850-7854.
36 Amundsen, S.K., Taylor, A.F., Chaudhury, A.M. and Smith, G.R. (1986) Proc. Nat. Acad. Sci. U.S.A. 83, 5558-5562.
37 Biek, D.P. and Cohen, S.N. (1986) J. Bacteriol. 167, 594-603.
38 Leach, D.R.F., Lindsey, J.C. and Okely, E.A. (1987) J. Cell Sci. Suppl. 7, 33-40.
39 Kulkarni, S.K. and Stahl, F.W. (1989) Genetics 123, 249-253.
40 Cohen, A. and Clark, A.J. (1986) J. Bacteriol. 45, 409-416.
41 Silberstein, Z. and Cohen, A. (1987) J. Bacteriol. 169, 3131-3137.
42 Kaiser, K., Murray, N.E. and Whittaker, P.A. (1995) in DNA Cloning, Vol. 1, 2nd edition (D.M. Glover, ed.), pp. 37-84, IRL Press, Oxford.
43 Sanger, F., Nicklen, S. and Coulson, A.R. (1977) Proc. Nat. Acad. Sci. U.S.A. 74, 5463-5467.
44 Maxam, A.M. and Gilbert, W. (1977) Methods Enzymol. 65, 499-560.
45 Waterston, R. and Sulston, J. (1995) Proc. Nat. Acad. Sci. U.S.A. 92, 10836-10840.
46 Fangman, W. L. and Novick, A. (1966) J. Bacteriol. 91, 2390-2391.

DNA ISOLATION, MANIPULATION AND CHARACTERIZATION FROM OLD TISSUES

Rob DeSalle and Elizabeth Bonwich

Department of Entomology
American Museum of Natural History
79th Street at Central Park West
New York, NY 10024

INTRODUCTION

Recent advances in molecular techniques have revolutionized our approaches and expanded our vistas in the biomedical and organismal sciences. The development of the Polymerase Chain Reaction (PCR) has been the major reason for this revolution. Nearly every field of biology now owes a great deal to the development of these molecular techniques. The historically oriented sciences are an area of biology that has especially benefitted from PCR technology. This is because deoxyribonucleic acids are molecules that can carry genealogical information. Its double helical complementary nature and method of replication make it a perfect storage receptacle for historical information, whether the history be ancient or recent. In particular, in the organismal sciences, large advances have been made in the fields of evolutionary biology, population biology and systematics.

One area of the organismal sciences that has seen great advances is approaches with the use of ancient DNA (aDNA). Without PCR technology, analysis of ancient tissues would be greatly limited. With the benefits of PCR, however, have come novel and extremely important aspects of data scrutiny and analysis. As any researcher who has worked with PCR knows, the technique is extremely sensitive and contamination from nearly any extraneous source will potentially damage results. It is the purpose of this chapter to detail the methods available for the extraction, PCR amplification, characterization and diagnosis of nucleic acids from old tissue sources, and to make clear the problems and pitfalls that accompany this approach to organismal and historical biology.

At the outset we shall define any old or ancient tissue source as a tissue not stored with the initial intent of recovery of nucleic acids (or other biomolecules) from the tissue and restrict our discussion of such tissues to exclude tissues used in forensic studies (1). This definition would exclude frozen tissue collections or ethanol-preserved specimens where the tissues are stored specifically for the isolation of DNA or proteins. Although this definition is somewhat arbitrary, it does demarcate tissues where DNA isolation will be rather trivial (frozen collections or ethanol-preserved collections where the explicit intention of storage is for later retrieval of nucleic acids or proteins) and tissues where

Genetic Engineering, Vol. 18, Edited by J.K. Setlow
Plenum Press, New York, 1996

nucleic acid isolation poses a much greater challenge. Although there are several reports of the characterization of amino acids in ancient tissues (2-7), we will concentrate entirely on the techniques and approaches developed for nucleic acids. In the course of this chapter, we will discuss the techniques that are most commonly used for DNA isolation, the precautions taken during the isolation and the protocols for PCR amplification of nucleic acids isolated from ancient tissue sources. We will also discuss the methods necessary for verification of the characterized DNA. Finally, the scope of this new field will be summarized and the utility of aDNA studies will be discussed.

We start the description of the aDNA techniques with the caveats expressed by Lindahl (8,9) in his discussion of the stability of the primary structure of DNA. His excellent review pointed out that DNA in aqueous solution "decays" at an extremely rapid rate. He calibrated this decay rate of nuclear DNA and suggests that the typical mammalian genome in aqueous solution would be completely decomposed in 30,000 years. Such a rapid decay rate would preclude the isolation of any intact nucleic acids from tissues much older than 30,000 years and equally importantly implies that any DNA isolated from ancient tissues will have experienced extreme damage to the DNA primary structure and will most likely be degraded into very small fragments. There are three points to be considered as a result of Lindahl's critique (9). The first two concern the quality and quantity of the DNA isolated from ancient tissues and the third concerns the outright disbelief that many scientists will hold from the outset of any aDNA study on tissues older than 30,000 years.

Lindahl (9) suggests four areas where aDNA studies should follow certain guidelines. These four suggestions were made in order that future aDNA studies, no matter what the source of tissue, could be evenly scrutinized. We feel that these suggestions are most important and repeat them here, but would include a fifth and equally important suggestion concerning the diagnosis of the aDNA after its isolation, amplification and sequencing. Diagnosis is the only way that certain studies can gain credibility, and most diagnosis procedures involve phylogenetic analysis. The recent controversy surrounding the recent report of the isolation and characterization of dinosaur DNA (10-14) only accentuates the importance of rigor and proper procedure for this stage of analysis. The advantages and pitfalls of the phylogenetic analyses used to diagnose DNA sequences will be discussed in the diagnosis section of this chapter.

The four suggestions made by Lindahl (9) are: 1) The presentation and publication of failed or negative results would greatly facilitate progress in the field. Currently, only those studies that have been successful are published, but for every successful aDNA study there is a great deal of failure. Lindahl (9) suggests that the difficulty of using the technique would become more clear and apparent to researchers if these negative results were made part of the published literature. 2) Corroboration of aDNA studies in at least two independent laboratories is essential for the credibility of any aDNA study. Lindahl (9) suggests that, if possible, aDNA studies should be conducted simultaneously in two laboratories preferably at two different institutions. 3) Controls for the majority of aDNA studies have been poorly documented and hence more complete documentation of control experiments in aDNA studies is required. Specific controls are suggested by Lindahl (9), including "mock" isolation of DNA from the medium surrounding the target tissue. If upon amplification products are found from this "mock" isolation, then contamination can be inferred as a factor in the experiment. 4) More detailed and probing examination by chemists and biochemists on the biomolecular composition of ancient tissues is needed before many of the studies that claim success can be credible. Lindahl (9) therefore suggests a stronger degree of collaboration among physical chemists and analytical chemists and those conducting aDNA studies in order to further our understanding of ancient tissues.

DNA EXTRACTION PROTOCOLS

Destructive vs. Non-Destructive Sampling

The first thing which must be considered before any extraction of "ancient" materials takes place is whether a less destructive or non-destructive sampling method can be used. Destructive sampling, where the entire fossil is given up for DNA extraction, leaving none of the original fossil behind, is not a sound practice for two major reasons. The first reason concerns replication of results. Since the first reaction to a positive result with aDNA should be skepticism concerning contamination and authenticity of the result, to be certain of the result it is prudent to try to replicate the experiment to as great a degree as possible in another laboratory. If destructive sampling is used then there is no way to verify a result and as such, that result can always be held in question. The second reason concerns the loss of morphological information by destructive sampling. The data contained in the morphology of a fossil organism are extremely important and informative. Most often fossil information is found in external structures such as on the surface of bones, the shape of bones, external morphology of invertebrates or of skins. In addition, there may be extreme fine detail internal to fossils as recently demonstrated by SEM studies of fossil insects (15-17). Therefore, before losing the morphological information from an entire tooth, bone or amber fossil for 200 to 300 base pairs of DNA sequence, alternatives must be considered.

The targets of aDNA studies have typically been bone, teeth, dried skins, pinned insects, some mummified soft tissues such as brain tissue and amber-embedded invertebrate, plant and bacterial material. Each of these sources of aDNA presents unique problems with respect to destructive sampling. Most bone and teeth preps require powdering and are extremely destructive. However, teeth can often be hollowed out (18), thereby leaving most of the external morphology intact. For bones, Meijer et al. (19) describe methods whereby the patella, tibia and caput of the femur are cleaved with an iron saw and the spongiosum is used as a source for DNA extraction. If the actual cleavage of the bone is done carefully it is conceivable that the two halves of these bones can be placed together, preserving the original morphology of the bone. Alternatively, a small wedge from larger bones can be "hacksawed," yielding a few grams of material without substantially altering the morphology of the bone (20). Hummel and Hermann (21) suggest that all outer surfaces of bones used in aDNA studies should either be cut away or extensively UV irradiated as a precaution against contamination.

The very nature of work with amber aDNA requires a destructive sampling technique. This being the case, it is imperative that the fossil taxon chosen for aDNA analysis be one that is commonly found. If the fossil taxon is rare, then it is inappropriate, if not unethical, to destroy it for aDNA analysis. Several of the amber aDNA studies and the ancient plant studies have followed this guideline. Cano et al. (22, 23) have done most of their amber insect work with *Proplebeia dominicana*, a commonly-found stingless bee in Dominican amber. Likewise, DeSalle et al. (24) used *Mastotermes*, a commonly-found termite in Dominican amber, in their analysis of termite phylogeny. Both Golenberg et al. (25,26) and Soltis, Soltis and Smiley (27) have taken advantage of the high abundance of fossilized plant material in the Clarkia fossil beds for their studies of Miocene plant genes.

Non-destructive methods for amber-preserved specimens have been discussed and there is one report of drilling into amber fossils and removing tissue without destruction of the exoskeleton (28). However, given the wealth of morphological information from SEM studies of the internal organs of amber-embedded insects (15-17), it may be more appropriate to be extremely conservative when working with amber specimens.

Museum and herbarium specimens are an excellent and abundant source of material for aDNA studies (29). However, sampling of specimens from collections is often in direct conflict with the goals of collection-oriented institutions and consequently museum and herbaria curators are rightfully reluctant to "loan" specimens to aDNA researchers. In addition, the cost of collection of many specimens from exotic locales far exceeds the benefits of obtaining one kilobase of sequence from the organism if the specimen is sampled destructively. These problems are more acute the smaller and the rarer the specimen. For instance, large mammal skins can be destructively sampled by removal of hair or small patches of skin without great loss of morphological information. Herbarium samples can also have small patches of leaf material removed for destructive sampling without much loss of morphological information. Insect sampling is more complicated, however, and non-destructive sampling techniques have been developed so that large-sized pinned museum specimens, such as crickets (but not as small as *Drosophila*) can be used for DNA extraction without ever being removed from the pin (30) (Table 5), and with virtually no alteration of surface morphology of the insect.

As a final statement on sampling, work with aDNA is labor intensive, time and money consuming with very little product to show for the effort (typically ~200 to 1000 base pairs). Because of this poor cost-to-benefit ratio, aDNA studies should not be entered on a whim, or for the sole purpose of seeing if they can be done. Failed attempts and contaminants far outnumber the successes.

Voucher Specimens and Experimental Replication

Voucher specimens are an essential part of many organismal studies and should be maintained and stored for aDNA studies. The role of voucher specimens in systematic studies with molecular approaches has been discussed and specific recommendations have been made concerning the storage and maintenance of such specimens (31-36). Lee et al. (37) clearly state that voucher specimens "physically and permanently document data in an archival report by 1) verifying the identity of the organism(s) used in the study; and 2) by so doing, ensure that a study which otherwise could not be repeated can be accurately reviewed or reassessed" (34). Consequently, we suggest that the most important reason for the maintenance of voucher specimens is to allow replication of the experiment. Obviously, if the entire ancient specimen is used in the isolation of DNA there will be no opportunity for corroboration with that tissue. If information concerning the ancient specimen is not archived properly then verification of work on that tissue is impossible. Storage of a portion of the isolated nucleic acids is also suggested for possible use by other investigators.

Preliminary Precautionary Measures - DNA Isolation

Certain precautions must be taken prior to extraction to reduce the possibility of contamination. All equipment (pipettors, tips, tubes, forceps, gloves and any other equipment used for maceration of the sample) should be exposed to ultraviolet (UV) radiation for a time and intensity expected to kill bacteria prior to use. Ideally, equipment should be kept under constant UV in a laminar flow hood when not being used. This is especially true for work with human aDNA (18). Extraction reagents should also be UV treated prior to extraction. The relative efficiency of washing pipettors and bench areas with sodium hypochloride (Chlorox) and hydrogen chloride has been examined (38). Even relatively low concentrations of Clorox (10%) appear to be much more efficient at removing contaminants than strong concentrations (2.0 N) of hydrogen chloride. Consequently, wiping down of bench areas and cleansing of pipettors suspected of contamination with 10% Chlorox may be conducted prior to UV treatment.

The extraction should be performed in an area with no possible PCR products of a similar organism or *Taq* polymerase present. Access to this area should be strictly limited. Protective clothing should be worn at all times when in the room. This protective clothing includes gloves, face masks and some form of protective outergarment which is either disposable or used and stored only in the aDNA room. The pipettors and other equipment used for the extraction should never be used interchangeably for PCR product manipulation.

Contamination on bone material can often be reduced by brushing the surface clean (39), cleansing the outer surface of the bone with 10% Chlorox (18) or removing a substantial layer from the outside of the bone by sand blasting (20). To reduce contamination in working with human remains, it is advisable to "sequence" the gene of interest from the individual(s) who are working on the project so that it is easy to identify a "positive" result which is actually amplified DNA from a laboratory worker (40, 41). Decalcification of bone material prior to DNA isolation has been a common precautionary method to prevent co-isolation of PCR inhibitors. However, Fisher et al. (42), using Civil War era bone, have demonstrated that decalcification is not a necessary precaution for the isolation of DNA from bone.

Specific Extraction Protocols

Several standard protocols for DNA preparation exist and can be consulted for the basics of DNA isolation. Most of these protocols were established prior to the development of PCR and are used on relatively large amounts of starting material for such manipulations as Southern blotting and genomic library construction. There are basic variations in standard preps for the wide range of organismal tissue sources due to the obvious biological differences of these tissues. For instance, tissue sources from bone, plants, invertebrate materials, blood and soft materials such as internal organs and skins are treated differently. Some of these standard preparations can and have been used on ancient tissues. Methods in Enzymology, Vol. 224 contains several chapters that describe in detail these standard methods [plants and algae (32,43-45); animal (33,46,47)]. Hillis and Moritz (48) is also an excellent source for protocols for standard DNA preparations. The Hermann and Hummel (49) volume has several chapters that describe specific ancient tissue types and the methods most commonly used on them [wet samples (50); frozen samples (51); dried samples (52); feathers (53); herbarium specimens (54); bones (20,55,56); plant material (57)].

The development of PCR has resulted in a proliferation of protocols for the isolation of DNA, specifically for amplification from non-conventional tissue sources. Table 1 lists several of these techniques that have been developed and these references can be consulted as alternative methods for difficult-to-manipulate tissue sources. Rather than give exhaustive descriptions of the techniques available we discuss four of the most common methods used to extract DNA from old or non-conventional tissues (Tables 2-5). There are also a number of extraction kits on the market (Table 6).

The issues to be taken into account when a protocol is to be chosen include potential for PCR inhibition (58), known effectiveness of method on relatively small starting amounts of material, and amount of sample loss during extraction. The Chelex preparation (Table 2) (59) has been used successfully on a variety of tissues. Its major advantage is that there

Table 1. A Compilation of More Recent DNA Isolation Techniques for Non-Conventional Tissue Sources

Tissue Source	Comment	Reference
bone powder	water elution	(98)
bone powder	heated and treated with detergent, chelating agents and reducing agents	(19)
teeth and soft tissue	chelating reagents	(99)
teeth	DNA isolation and diagnosis	(100)
archival celloidal bone sections	originally used in pathology tests	(101)
fingernail		(102)
human hair shafts	extracted in organic solvent and purified by filtration methods	(103)
histological samples	originally used in pathology tests	(88)
paraffin-embedded samples	microwave-based isolation procedure	(104)
cell samples affixed to slides	slides originally used for microscopy	(105)
highly degraded DNA	gel filtration used as technique	(106)
plant tissues	treated with alkali, neutralized and then heat treated prior to PCR in a single tube	(107)
teeth	rapid CTAB DNA isolation prep	(108)
contemporary tissues	sephacryl-500HR used as a filtration media in microtiter plates for multiple DNA isolations	(109)
whole blood	treated with formamide at 95°C followed by PCR (Fol T) with reduced stringency thermal cycling	(110)
blood clots	guanidine thiocyanate-silica particle preparation	(111)
paraffin-embedded tissue	tissues prepared in formalin prior to embedding in paraffin	(112)
Guthrie cards	originally saved for blood typing	(113)
paraffin-embedded tissues	originally for clinical pathology studies	(114)
sperm	single-cell typing with the use of microsatellites	(115)
specimens originally extracted and stored for allozyme analysis	common storage practice of systematists and population geneticists in 1970s	(116)
retail whale meat	forensic conservation and analysis	(188)

Table 2. Chelex Preparations (taken primarily from refs. 53, 59)

1 Use Chelex 100.
2 Make up stock solution of 5% Chelex and aliquot 500 ul of the stock per tube.
3 With sterile UV-treated forceps or spatula, add a small amount of sample to a tube containing 500 ul of Chelex.
4 Heat sample tube and two blank Chelex tubes at 56°C on a rotating platform for 45 min to overnight, until most tissue has disintegrated.
5 Vortex at maximum speed for 10-15 sec.
6 Heat at 95° - 100°C for 15 min in a heat block.
7 Vortex at maximum speed for 10-15 sec.
8 Store at either 4°C or -20°C.
9 Prior to PCR spin down tubes and use ~ 1 ul of supernatant per PCR reaction.

Table 3. Phenol/Chloroform with Centricon Purification

1 All reagents must be sterile.
2 Begin with standard homogenization (lysis) buffer. Per extraction, you will need a total of 1 ml of solution consisting of 900 ul Tris/EDTA, and 100 ul 10% SDS [100 mM EDTA, 10 mM Tris (pH 7.5)].
3 Mix these first to use with grinding. Add approximately 20 ug/ul proteinase K.
4 Depending on how the specimen was preserved, grind it up.
   a. If in alcohol, let dry out on kimwipe and then dry some more in speed vac before grinding. Then proceed as below.
   b. If amber, then after cracking open the fossil scrape out the inside with a buffer-dampened needle and place into buffer tube. Add 100 ul of SDS and continue at step three.
   c. If pinned or fresh grind in 1.5 ml microfuge tube set in liquid nitrogen. Put some liquid nitrogen inside the tube as well. Once ground thoroughly, add Tris/EDTA buffer and continue grinding. Add 100 ul of SDS.
5 Cap tubes and wrap tops with parafilm. Place in 55°C incubator and a rotating plate overnight.
6 Spin down and take off supernatant for phenol and chloroform extractions.
7 Proceed with phenol and chloroform extractions:
   a. Add equilibrated phenol to top of 1.5 ml tube. Vortex. Spin 5 min at top speed.
   b. Remove aqueous (top) layer to new tube but save old tubes just in case a need to back extract arises.
   c. Repeat phenol extractions until the interface between the organic and aqueous layer is clean.

   CAUTION! Be careful that you don't lose too much sample when you're pulling off the aqueous layer.
8 When you have clean interface, remove the aqueous layer and add lots of chloroform. Vortex.
9 Spin at top speed 5 min and remove supernatant to new tube.
10 Ethanol precipitate or Centricon purify:
   a. Centricon 30 is the preferred column for spin purification.

b. Otherwise, add 2-3x the volume of your product of cold 100% EtOH. Can also add 3 M NaAc and chill 15 min in -70°C. Spin 30 min at max speed. Wash pellet twice in 80% EtOH.

11 Resuspend pellet in H_2O. For a small pellet use 300-400 ul of water. For a large pellet use ~900 ul.

Table 4. Silica Method (Adapted from refs. 117-119)

1 Prepare buffers:

a. Silica suspension aliquots:
Add 100 ml water to 12 g silica particles. Mix until completely suspended. Allow silica to settle for 24 hr at room temperature. Then remove the upper 86 ml of liquid and refill with water to 100 ml final volume. Allow settling again about five hr, remove upper 88 ml of liquid, and then add 120 ul 10 M HCl. Resuspend to the 100 ml volume and aliquot into 1.5 ml microtubes. Store in the dark at room temperature.

b. Extraction buffer:
Dissolve 24 g of guanidinium thiocyanate (GuSCN) in 20 ml of 0.1 M Tris- HCl (pH 6.4). Heat if necessary to dissolve. Then add 4.4 ml of 0.2 M EDTA (pH 8.0) and 0.5 ml of Triton X-100. Mix by inversion.

c. Wash buffer:
Same as extraction buffer but without the EDTA or Triton.

2 Add a small amount (~0.5 g) of macerated sample to ~1 ml extraction buffer. Incubate at 60°C for at least 1 hr with occasional agitation.

3 Centrifuge at 5,000 rpm for 5 min. Remove 500 ul of the supernatant to a fresh tube and add 500 ul of the extraction buffer and 40 ul of the silica suspension. Mix and incubate at room temperature for 10 min. Centrifuge.

4 Wash pellet twice with the wash buffer and then twice with 70% ethanol. Allow to dry and resuspend in water or TE at 56°C. Store at -20°C.

Table 5. DTAB Method for Medium- to Large-Sized Pinned Insect Specimens (30)

1 Make DTAB extraction solution:
8% dodecyltrimethylammonium bromide (DTAB)
1.5 M NaCl
100 mM Tris-Cl pH 8.8
50 mM EDTA

2 Pierce exoskeleton several times with a fine needle.

3 Without removing from pin, submerse insect abdomen in 600 ul of extraction solution.

4 Leave overnight in a 68°C water bath.

5 The pinned specimen can now be removed, washed with chloroform, air dried and placed back into the collection.

6 Agitate to mix. Then add 600 ul of chloroform. Mix by inversion and centrifuge at 10,000 rpm for 2 min.

7 Transfer upper aqueous layer, chloroform extract again and transfer that aqueous layer to another tube.

8 To this add a mixture of 900 ul water and 100 ul cetyltrimethylammonium bromide (CTAB) solution (5% CTAB, 0.4 M NaCl).
9 Invert to mix and let sit at room temperature for 2 min.
10 Centrifuge at 10,000 rpm for 10 min. Discard supernatant and resuspend pellet in 300 ul 1.2 M NaCl.
11 Follow standard ethanol precipitation protocol. Resuspend DNA pellet in 50 ul TE. Use this DNA solution for PCR amplification.

Table 6. Commercially Available Kits for the Isolation of DNA

Kit Name	Company	Comments
EasyDNA	Invitrogen	50 isolations per kit
DNA Isolator	Genosys	Phenol-free kit
GeneClean for aDNA	Bio 101	New on market; all components gamma irradiated
GeneClean	Bio 101	

Note that many of this type of kit call for an ethanol precipitation which may give only a very small (~10 ul) amount of aDNA to work with. If such a kit is used, it is wise to ask about the effectiveness of Centricon purification instead of ethanol precipitation.

is almost no chance of sample loss in the course of the preparation since the protocol involves only one solution, whereas a phenol/chloroform prep, for instance, includes many steps where the sample can be lost. One of the drawbacks of the Chelex method is the apparent degradation of DNA if left too long in Chelex. Another problem that is encountered with the Chelex method involves PCR inhibition by the Chelex solution. Consequently, several dilutions of the Chelex-isolated DNA are sometimes required for successful amplification of DNA. The phenol/chloroform with Centricon-30 (Amicon, Inc.) prep (Table 3) overcomes this problem as most phenol preparations produce extremely "clean" DNA. This protocol was adapted for use with rare and extremely tiny insect samples. As it is very successful with those samples, it can be used on rare and small "ancient" samples. The main reason behind using this protocol is that it will give very clean nucleic acids, even if there is less final sample for use in PCR. The very "clean" prep is desired over a "dirty" prep that will give problems during PCR. With respect to PCR inhibition, some researchers have found that the silica method (Table 4) gives the least PCR inhibition of the standard protocols.

Choice of DNA precipitation method, generally ethanol or Centricon concentration, has similar guidelines. Less product is obtained and there is a higher risk for sample loss by ethanol precipitation. Centricon treatment will retain more product, although if the pre-Centricon DNA prep is not "clean" then the final product will cause problems later in the PCR. Goodyear et al. (58) have suggested a protocol for removal of co-purifying inhibitor in ancient DNA studies that is both reliable and results in minimal loss of sample DNA. In addition, Hanni et al. (60) suggest that isopropanol precipitation can remove extraneous PCR inhibitors from ancient bone extracts without extreme loss of sample. Another method of concentrating aDNA isolations and producing cleaner preparations is through the use of the GeneClean (Bio 101; Table 6) method after the initial isolation of DNA with any of the above-mentioned methods.

aDNA PCR

Due to the sensitivity of the PCR technique, precautions must be taken to avoid undesired DNA being amplified along with the "ancient" template. For this reason, all equipment, e.g., tubes, tips, pipettors, etc., and most reagents, must be sterilized and UV treated prior to use. PCRs should be set up in an area other than the extraction room. Of course, template DNA as well as dNTPs and *Taq* polymerase should not be UV treated.

Hummel and Hermann (21) suggest that there are two pitfalls to PCR in aDNA studies. The first concerns contamination and the second inhibition. Control PCR reactions are the most obvious solution to the problem of contamination which will cause the production of false positives. False positives are most commonly caused by product carry-over of a previously amplified product. The occurrence of product carry-over is very common and proceeds through direct contact by gloves or indirect contact by aerosols.

There are several control reactions that must be run with every aDNA PCR experiment. As with all types of PCR, a positive control must be used to test the efficacy of the PCR reagents in any aDNA experiment. Hummel and Hermann (21) recognize three different classes of PCR controls for the detection of false positives. The first is the *negative control* and is essentially the same control suggested by Lindahl (9), where a "mock" isolation from some material similar in composition to the actual target tissue is performed and the resultant solution is used as template for PCR. As an example, in the case of an amber aDNA study, a piece of surrounding amber without an inclusion could be used. If a positive PCR reaction occurs with this control then the experiment must be discarded. The second control is the *blind control* which is an extraction without sample material. If any of the isolation reagents are contaminated, a positive PCR reaction will be obtained and the experiment can be designated as invalid. The final control suggested by Hummel and Hermann (21) is the *no-template control*, which can only help determine if the PCR reagents and conditions are producing a false positive.

Another control that should be used is a negative PCR control with primers that should not allow for amplification. As Handt et al. (40) point out, "Any indication that it [aDNA] fails to show signs of degradation should be regarded as a serious warning signal". DeSalle et al. (24) used what they termed "large DNA" controls in their amber termite study. The rationale behind this control is that aDNA should be highly degraded (less than 500 bases in length). Contaminating DNA from contemporary sources will not be so highly degraded. If a primer pair is used that amplifies a large (1kb or larger) fragment of DNA, then if there is contamination a 1 kilobase PCR product will be observed. If the large DNA control does not produce a PCR product, then the presence of high molecular weight contaminating DNA can be ruled out. This control does not rule out low molecular weight contaminants, however.

The second pitfall concerns the inhibition of the PCR reaction by the solution isolated from the specimen. Such inhibition will give a false negative for a PCR experiment. Inhibition is most likely produced by co-isolation of inhibiting materials from the ancient tissue source or from reagents used in the isolation (40). In addition, it is possible that contaminating microorganism DNA could be co-isolated with the target aDNA and the large amounts of this contaminant DNA could inhibit the PCR (21). In addition to the positive controls mentioned above for detecting contamination, an aDNA sample (if available) that has worked previously can be run as a positive control. Another set of controls that can be run are "mixing controls" where aDNA suspected of inhibition is added to a template that is known to amplify easily. A reaction with the easily amplified template alone is also performed. If no PCR product appears in the "mixing control" and does appear in the reaction without the suspect aDNA, then inhibition of the PCR reaction can be concluded. Different concentrations of the suspect aDNA can be added and the critical concentration of the suspect aDNA for inhibition can also be determined.

The major gene targets for non-human aDNA studies (and to a certain extent for human aDNA studies too) are mitochondrial genes and nuclear ribosomal genes. Designing primers for amplifying these regions has been facilitated by a proliferation of comparative studies with these genes. One general consideration that is required for designing primers for aDNA studies concerns the target size of the PCR product. In general aDNA is highly degraded and hence primer pairs should be designed to accommodate amplification products that are relatively small (less than 500 base pairs). Most aDNA primer pairs span 200 base pairs or less, and some studies have used suites of primer pairs less than 100 bases apart to obtain relatively large stretches of sequence from aDNA study organisms (40). For mitochondrial DNA primers, several good lists of conserved primer sequences exist and these should be consulted as a starting point (40,46,55,61-63). For nuclear ribosomal primers, several publications can be consulted as starting points (64-68).

The primer systems and cycling conditions for most aDNA studies follow typical parameters. Recent developments of different primer systems and cycling conditions may also become useful for aDNA studies. Semi-nested (69) or nested (70,71) primer pairs have been used in many PCR experiments on standard DNAs of extremely low concentration with good success. For instance, Honda et al. (70) estimate that a nested primer system with 47 cycles is 2000 times more sensitive than standard PCR with 27 cycles. Meyerhans et al. (72) and Weiss et al. (73) describe an adapter attachment primer system that allows for the amplification of DNA from small amounts of starting material. This adapter attachment system requires a primer specific for the target that has an adapter tagged onto its five prime end. This primer is used in a first round of PCR reaction. A primer identical to the adapter is then used in second round PCR. Another new PCR method that may be useful in aDNA studies is the whole genome amplification procedure developed by Zhang et al. (74). In this procedure, DNA from a single cell is the starting template in combination with a mixture of 15-base random oligonucleotides. The random oligonucleotides implement the amplification of over 75% of the genome which can then be used in second round amplifications that are specific for particular gene regions.

Sequencing aDNA PCR Products

Once a PCR product has been produced that has passed the scrutiny of the negative and positive controls mentioned above, there are two ways that the sequence information can be obtained from the fragment. The first is to sequence the PCR product directly by standard procedures (47,75). The second is to clone the PCR product (76,77; TA cloning; Invitrogen) and to sequence the clones. The disadvantage to cloning is that errors are introduced into the clones at a very high rate (47,78). When clones are sequenced, at least three per mtDNA or per nuclear allele should be sequenced to obtain the data. The disadvantage to direct sequencing is that the presence of different alleles or chimeric PCR products will produce unresolved regions in the sequence. The problem of chimeric PCR products for damaged DNA has been recognized for some time and has been termed PCR jumping (79-81). Handt et al. (40) suggest that an advantage of cloning is that the chimeric nature of PCR products can only be discovered by sequences from clones.

METHODS FOR VERIFICATION OF aDNA

Lindahl (9) makes the pointed statement that, "DNA might be obtained from pulverized material from the Burgess shale after a sufficient number of PCR cycles, and the DNA might not be identical to any known sequence that has been deposited in data banks, but this does not prove that such DNA was of Cambrian rather than of contemporary origin". This skepticism is an important and healthy starting point for all aDNA studies. Lindahl (9) made the four suggestions mentioned earlier as guidelines for preventing misdiagnosis. While it should be apparent that we agree with using these four guidelines and that the controls that Lindahl (9) suggests are essential, we suggest that a fifth requirement be added to this list. When one is working with fairly divergent specimens, the phylogenetic approach can be used to verify the DNA sequences as authentic (1,40). The phylogenetic verification procedure is a hypothetico-deductive process where the more rigorous challenge to the hypothesis, the less doubt can be placed on the rejection of the hypothesis. This approach implies two important points concerning the components involved in the phylogenetic tests. The first point concerns taxon sampling and implies that very close relatives of the fossil taxon from which the DNA is extracted be used in the phylogenetic analysis. The second concerns the amount of evidence needed and suggests that as many characters as possible be used in the diagnosis. To demonstrate the nuances of verification of aDNA studies, we examine two case studies. The first concerns a situation where the aDNA sequences are identical or extremely similar to contemporary sequences. The second case study concerns an apparent misdiagnosis due to lack of phylogenetic analysis.

Case Study 1. The Tyrolean "Ice-Man"

An excellent example of following the guidelines set out by Lindahl (9) and verification of ancient DNA can be found in the Handt et al. (82) analysis of the Tyrolean "ice man" (Oetzi). DNA was isolated from this 4000-year-old human corpse with ancient DNA techniques. The D-Loop of the mtDNA was targeted as a source for diagnostic characters. Handt et al. (82) suspected that the DNA sequences from available primer pairs in this mitochondrial region would be extremely similar to contemporary human mtDNA sequences and so at the outset of their experiments they were faced with the problem of verifying a DNA sequence that would be similar to sequences of researchers in their laboratory. They also used primer pairs that spanned only short stretches of DNA and characterized the DNA isolated from the "ice man" as highly degraded.

In this example diagnosis by phylogenetic or genealogical analysis gave the result that the "ice man" had mtDNA D-Loop sequences that are similar to those of contemporary Europeans. In certain respects, this diagnosis step is needed to determine that the sequences these researchers obtained are human and not some fungal contaminant as was observed for the Mylodon in a study discussed in Handt et al. (82). However, the strongest evidence for the authenticity of the "ice man's" DNA rests on the careful controls and corroboration of the results in two independent laboratories.

Case Study 2. Dinosaur DNA

Woodward et al. (10) reported the isolation and characterization of DNA from 80-million-year-old bone, presumably dinosaur in origin. Several PCR products were generated from 42 separate extracts and DNA was obtained from two fossil bones. The control experiments for these PCR experiments were adequately documented in the publication, but were not replicated in another laboratory. Woodward et al. (10) then sequenced nine different PCR products generated from the isolated DNA and obtained 133 bases from the mitochondrial cytochrome b gene. No two sequences from these nine PCR products were the same; additionally, a search of the database indicated that these nine sequences were quite different from sequences deposited in the database. Sequences in the database represent "mammals, birds, reptiles, amphibians and insects. In addition, other partial sequences not represented in these databases were obtained in our laboratory for ostrich, rhea, emu, turkey, box turtle, garter snake and alligator". The results of these comparisons suggested that the sequences obtained from these two bones were equally divergent from the existing mammalian sequences, reptilian sequences and bird sequences. The authors used this similarity approach to suggest that their sequences were authentic Cretaceous period sequences.

There are three points to be made concerning phylogenetic analysis that are pertinent to these claims that were discussed at length in a Technical Comment in Science (11-14, and see also Woodward's reply, 83). The first point concerns the fact that the similarity comparisons made by Woodward et al. (10) are not phylogenetic. In fact, when phylogenetic analyses with the use of parsimony, maximum likelihood and neighbor joining are performed, the relationship of the two Cretaceous bone sequences are unambiguous. Hedges and Schweitzer (13) show that the Cretaceous bone sequences are most closely related to human cytochrome b sequences. Thus the first point to be made is that careful phylogenetic analysis of these sequences leaves us with the suggestion that they are most likely related to human and perhaps the result of a contaminant.

The last two of the three points are directly related to taxon sampling and to the number of characters used in the phylogenetic analysis that were discussed in the beginning of this section. The first of these concerns the strength of the phylogenetic inference which, as suggested by Hedges and Schweitzer (13), "is low as a result of the short length of this region (133 base pairs)". The final point concerns more complete sampling in the phylogenetic diagnosis and is addressed by Zischler et al. (14). They point out that Woodward et al. (10) failed to include critical sequences that might have aided in the diagnosis of the Cretaceous bone sequences. Zischler et al. (14) have isolated several clones of cytochrome b sequences that originate from the human nuclear genome. They show that these sequences have somehow been transposed or inserted into the human nuclear genome and are nonfunctional. When a phylogenetic analysis is performed of the Cretaceous bone sequences, the human nuclear inserted cytochrome b sequences and several other mammalian and bird cytochrome b sequences, the Cretaceous bone sequences are found to be most closely related to the nuclear inserted cytochrome b sequences. These results suggest that even though extreme precautions are taken in the PCR of ancient DNA studies, contamination from human sources in this case is a possibility. The Woodward et al. (10) study and its criticisms are an excellent example of the necessity of phylogenetic analysis in aDNA studies.

THE UTILITY OF aDNA STUDIES

Recent advances in ancient DNA have been reviewed in several articles (1,48,84-87). There have also been three international ancient DNA conferences held in the last five years where results and techniques are discussed and debated. These reviews and conferences demonstrate a proliferation of techniques for the isolation of nucleic acids and the application of nucleic acid sequence information from ancient tissue sources in a wide variety of disciplines. Table 7 summarizes the extent of these studies and places them in a temporal and organismal context.

The application of aDNA techniques in forensics is obvious and highly visible. In addition the development of aDNA techniques for embedded tissues, tissues on slides and other sources has opened new areas in the medical sciences, in particular in medical pathology (88). The more sedate organismal sciences have perhaps seen the greatest benefit from development of this technology. Systematics, population genetics, ecology,

Table 7. Summary of Selected aDNA Studies

Age (Years)	Humans	Other vertebrates	Plants	Insects	Bacteria
10^1-10^2	19,42,120-127 139-141,154,155 157,176,177	41,92,93,97,142-151	32,54 166 167 180	30,180	184,185
10^2-10^3	18,19,56,78 123-133,156	55,134,152,153 159,161-163	168-172		
10^3-10^4	82,134-138,158 160	152-154,161,164 165	173-175 178		
10^4-10^5					
10^5-10^6					
10^6-10^7					
10^7-10^8			25-27	22-24,81	
186,187					
			179	180-183	
10^8-10^9					

The abundance of strictly forensic aDNA studies precludes the inclusion of such studies in this table. A few highly visible representative forensic studies are included.

conservation biology archeology and anthropology all have a need for incorporating information from long-dead organisms. The importance of incorporating fossil information in systematic studies has been discussed at length at the morphological level (89,90) and at the molecular level (91). Population genetics studies and conservation studies now routinely incorporate genotypes of long-dead organisms or populations of organisms. The importance of aDNA in population biology has been discussed by Thomas et al. (92) and Villablanca (93). The arguments for the utility of aDNA in conservation genetics are similar to those engendered for population genetics and are articulated by Roy et al. (94), Thomas (29) and Vogler (95). Anthropological applications of aDNA were first discussed by Pääbo et al. (96) and have been expanded recently by Hagelberg (85) and Brown and Brown (87).

The test of time will tell the true utility of aDNA in all of these fields. Whether the use of sequence information from long-dead organisms will remain useful in these fields depends on the ability of these studies to meet and deal with the skepticism that is leveled at aDNA sequences. We feel that the techniques for aDNA isolation are at an advanced stage and that recent improvements in PCR technology have also resulted in great advances. The experience of over ten years of aDNA success (97) has led researchers to develop adequate and sometimes exquisite controls for their experiments. Consequently, the technical aspects of aDNA work are at this time quite adequate. We suggest, however, that since the phylogenetic diagnosis step in aDNA studies is critical, the two points discussed earlier concerning character support and taxonomic sampling are an important part of dealing with skepticism. It is therefore extremely important that we expand our taxonomic base in aDNA studies and begin to develop newer more sensitive probes for gene regions that are suitable as sources for characters in the diagnosis step.

REFERENCES

1 DeSalle, R. and Grimaldi, D.A. (1994) Curr. Opin. Gen. and Dev. 4, 810-815.
2 Lowenstein, J.M., Sarich, V. and Richardson, B.J. (1981) Nature 291, 409-411.
3 Tuross, N. (1993) in Prehistoric Human Bone (Lambert, J.B. and Grupe, G.), pp. 275-292, Springer-Verlag, Berlin.
4 Tuross, N. (1994) Experientia 50, 530-535.
5 Bada, J., Wang, X.S., Poinar, H.N., Pääbo, S. and Poinar, G.O. (1994) Geochimica Cosmochimica Acta 14, 3131-3135.
6 Logan, G.A., Boon, J. and Eglinton, G. (1993) Proc. Nat. Acad. Sci. U.S.A. 90, 2246-2250.
7 Gurley, L.R, Valdez, J.G., Spall, W.D., Smith, B.F. and Gillette, D.D. (1991) J. Prot. Chem. 10, 75-90.
8 Lindahl, T. (1993) Nature 362, 709-715.
9 Lindahl, T. (1993) Nature 365, 700.
10 Woodward, S.R., Weyand, N.J. and Bunnell, M. (1994) Science 266, 1229-1232.
11 Henikoff, S. (1995) Science 268, 1192.
12 Allard, M.W., Young, D. and Huyen, Y. (1995) Science 268, 1192.
13 Hedges, S.B. and Schweitzer, M.H. (1995) Science 268, 1191.
14 Zischler, H., Höss, M., Handt, O., von Haeseler, A., van der Kuyl, A.C., Goudsmit, J. and Pääbo, S. (1995) Science 268, 1192-1193.
15 Henwood, A. (1992) Palaios 7, 203-212.
16 Henwood, A. (1992) Paleontology 35, 901-912.
17 Grimaldi, D., Bonwich, E., Delannoy, M. and Doberstein, S. (1994) Amer. Mus. Nov. 3097, 1-31.
18 Merriwether, D.A., Rothhammer, F. and Ferrell, R.E. (1994) Experientia 50, 592-601.
19 Meijer, H., Perizonius, W.R. and Geraedts, J.P. (1992) Biochem. Biophys. Res. Commun. 183, 367-74.
20 Hagelberg, E. (1994) in Ancient DNA (Herrmann, B. and Hummel, S., eds.), pp. 195-204, Springer-Verlag, New York, NY.
21 Hummel, S. and Herrmann, B. (1994) in Ancient DNA (Herrmann, B. and Hummel, S.), pp. 205-210, Springer-Verlag, New York, NY.
22 Cano, R.J., Poinar, H.N. and Poinar, G.O. (1992) Med. Sci. Res. 20, 249-251.
23 Cano, R.J., Poinar, H.N., Roubik, D. and Poinar, G.O. (1992) Med. Sci. Res. 20, 619-623.
24 DeSalle, R., Gatesy, J., Wheeler, W. and Grimaldi, D. (1992) Science 257, 1933-1936.
25 Golenberg, E.M., Giannasi, D.E., Clegg, M.T., Smiley, C.J., Durbin, M., Henderson, D. and Zurawski, G. (1990) Nature 344, 656-658.
26 Golenberg, E.M. (1991) Phil. Trans. Roy. Soc. Lond. [Biol] 333, 419-426.
27 Soltis, P.S., Soltis, D.E. and Smiley, C.J. (1992) Proc. Nat. Acad. Sci. U.S.A. 89, 449-51.
28 Poinar, G.O., Poinar, H.N. and Cano, R.J. (1994) in Ancient DNA (Herrmann, B. and Hummel, S.), pp. 92-103, Springer-Verlag, New York, NY.
29 Thomas, R.H. (1994) in Molecular Ecology and Evolution: Approaches and Applications (Schierwater, B., Streit, B., Wagner, G.P. and DeSalle, R.) pp. 311-321, Birkhauser-Verlag, Basel.
30 Phillips, A.J. and Simon, C. (1995) Ann. Entom. Soc. 88, 281-283.
31 Dessauer, H.C., Cole, C.J. and Hafner, M.S. (1992) in Molecular Systematics

(Hillis, D.M. and Moritz, C., eds.) pp. 25-42, Sinauer, Sunderland, MA.
32 Sytsma, K., Givnish, T.J., Smith, J.F. and Hahn, W. (1993) in Molecular Evolution: Producing the Biochemical Data (Zimmer, E.A., White, T.J., Cann, R.L. and Wilson, A.C.) pp. 23-38, Academic Press, San Diego, CA.
33 Dick, M., Bridge, D., Wheeler, W.C. and DeSalle, R. (1993) Methods Enzymol. 224, 51-64.
34 Cann, R.L., Feldman, R.A., Freed, L.A., Lum, J.K. and Reeb, C.A. (1993) Methods Enzymol. 224, 38-51.
35 Blackwell, M. and Chapman, R.L. (1993) in Molecular Evolution: Producing the Biochemical Data (Zimmer, E.A., White, T.J., Cann, R.L. and Wilson, A.C., eds.) pp. 65-77, Academic Press, San Diego, CA.
36 Whitfield, J. and Cameron, S. (1994) Mol. Phyl. Evol. 3, 268-271.
37 Lee, W.L., Bell, B.M. and Sutton, J.F. (1982) Assoc. of Syst. Coll., Lawrence, Univ. of Kansas Press.
38 Prince, A.M. and Andrus, L. (1992) BioTechniques 12, 358-60.
39 Tuross, N. and Stathoplos, L. (1993) Methods Enzymol. 224, 121-129.
40 Handt, O., Höss, M., Krings, M. and Pääbo, S. (1994) Experientia 50, 524-529.
41 Höss, M., Kohn, M., Knauer, F., Pääbo, S. and Schroder, W. (1992) Nature 359, 199.
42 Fisher, D.L., Holland, M.M., Mitchell, L., Sledzik, P.S., Wilcox, A.W., Wadhams, M. and Weedn, V.W. (1993) J. Forensic Sci. 38, 60-68.
43 Jupe, E.R. and Zimmer, E.A. (1993) Methods Enzymol. 224, 541-551.
44 Stein, D.B. (1993) Methods Enzymol. 224, 153-167.
45 Chesnik, J.M. and Cattolico, R.A. (1993) Methods Enzymol. 224, 168-175.
46 DeSalle, R., Williams, A.K. and George, M. (1993) Methods Enzymol. 224, 176-203.
47 Thomas, W.K. and Pääbo, S. (1993) Methods Enzymol. 224, 406-418.
48 Hillis, D.M. and Moritz, C. (1992) Molecular Systematics Sinauer, Sunderland, MA.
49 Herrmann, B. and Hummel, S. (1994) Ancient DNA, Springer-Verlag, New York, NY.
50 Hauswirth, W.W., Dickel, C.D. and Lawlor, D.A. (1994) in Ancient DNA (Herrmann, B. and Hummel, S., eds.) pp. 104-121, Springer-Verlag, New York, NY.
51 Nielsen, H., Engberg, J. and Thuesen, I. (1994) in Ancient DNA (Herrmann, B. and Hummel, S., eds.) pp. 122-140, Springer-Verlag, New York, NY.
52 Sensabaugh, G.F. (1994) in Ancient DNA (Herrmann, B. and Hummel, S., eds.) pp. 141-148, Springer-Verlag, New York, NY.
53 Ellgren, H. (1994) in Ancient DNA (Herrmann, B. and Hummel, S., eds.) pp. 211-217, Springer-Verlag, New York, NY.
54 Taylor, J.W. and Swann, E.C. (1994) in Ancient DNA (Herrmann, B. and Hummel, S., eds.) pp. 166-181, Springer-Verlag, New York, NY.
55 Cooper, A. (1994) in Ancient DNA (Herrmann, B. and Hummel, S., eds.) pp. 149-165, Springer-Verlag, New York, NY.
56 Rogan, Salvo, J.J. (1994) in Ancient DNA (Herrmann, B. and Hummel, S., eds.) pp. 182-194, Springer-Verlag, New York, NY.
57 Rollo, F., Venanzi, F.M. and Amici, A. (1994) in Ancient DNA (Herrmann, B. and Hummel, S., eds.) pp. 31-58, Springer-Verlag, New York, NY.
58 Goodyear, P.D., MacLaughlin-Black, S. and Mason, I.J. (1994) BioTechniques 16, 232-235.
59 Walsh, P.S., Metzger, D.A. and Higuchi, R. (1991) BioTechniques 10, 506-513.

60 Hanni, C., Brousseau, T., Laudet, V. and Stehelin, D. (1995) Nucl. Acids Res. 2379, 881-882.
61 Kocher, T.D., Thomas, W.K., Meyer, A., Edwards, S.V., Pääbo, S., Villablanca, F.X. and Wilson, A.C. (1989) Proc. Nat. Acad. Sci. U.S.A. 86, 6196-6200.
62 Kessing, B., Croom, H., Martin, A., McIntosh, C., McMillan, W.O. and Palumbi, S. (1989) Simple Fools Guide to PCR. Univ. of Hawaii.
63 Simon, C., Frati, F., Beckenbach, A., Crespi, B., Liu, H. and Flook, P. (1994) Ann. Entom. Soc. 87, 651-701.
64 Bachellerie, J. and Qu, L. (1993) Methods Enzymol. 224, 349-356.
65 Wheeler, W.C., Cartwright, P. and Hayashi, C.Y. (1993) Cladistics 9, 1-39.
66 Brower, A.V.Z. and DeSalle, R. (1994) Ann. Entom. Soc. 87, 702-716.
67 Wray, C.G. and DeSalle, R. (1994) Mol. Mar. Biol. Biotechnol. 3, 13-22.
68 Hillis, D.M. and Dixon, M.T. (1992) Quant. Rev. Bio. 6, 411-53.
69 Uchihi, R., Yamamoto, T., Kojima, T., Tamaki, K. and Katsumata, Y. (1994) Nippon Hoigaku Zasshi 48, 329-335.
70 Honda, K., Sugiyama, K.E., Tsuchikane, A., Katsuyama, Y., Harashima, N., Ota, M. and Fukushima, H. (1994) Nippon Hoigaku Zasshi 48, 156-160.
71 Chen, F. and Suttle, C.A. (1995) BioTechniques 18, 609-612.
72 Meyerhans, A., Vartanian, J.P. and Wain-Hobson, S. (1992) Nucl. Acids Res. 20, 521-523.
73 Weiss, K.M., Buchanan, A.V., Daniel, C. and Stoneking, M. (1994) Hum. Biol. 66, 789-804.
74 Zhang, L., Cui, X., Schmitt, K., Hubert, R., Navidi, W. and Arnheim, N. (1992) Proc. Nat. Acad. Sci. U.S.A. 89, 5847-5851.
75 Thomas, W.K. and Kocher, T.D. (1993) Methods Enzymol. 224, 391-398.
76 Maniatis, T., Frisch, E. and Sambrook, J. (1982) Molecular Cloning. Cold Spring Harbor Laboratory, Cold Spring Harbor, NY.
77 Mitchell, D.B., Ruggli, N. and Tratschin, J.D. (1992) PCR Methods Appl. 2, 81-82.
78 Pääbo, S. (1989) Proc. Nat. Acad. Sci. U.S.A. 86, 1939-1943.
79 Pääbo, S., Irwin, D.M. and Wilson, A.C. (1990) J. Biol. Chem. 265, 4718-4721.
80 Lawlor, D.A., Dickel, C.D., Hauswirth, W.W. and Parham, P. (1991) Nature 349, 785-788.
81 DeSalle, R., Barcia, M. and Wray, C. (1993) Experientia 49, 906-909.
82 Handt, O., Richards, M., Trommsdorff, M., Kilger, C., Simanainen, J., Georgiev, O., Bauer, K., Stone, A., Hedges, R., Schaffner, W., Utterman, G., Sykes, B., and Pääbo, S. (1994) Science 264, 1775-1778.
83 Woodward, S.R. (1995) Science 268, 1194.
84 Pääbo, S. (1993) Sci. Amer. 269, 86-92.
85 Hagelberg, E. (1993) The Biochemist August, 17-22.
86 Hauswirth, W.W. (1994) Experientia 50, 521-524.
87 Brown, T.A. and Brown, K.A. (1994) BioEssays 16, 719-726.
88 Pan, L.X., Diss, T.C. and Isaacson, P.G. (1995) Histopathology 26, 201-217.
89 Donoghue, M., Doyle, J., Gauthier, J., Kluge, A. and Rowe, T. (1989) Annu. Rev. Ecol. Syst. 20, 431-460.
90 Novacek, M. (1992) in Extinction and Phylogeny (Novacek, M. and Wheeler, Q., eds.) pp. 46-88, Columbia University Press, New York, NY.
91 DeSalle, R. (1994) Experientia 50, 543-550.
92 Thomas, W.K., Pääbo, S., Villablanca, F.X. and Wilson, A.C. (1990) J. Mol. Evol. 31, 101-112.
93 Villablanca, F. (1994) in Ancient DNA (Herrmann, B. and Hummel, S., eds.) pp. 31-58, Springer-Verlag, New York, NY.

94 Roy, M.S., Girman, D.J., Taylor, A.C. and Wayne, R.K. (1994) Experientia, 50, 551-557.
95 Vogler, A.P. (1994) in Molecular Ecology and Evolution: Approaches and Applications (Schierwater, B., Streit, B., Wagner, G.P. and DeSalle, R., eds.) pp. 261-274, Birkhauser-Verlag, Basel.
96 Pääbo, S., Higuchi, R.G. and Wilson, A.C. (1989) J. Biol. Chem. 264, 9709-9712.
97 Higuchi, R., Bowman, B., Freiberger, M., Ryder, O.A. and Wilson, A.C. (1984) Nature 312, 282-284.
98 Petrishchev, V.N., Kutueva, A.B. and Rychkov, V. (1993) Genetika 29, 690-693.
99 Woodward, S.R., King, M.J., Chiu, N.M., Kuchar, M.J. and Griggs, C.W. (1994) PCR Methods Appl. 3, 244-247.
100 Ginther, C., Issel-Tarver, L. and King, M.C. (1992) Nature Genet. 2, 135-138.
101 Simpson, T.A. and Smith, R.J. (1995) Laryngoscope 105, 28-34.
102 Kaneshige, T., Takagi, K., Nakamura, S., Hirasawa, T., Sada, M. and Uchida, K. (1992) Nucl. Acids Res. 20, 5489-5490.
103 Wilson, M.R., Polanskey, D., Butler, J., DiZinno, J.A., Replogle, J. and Budowle, B. (1995) BioTechniques 18, 662-669.
104 Banerjee, S.K., Makdisi, W.F., Weston, A.P., Mitchell, S.M. and Campbell, D.R. (1995) BioTechniques 18, 768-770.
105 Yap, E.P.H. and McGee, J.O'D. (1991) Nucl. Acids Res. 19, 4294.
106 Akane, A., Matsubara, K., Nakamura, H., Takahashi, S. and Kimura, K. (1994) BioTechniques 16, 235-238.
107 Klimyuk, V.I., Carroll, B.J., Thomas, C.M. and Jones, J.D. (1993) Plant J. 3, 493-494.
108 Stewart, C.N. and Via, L.E. (1993) BioTechniques 14, 748-750.
109 Wang, K., Gan, L., Boysen, C. and Hood, L. (1995) Anal. Biochem. 226, 85-90.
110 Panaccio, M., Georgesz, M. and Lew, A.M. (1993) BioTechniques 14, 238-243.
111 Zeillinger, R., Schneeberger, C., Speiser, P. and Kury, F. (1993) BioTechniques 14, 202-203.
112 Greer, C.E., Wheeler, C.M. and Manos, M.M. (1994) PCR Methods Appl. 3, S113-122.
113 Gregory, C.A., Myal, Y. and Shiu, R.P. (1995) BioTechniques 18, 758-760.
114 Shibata, D., Martin, W.J. and Arnheim, N. (1988) Cancer Res. 48, 4564-4566.
115 Hubert, R., Weber, J.L., Schmitt, K., Zhang, L. and Arnheim, N. (1992) Amer. J. Hum. Genet. 51, 985-991.
116 Tan, A.M. and Orrego, C. (1992) Mol. Ecol. 1, 195-197.
117 Boom, R., Sol, C.J.A., Salimans, M.M.M., Jansen, C.L., Wertheim-VanDillen, P.M.E. and Van Der Noordaa, J. (1990) J. Clin. Microbiol. 28, 495-503.
118 Höss, M. and Pääbo, S. (1993) Nucl. Acids Res. 21, 3913-3914.
119 Höss, M. (1994) Ancient DNA Newsletter 2, 10-12.
120 Hochmeister, M.N., Budowle, B., Borer, U.V., Eggmann, U., Comey, C.T. and Dirnhofer, R. (1991) J. Forensic Sci. 36, 1649-1661.
121 Yokoi, T., Aoki, Y. and Sagisaka, K. (1991) Z. Rechtsmed. 102, 323-330.
122 Hanni, C., Laudet, V., Sakka, M., Begue, A. and Stehelin, D. (1990) C. R. Acad. Sci. III 310, 365-370.
123 Kurosaki, K., Matsushita, T. and Ueda, S. (1993) Amer. J. Hum. Genet. 53, 638-643.
124 Hummel, S. and Herrmann, B. (1991) Naturwissenschaften 78, 266-267.
125 Hummel, S., Nordsiek, G. and Herrmann, B. (1992) Naturwissenschaften 79, 359-360.

126 Hagelberg, E. and Clegg, J.B. (1991) Proc. Roy. Soc. Lond. B 244, 45-50.
127 Stone, A.C. and Stoneking, M. (1993) Amer. J. Phys. Anthropol. 92, 463-471.
128 Rogan, P. and Salvo, J. (1991) Yrbk. Phys. Anthrop. 33, 195-214.
129 Hagelberg, E. and Sykes, B. (1989) Nature 342, 485.
130 Pääbo, S. (1985) Nature 314, 644-645.
131 Horai, S., Hayasaka, K., Muryama, K., Wate, N., Koike, H. and Nakai, N. (1989) Proc. Japan. Acad. B 65, 229-233.
132 Horai, S., Kondo, R., Murayama, K., Hayashi, S., Koike, H. and Nakai, N. (1991) Phil. Trans. Roy. Soc. Lond. [Biol] 333, 409-417.
133 Hagelberg, E. and Clegg, J.B. (1993) Proc. Roy. Soc. Lond. B Biol. Sci. 252, 163-170.
134 Lister, A.M. (1994) Trends Ecol. Evol. 9, 82-84.
135 Pääbo, S., Gifford, J.A. and Wilson, A.C. (1988) Nucl. Acids Res. 16, 9775-9787.
136 Doran, G.H., Dickel, D.N., Ballinger, W.E., Agee, O.F., Laipis, P.J. and Hauswirth, W.W. (1986) Nature 323, 803-806.
137 Hauswirth, W.W., Dickel, C.D., Rowald, D.J. and Hauswirth, M.A. (1994) Experientia 50, 585-591.
138 Merriwether, D.A., Rothammer, F. and Ferell, R.E. (1992) Amer. J. Hum. Genet. 51, A13.
139 Hagelberg, E., Quevedo, S., Turbon, D. and Clegg, J.B. (1994) Nature 369, 25-26.
140 Gibbons, A. (1994) Science 263, 32-33.
141 Holland, M.M., Fisher, D.L., Mitchell, L.G., Rodriquez, W.C., Canik, J.J., Merril, C.R. and Weedn, V.W. (1993) J. Forensic Sci. 38, 542-553.
142 Wayne, R. and Jenks, S.M. (1991) Nature 351, 565-568.
143 Pääbo, S. and Wilson, A.C. (1988) Nature 338, 774.
144 Wayne, R.K., Lehman, N., Allard, M.W. and Honeycutt, R.L. (1992) Conserv. Biol. 6, 559-569.
145 Roy, M.S., Geffen, E., Smith, D., Ostrander, E.A. and Wayne, R.K. (1994) Mol. Biol. Evol. 11, 553-570.
146 Wayne, R.K., Meyer, A., Lehman, N., Van Valkenburgh, B. and Kat, P.W. (1990) Proc. Nat. Acad. Sci. U.S.A. 87, 1772-1776.
147 Taylor, A.C., Sherwin, W.B. and Wayne, R.K. (1994) Mol. Ecol. 3, 277-290.
148 Houde, P. and Braun, M.J. (1988) Auk 105, 773-776.
149 Ellgren, H. (1991) Nature 354, 113.
150 Krajewski, C., Driskell, A.C., Baverstock, P.R. and Braun, M.J. (1992) Proc. Roy. Soc. Lond. B Biol. Sci. 250, 19-27.
151 Thomas, R.H., Schaffner, W., Wilson, A.C. and Pääbo, S. (1989) Nature 340, 465-467.
152 Hardy, C., Casane, D., Vigne, J.D., Callou, C., Dennebouy, N., Mounolou, J.C. and Monnerot, M. (1994) Experientia 50, 564-570.
153 Hardy, C., Callou, C., Vigne, J.D., Casane, D., Dennebouy, N., Mounolou, J.C. and Monnerot, M. (1995) J. Mol. Evol. 40, 227-237.
154 Hagelberg, E., Thomas, M.G., Cook, C.E., Sher, A.V., Baryshnikov, G.F. and Lister, A.M. (1994) Nature 370, 333-334.
155 King, M.C. (1991) Mol. Genet. Med. 1, 117-131.
156 Gaensslen, R.E., Berka, K.M., Herrin, G., Mishler, R.E., Sveum, L.K. and Lee, H.C. (1993) Naturwissenschaften 80, 80-81.
157 Schweitzer, R.R. (1995) Nature Genet. 9, 345.
158 Benditt, J. (1989) Sci. Amer. 261, 25-26.

159 Hanni, C., Laudet, V., Stehelin, D. and Taberlet, P. (1994) Proc. Nat. Acad. Sci. U.S.A. 91, 12336-12340.
160 Del Pozzo, G. and Guardiola, J. (1989) Nature 339, 431-432.
161 Hardy, C., Vigne, J.D., Casane, D., Dennebouy, N., Mounolou, J.C., Monnerot, M. (1994) J. Evol. Biol. 7, 217-226.
162 Cooper, A., Mourer-Chauvire, C., Chambers, G.K., von Haeseler, A., Wilson, A.C. and Pääbo, S. (1992) Proc. Nat. Acad. Sci. U.S.A. 89, 8741-8744.
163 Cooper, A. (1994) Experientia 50, 558-563.
164 Johnson, P.H., Olson, C.B. and Goodman, M. (1985) Comp. Biochem. Physiol. [B] 81, 1045-1051.
165 Janczewski, D.N., Yuhki, N., Gilbert, D.A., Jefferson, G.T. and O'Brien, S.J. (1992) Proc. Nat. Acad. Sci. U.S.A. 89, 9769-9773.
166 Bruns, T.D., Fogel, R. and Taylor, J.W. (1990) Mycologia 82, 175-184.
167 Miller, O.K. (1990) Mycol. Soc. Amer. News 41, 29-30.
168 Rollo, F., Venanzi, F.M. and Amici, A. (1991) Genet. Res. 58, 193-201.
169 Goloubinoff, P., Pääbo, S. and Wilson, A.C. (1993) Proc. Nat. Acad. Sci. U.S.A. 90, 1997-2001.
170 Brown, T.A., Allaby, R.G., Brown, K.A. and Jones, M.K. (1993) World Archaeol. 25, 64-73.
171 Brown, T.A., Allaby, R.G., Brown, K.A., O'Donoghue, K. and Sallares, R. (1994) Experientia 50, 571-575.
172 Venanzi, F.M. and Rollo, F. (1990) Nature 343, 25-26.
173 Rogers, S.O. and Bendich, A.J. (1985) Plant Molec. Biol. 5, 69-76.
174 O'Donoghue, K., Brown, T.A., Carter, J.F. and Evershed, R.P. (1994) Rapid Commun. Mass Spectrometry 8, 503-508.
175 Allaby, R.G., Jones, M.K. and Brown, T.A. (1994) Antiquity 68, 126-132.
176 Gill, P., Ivanov, P.L., Kimpton, C., Piercy, R., Benson, N., Tully, G., Evett, I., Hagelberg, E. and Sullivan, K. (1994) Nature Genet. 6, 130-135.
177 Jeffreys, A.J., Allen, M.J., Hagelberg, E. and Sonnberg, A. (1992) Forensic Sci. Internat. 56, 65-76.
178 Rollo, F., Asci, W., Antonini, A., Marota, I. and Ubaldi, A. (1994) Experientia 50, 576-584.
179 Poinar, H.N., Poinar, G.O. and Cano, R.J. (1993) Nature 363, 677.
180 Cano, R.J. and Poinar, H.N. (1993) BioTechniques 15, 432-434, 436.
181 Poinar, G.O. (1994) Experientia 50, 536-542.
182 Hoelzel, R. (1992) PhD thesis, University of Cambridge, England.
183 Cano, R.J., Poinar, H.N., Pieniazek, N.J., Acra, A. and Poinar, G.O. (1993) Nature 363, 536-538.
184 Persing, D.H., Telford, S.R., Rys, P.N., Dodge, D.E., White, T.J., Malawista, S.E. and Spielman, A. (1991) Science 249, 1420-1423.
185 Salo, W.L., Aufderheide, A.C., Buikstra, J. and Holcomb, T.A. (1994) Proc. Nat. Acad. Sci. U.S.A. 91, 2091-2094.
186 Cano, R.J., Borucki, M. and Higby-Schweizer, M. (1994) Appl. Environ. Microbiol. 60, 2164-2168.
187 Cano, R.J. and Borucki, M.K. (1995) Science 268, 1060-1064.
188 Baker, C.S. and Palumbi, S.R. (1994) Science 265, 1538-1539.

GROWTH FACTORS AND NEURAL CONNECTIVITY

Sarah McFarlane and Christine E. Holt

Department of Biology, 0366
University of California, San Diego
La Jolla, CA 92093-0366

INTRODUCTION

The formation of connections in the developing nervous system requires axons to navigate through complex environments to reach their targets. It is the growth cone, a sensory and motor apparatus located at the tip of the developing axon, that senses and responds to the environment. In so doing the growth cone makes decisions about where to grow and when to stop and innervate. Molecular cues in the local milieu of the growth cone control these behaviors. A number of different classes of signaling molecules are involved in axon guidance including cell-cell and cell-substrate adhesion molecules, extracellular matrix (ECM) molecules and their receptors, and soluble factors (see reviews 1-4). Growth factors are one group of soluble molecules that promote neurite outgrowth *in vivo* and *in vitro*. Research on growth factors has focused primarily on their role in promoting neuronal survival and differentiation. Only recently have studies suggested that they play a much broader role in patterning neuronal connections during development. This review will discuss recent findings that implicate growth factors in axon guidance and target recognition.

Understanding the cues controlling axonal navigation and target recognition is not as simple as listing molecules and the effects they have. Rather, integrated influences of numerous molecules impinging on the growth cone direct the behavior of the axon tip. Recently, several diffusible molecules have been shown to affect growth cone behavior by attracting or repulsing axons (see reviews 5-7). Because the combined action of attractive and repulsive signals decide the growth cone's actual response, the nature of a particular cue is critically dependent on the "cast" of molecules with which the molecule is presented. An axon growing towards its target may encounter the same molecule at several points but may respond differently to the cue depending on where along the pathway it is sampled.

An additional consideration is that growth cones of different neurons will not necessarily respond in the same manner to a given molecular cue. For example, netrin-1, a diffusible guidance factor expressed by the spinal cord floor plate, attracts spinal commissural axons *in vitro* (8,9) but repels trochlear motor axons (10). Thus, it is not the particular signaling molecule but how the growth cone responds to it that is important for axon guidance. Proper connectivity could not arise if growth cones in overlapping or

Genetic Engineering, Vol. 18, Edited by J.K. Setlow
Plenum Press, New York, 1996

intersecting pathways responded similarly to the cues in their environment. Differential responsiveness provides specificity critical in producing the precise network of neuronal connections that comprise a functional nervous system.

GROWTH FACTORS AND THEIR RECEPTORS

The main growth factor families that appear to be involved in establishing and maintaining connections in the developing nervous system are the fibroblast growth factor (FGF) family and the neurotrophin family. A number of parallels exist between the two families (see reviews 11-14). First, both families consist of several members with distinct but related functions. Second, they act *via* high affinity membrane bound receptor tyrosine kinases (RTKs) to ultimately alter the phosphorylation state of key cellular proteins. Third, they are promiscuous. A certain degree of cross-talk exists between the different growth factors and their receptors, indicating that specificity of function is likely a result of precise temporal and spatial regulation of the expression of the growth factor and its receptor. Fourth, the high affinity tyrosine kinase receptors require a low affinity co-receptor for function. The co-receptor is usually membrane bound and has no tyrosine kinase activity.

Fibroblast Growth Factors

At present, the FGF family consists of 9 members, of which acidic FGF (aFGF or FGF-1) and basic FGF (bFGF or FGF-2) are best characterized. These two FGFs promote proliferation, survival and differentiation of numerous neuronal types (11,15). Three different types of FGF receptor (FGFR) have been identified (14): a family of membrane RTKs, which serve as high affinity FGF receptors, the low affinity heparan sulfate proteoglycans (HSPGs) and a cysteine rich FGFR (16). The activated FGFR is thought to exist as a ternary complex of an RTK, an HSPG and an FGF. Thus the biological activity of bFGF depends on its ability to bind heparan sulfate on the cell surface or in the ECM (17-20). The high affinity family of RTKs consists of at least four receptor populations, several of which can be activated *in vitro* by different FGFs (14,17).

Neurotrophins

The neurotrophin family consists of nerve growth factor (NGF), brain-derived neurotrophic factor (BDNF), and neurotrophins 3 and 4/5 (NT-3 and NT-4/5) (reviewed in 12,21). Neurotrophins and their receptors have been implicated in many aspects of the establishment of synaptic connectivity, and all promote the survival and differentiation of sensory neurons (12,21,22). These growth factors act *via* high affinity membrane RTKs called trks (13,23). For the most part, trks are selective for different neurotrophins, with BDNF binding trk B, NGF trk A, and NT-3 trk C. However, the picture is complicated by NT-3 which can also activate trk A and trk B, and NT-4/5 which can activate trk B and trk C. The neurotrophin-trk interaction involves a co-receptor, $p75^{NGFR}$, a transmembrane glycoprotein which can act as a low affinity receptor on its own. $p75^{NGFR}$ potentiates the binding of all the neurotrophins to their receptors (24).

APPROACHES TO THE STUDY OF GROWTH FACTORS IN AXONAL PATHFINDING AND TARGETING

For a growth factor to be said to play a role in neuronal connectivity it must meet a number of criteria. The growth factor must be expressed in a temporally and spatially meaningful manner as demonstrated by immunocytochemistry or *in situ* hybridization. In addition, the growth factor must have a biological effect on the neurons in question. Experimentally, this involves showing either that the growth factor promotes neurite outgrowth in culture, or that it acts as a chemotropic molecule in a chemoaffinity assay. One of the most successful chemoaffinity assays was pioneered by Lumsden and Davies (25), and involves placing beside each other in a collagen matrix a neuronal tissue explant and an explant of its target tissue. The collagen provides a permissive substrate for axon growth and stabilizes any gradient of a molecule secreted by the target. This assay has been used to identify target-derived chemoattractant and chemorepulsive molecules (see reviews 4,6,7).

However, the strictest criterion requires that guidance or target innervation be affected by disturbing growth factor signaling. *In vivo* technical advances have started to yield interesting experimental results. One powerful approach is genetic manipulation with the use of transgenic mice in which the expression of either the growth factor or its receptor is altered. Current experiments are utilizing a knock-out approach, in which a null mutation is generated, thereby eliminating either the growth factor or the receptor gene product. Knock-out mice for all of the neurotrophins and their receptors have been generated (26-34), and certain phenotypes observed in these mice suggest a role for neurotrophins in axonal navigation (see section on axon pathfinding).

One serious problem with such knock-out experiments is that the growth factors are often necessary for the survival of the neurons in question. One can hardly study a role for a growth factor in neuronal connectivity if the cells die. An alternative approach is to overexpress or misexpress the growth factor or the receptor in a novel cell type. Since growth factors are involved in many aspects of development, the most elegant and useful experiments are those where expression of the transgene is under the control of a tissue or cell type-specific heterologous promoter. For instance, NGF has been expressed in sympathetic neurons and the effects on pathfinding and target recognition analyzed (see section on target recognition and 35).

Non-genetic approaches to studying the role of growth factor signaling in neural connectivity are also available. The most common is a form of overexpression, in which growth factors are injected or infused *in vivo*, and the effects on neurite growth analyzed (36-38). Our laboratory developed two additional approaches to investigate guidance and targeting cues, which we are using to examine a role for growth factor signaling in the establishment of the retinotectal pathway in the frog *Xenopus laevis*. The "exposed brain preparation", a twist on the injection approach, was developed to examine the effects of exogenous application of interesting molecules to the developing optic projection (39-41). Retinal ganglion cell (RGC) axons grow 2 to 10 μm below the surface of the brain. Therefore, by removing the skin and dura from one side of the embryonic *Xenopus* brain, retinal axons growing *in vivo* can be exposed to a relevant drug. The exposed brain approach serves as a useful screen for molecules involved in axon pathfinding and/or target recognition, and has been used to implicate bFGF in axonal targeting (see section on target recognition and 42).

A complementary and very powerful approach involves transfecting individual RGCs *in vivo* with DNA constructs encoding mutant proteins (43). At stage 19 (neurula stage), DNA and a lipofection agent (DOTAP) are injected into the developing neuroepithelium which gives rise to the eye. Embryos develop until stage 40 (tailbud stage) when many RGC axons have innervated their target. Transfected cells are then identified by immunostaining. This transfection approach has been used to investigate the roles of ß1-integrin (43) and the cell-adhesion molecule N-cadherin (44) in axonal guidance *in vivo*.

In these studies, the functions of the endogenous molecules were perturbed or inhibited with the use of mutant constructs. Alternatively, it is possible to overexpress molecules in retinal axons with this technique. We have started to use such approaches to manipulate FGF signaling and investigate effects on axonal guidance and targeting (45).

THE FORMATION OF NEURONAL CONNECTIONS

The journey an axon makes from its birthplace to its target has many steps, and growth factors may influence the growth cone machinery at any point along this journey. The first step is the initiation and subsequent outgrowth of an axon to the appropriate part of the nervous system. En route the axon reaches points where it must choose in which direction to grow. There is evidence that the morphology of the growth cone becomes more complex at these choice points as it feels out its environment (46-48). The axon is typically guided to the target along a highly stereotyped route. Once the growth cone arrives at the target, it must then recognize and innervate this region. Here we want to distinguish between axonal pathfinding, which entails guiding the axon along the correct pathway to the proper target, and target recognition, which involves the axon growing into and innervating its target. In addition, we limit the definition of target recognition to those early signaling events which trigger an axon to slow and arborize in the target tissue. Later innervation events include synaptogenesis, followed by activity-dependent rearrangement of these initial connections.

We will now discuss examples where growth factor signaling has been perturbed *in vivo*, and the insights such experiments provide towards an understanding of the roles growth factors play in each of the events of neuronal connectivity.

AXON INITIATION AND GROWTH

In vitro

A hallmark of growth factors is their ability to promote neurite outgrowth. Most of the work on growth factor-stimulated neurite outgrowth has been conducted in tissue culture. These experiments have produced a list of neurons for which each growth factor promotes neurite extension. However, it is difficult to separate effects on neurite outgrowth from effects secondary to the growth factor's ability to promote neuronal survival (49). With this caveat, both aFGF and bFGF have neurite-promoting activities for several distinct populations of CNS neurons (49-54). In some cases FGF's effect on neurite outgrowth can be separated from its effects on neuronal survival (50). Neurotrophins can also stimulate neurite outgrowth in a non-trophic fashion (55-58). In fact, NGF was the first molecule described that clearly promoted neurite extension (reviewed in 59). Other growth factor families also promote neurite extension *via* a non-trophic mechanism. Transforming growth factor-ß1 and -ß2 (TGF-ß1; TGF-ß2) induce neurite sprouting and increase process length of cultured rat hippocampal neurons (60). Similarly, interleukin-2 (IL-2) enhances neurite outgrowth in cultures of rat hippocampal neurons (60) and chick and rat sympathetic neurons (61).

In vivo

In vivo studies in rats have shown that exogenous NGF promotes neurite outgrowth in both the developing central (CNS) and peripheral (PNS) nervous systems (36-38,62). For instance, NGF injected into the rat spinal cord *in utero* appears to enhance the rate of extension of dorsal root ganglion (DRG) axons, with collaterals extending almost twice as far as normal in the gray matter (38). In the adult rat CNS, NGF supplied exogenously after nerve transection results in enhanced axon regeneration (63,64). It is important to note that in both the injection and the regeneration studies effects on axon extension may be secondary to enhancement of cell survival.

FGFR May Transduce CAM-Stimulated Neurite Outgrowth

Doherty and colleagues have put forth the intriguing hypothesis that cell adhesion molecule (CAM)-stimulated neurite extension is mediated *via* the FGFR (62). The signaling pathways of both CAMs and the FGFR are blocked by the same drugs. In addition, a region of the extracellular portion of the FGFR called the CAM homology domain (65) has stretches which are very similar to extracellular regions of the different CAMs: L1, N-CAM and N-cadherin. To test the idea that CAM-stimulated neurite outgrowth requires the FGFR, antibodies were raised against the FGFR CAM homology domain. These antibodies were shown by ELISA not to recognize the different CAMs. However, they were able to almost completely block CAM-stimulated neurite outgrowth of P1 cortical neurons growing on 3T3 cells engineered to express a particular CAM. FGF does not appear to be necessary for CAM signaling through the FGFR. A critical test of FGFR involvement in CAM signaling remains to be done; a dominant-negative form of the FGFR (66), which blocks endogenous FGFR function, should also block CAM dependent outgrowth. Though not yet definitive, these studies raise the possibility that the FGFR may not only mediate growth factor signaling, but in certain circumstances CAM signaling as well.

AXON PATHFINDING

Axons take stereotyped routes to their targets, presumably by reading specific instructions on a molecular map. Steering could be accomplished by several means, including orientation along chemotropic gradients, selective fasciculation with other axons, or interactions between the growth cone and cell-surface or ECM molecules.

Chemotropism

The discovery of NGF provided the first candidate for a target-derived diffusible factor that could attract axons. Both sympathetic and sensory neurons require NGF for survival and innervate NGF-expressing targets (12,67). Additionally, the growth cones of chick DRG neurons *in vitro* turn towards a gradient of NGF, though high concentrations are required for this phenomenon (68), and intracerebral injections of NGF promote sympathetic nerve growth into the developing rat CNS (37). However, more recent studies have argued strongly against a role for NGF in chemoattraction towards target tissues; in the trigeminal sensory system NGF is not expressed in the target until after the first axons have arrived, and developing axons only respond to NGF after they reach their target (22,69).

Chemoattraction as a guidance mechanism has gained acceptance with the recent discovery and cloning of the first chemotropic molecules, netrin-1 and netrin-2, by Tessier-Lavigne and co-workers (8,9). Netrin-1 is secreted by the floor plate, an intermediate target of dorsal commissural interneurons, and is chemotropic for these interneurons in a chemoaffinity assay. A similar chemotropic function has, in preliminary experiments, been attributed to NT-3 by O'Leary and colleagues (70). They showed that projections of cortical layer 5 neurons to subcortical targets such as the basilar pons result from sprouting of branches from the primary axon shaft (71). By co-culturing basilar pons and cortical axons in collagen matrices, they have established that a factor secreted by the basilar pons increases branch initiation along the axon shaft, and directs the axons towards the pons (72,73). This group has gone on to show that NT-3 has a similar effect in an *in vitro* chemoaffinity assay (70). Furthermore, NT-3 injected into the neonatal rat spinal cord promotes collateral sprouting by corticospinal axons, suggesting that NT-3 could have a chemotropic role in other systems as well (74).

Because neurotrophins are often expressed by target cells, recently generated null mutant mice that lack neurotrophins or their receptors have significant potential as systems to investigate chemotropic roles for these growth factors (reviewed in 34,75). Analysis of these animals has concentrated on a survival role for neurotrophins, primarily because loss of neuronal populations is the main phenotype observed. Targeting of gene expression with the use of cell-specific promoters may be necessary to define roles of neurotrophins in axon guidance and targeting. Nonetheless, the BDNF knock-out mice provide tantalizing evidence that BDNF may attract axons of the vestibular ganglion to their targets in the sensory epithelium which normally express BDNF (28). In BDNF knock-out mice there is a massive reduction in the number of cells within the vestibular ganglion, yet some cells survive and send out axons (28). These axons pathfind correctly to the inner ear, but fail to innervate the vestibular sensory apparatus, and instead terminate prematurely in the adjacent connective tissue.

Growth factors could also be acting as chemoattractants to steer growth cones to intermediate targets. In the insect nervous system, guidepost cells located at strategic sites act to direct growing axons to their ultimate targets (76). While there is no evidence for solitary guidepost cells in vertebrates, groups of cells expressing a particular growth factor may function similarly. For instance, mRNA for trk A, the high affinity NGF receptor, is expressed in sensory ganglia at the same time that NGF mRNA is expressed in the peripheral somites and spinal cord (77). Expression in the somites is transient, indicating that NGF-dependent sensory axons growing through these structures on the way to their target may use them as guideposts.

Finally, several growth factors have been shown to direct neurite outgrowth in studies where fibroblasts, genetically altered to express a growth factor, are transplanted into rat brains (78-80). NGF-expressing grafts placed in the brain lead to the sprouting of cholinergic neurons towards the transplanted cells (78), and in the spinal cord cause ingrowth of sensory neurites (79). In addition, BDNF-expressing grafts promote sustained and directed growth from embryonic retina that have been co-grafted with the fibroblasts to the cerebral cortex of neonatal rats (80).

While it is a satisfying idea that the target molecules necessary for neuronal survival (reviewed in 12) are those that attract their axons to the target, the evidence to date is only suggestive of a chemotropic guidance function for growth factors. To test critically whether a growth factor acts as a chemoattractant, future transgenic studies must show that elimination of the growth factor prevents axons from finding their target. Since the growth factor is often required for survival of the neurons, it will probably be necessary to study axon trajectories before the axons have reached their target. An alternative transgenic approach is to express the null mutant under the control of an inducible promoter, limiting the knock-out to the period of axon growth.

Substrate Preference

Thus far we have discussed the possibility that growth factors act on axon tips and direct their growth. Growth factors could also control axonal navigation by affecting the substrate preference of axons, influencing either ECM molecules or the receptors that recognize them. For instance, bFGF upregulates astrocytic tenascin expression (81,82), an ECM molecule that in the visual system is inhibitory to axon growth (83,84). The ability of neurotrophins to influence the substrate preference of DRG neurites was examined recently in a cryostat section choice assay (85). A DRG neuron is unique in that it sends one axon into peripheral target tissue and a second axon into the CNS. To test the possibility that growth factors are important in allowing DRG axons to view both areas as permissive, DRG explants were placed between sections of CNS (neonatal spinal cord) and PNS (sciatic nerve) tissue. In the absence of neurotrophins neurites grew almost exclusively on the CNS tissue. When NGF was present neurites grew equally well on both tissues, as *in vivo*. NT-3 and BDNF also permitted growth in the PNS, but had a much weaker influence. The authors demonstrated that NGF acted through a non-trophic mechanism, and not *via* survival of PNS-selective neurons. In cultures where NGF was added only after 24 to 36 hours, allowing time for the death of NGF-dependent neurons, neurites showed no tissue preference, just as when NGF was present from the beginning. The possibility that NGF was acting on the substrate itself was eliminated by experiments which showed that PNS sections, pre-treated with NGF, were unable to support neurite outgrowth. Thus, insensitivity of neurites to PNS inhibitors is dependent on NGF. This could involve *de novo* extension of neurites that are PNS selective, or NGF could switch the ability of an axon's growth cone to grow in a previously repellent area.

Additional evidence for growth factor modulation of substrate preference comes from studies with migrating cells, which face many of the same guidance problems as extending axons. For instance, bFGF regulates the expression of ß1-integrin by migrating neuroepithelial cells (86). Integrins are receptors for ECM molecules such as laminin and collagen, which stimulate axon extension (87-89). In addition, TGF-ß increases the substrate adhesive properties of neural crest cells (90). Remodeling of the ECM or alterations in the ability of an axon's growth cone to respond to its environment would both be powerful ways by which a growth factor might affect an axon's behavior.

Fasciculation

One mechanism for axonal steering is for ingrowing axons to fasciculate with axons already present in the pathway using them as guides. For example, in the developing insect CNS axons grow along fascicles, or bundles of axons, that have preceded them. If the axons of a particular bundle are eliminated by ablating the cell bodies, axons that normally use the fascicle become lost and wander aimlessly (91). In other systems, fasciculation may not be necessary for axon guidance, but may facilitate growth to the target. Growth factors have been shown to enhance fasciculation between growing axons. When leukemia inhibitory factor (LIF) is applied to the distal end of sympathetic neurites, one effect is to increase neurite bundling (92). A similar effect is seen in aggregate cultures of murine cerebellar neurons treated with NT-3 (93). The halo of neurites leaving the aggregate is more densely packed, suggesting that the neurites are fasciculating in response to NT-3 treatment.

TARGET RECOGNITION

Once axons reach their target they are somehow signaled by the target that they have arrived, triggering a switch in the behavior of their growth cones (89-91). Time-lapse videomicroscopy has revealed that RGC axons slow upon reaching their target, the optic tectum, and put out backbranches as they begin to arborize (94,95). Recent data from our laboratory and others suggest that one such target signal is a difference in the concentration of growth factor between the pathway and the target.

Hoyle et al. (35) provided an elegant demonstration that a gradient of NGF is needed for proper innervation of the target, by using the dopamine-ß-hydroxylase promoter to direct transgenic expression of NGF to developing sympathetic neurons. Normally in this system the target is the only source of NGF. Interestingly, sympathetic axons in these animals reached their peripheral targets, confirming that a target-derived NGF gradient is not involved in pathfinding. However, terminal branching of sympathetic axons within the target was greatly reduced. The authors postulated that transgenic sympathetic axons secrete NGF, eliminating a target-derived NGF concentration gradient that appears necessary for branching and proper innervation. Further support for this idea was provided by showing that when NGF expression was increased in one target (the pancreas), presumably restoring the NGF gradient, there was a concomitant restoration of the terminal branch formation at that target.

Increasing NGF expression in the target, and not the innervating axons, also disrupts normal target innervation. During development, sympathetic and sensory neurons innervate NGF-expressing target cells in the presumptive dermis and epidermis (69). To investigate an *in vivo* role for neurotrophins in target innervation, Albers and co-workers used an epidermal keratin promotor to generate transgenic mouse lines that overexpress either NGF, NT-3 or BDNF selectively in the skin (96-99). So far, preliminary results are available only for the overexpression of NGF, and indicate that the normal segregation of sympathetic and sensory innervation is altered. In these mice, in addition to the expected hyperinnervation of the skin (96), there is evidence for ectopic sympathetic innervation of primary sensory neuron somata (97); neurons in the trigeminal ganglia potentially transport NGF retrogradely from the skin, which could result in ectopic sympathetic innervation. Interestingly, there is also greatly reduced innervation of the sweat glands by cholinergic sympathetic and sensory fibers (98).

Additional evidence implicating a neurotrophin gradient in controlling target innervation comes from studies in the embryonic rat spinal cord where recognition of motor neurons by Ia afferents may be dependent on a gradient of NT-3 (38). NT-3 mRNA is normally expressed in motor neurons before Ia afferents arrive, suggesting that the peptide functions to attract sensory axons to their appropriate targets in the gray matter (ventral horn and dorsal column nuclei). Experiments where NT-3 is injected into the embryonic spinal cord support this idea; NT-3 injection inhibits axon collateral growth into the gray matter, possibly because disruption of the normal growth factor gradient between the white and gray matter prevents the Ia afferents from recognizing and innervating their target.

Recent experiments from our laboratory also suggest that a change in growth factor concentration at the target is critical for proper innervation, but with an interesting difference: the growth factor is at low levels in the target and at high levels in the pathway (42). In the developing *Xenopus* visual system, RGC axons grow out and innervate the optic tectum. The entire pathway through which the axons grow, with the exception of the optic tectum, expresses bFGF. The drastically lower level of bFGF in the target seems to serve as an initial recognition signal for ingrowing retinal fibers, triggering a change in growth behavior required for subsequent recognition of target-specific molecules. When the normal pattern of bFGF expression is disrupted by applying bFGF exogenously to the exposed brain preparation, axons fail to innervate their target. RGC axons pathfind correctly to the tectum, but then bypass their target, growing dorsally and ventrally around its borders. This exogenous application of bFGF introduces the growth factor to the entire optic pathway, including the tectum. One possible explanation for the mistargeting phenotype is that without the difference in bFGF concentration between the pathway and the optic tectum, retinal fibers cannot recognize their target and either stall or grow past it.

Perspective

Historically, a molecule classified as a growth factor has been growth promoting. Yet, several of the studies discussed in this section raise the intriguing possibility that growth factors may in certain situations act to inhibit (or at least not promote) axon growth. Since growth factors are known to stimulate axon extension, it seems unlikely that a growth factor acts as a chemorepellant molecule like the one secreted by rat midline septal cells that actively repels axon growth of olfactory bulb neurons (6,100). We and others have found that when the restricted expression pattern of a growth factor normally recognized by an axon's growth cone is disrupted, the growth factor can be in that axon's direct path and yet fail to promote its extension into the area. This appears to be the situation at the target when the growth factor signal is not presented to the growth cone as a difference in concentration. As suggested for other target recognition signals (101), growth cones may ratio the difference in growth factor concentration between its immediate location and the path ahead. This difference, whatever its directionality, appears to be necessary for proper target innervation, and in its absence axons will not enter previously attractive areas.

TARGET INNERVATION

After a growth cone finds its general target area, it must form synapses with the appropriate neurons. Growth factors may play both an early role in synaptogenesis and a later role in the remodeling of initial connections in both motor and sensory systems.

Branching

The arbor formed as a result of axonal branching determines the precise connectivity between an axon and its target cells. In neonatal rodents, exogenous neurotrophins promote axonal branching; NGF causes extensive branching of developing sympathetic axons (102), and collateral branching of the corticospinal tract is enhanced by NT-3 (74). In some cases, branching and neurite extension can be independent processes, as bFGF enhances the branching of neurites of hippocampal neurons in culture without stimulating process extension (103). Recently, it has been suggested that in neonatal mice a switch in responsiveness from one neurotrophin to another may correspond to a switch from axon elongation to axonal branching. In cerebellar granule cells there is a developmental change in neurotrophin responsiveness, from BDNF to NT-3, corresponding to a 15-fold change in the ratio of trk C to trk B mRNA (93). This *in vivo* change has been correlated with the differential effects of the two growth factors in cerebellar aggregate cultures; BDNF enhances axon elongation while NT-3 influences the pattern of neurite outgrowth, possibly by enhancing branching.

Synaptogenesis

Although the role of growth factors in CNS synapse formation appears limited, there is some evidence that growth factors are involved in synaptogenesis at the frog neuromuscular junction (NMJ). At this synapse, the contact between nerve and muscle induces both pre- and post-synaptic specializations. Heparan sulfate proteoglycan, the low affinity FGF co-receptor, is a major component of the basal lamina surrounding the muscle, which therefore may serve as a storage site for bFGF. In culture, bFGF-coated beads are able to induce pre- and post-synaptic specializations at the NMJ. In the muscle, bFGF beads cause acetylcholine receptor clustering at the site of bead contact (104). These beads are also able to mimic the muscle target by inducing certain pre-synaptic specializations, such as clustering of synaptic vesicles and a prolonged elevation in the intracellular Ca^{2+} level at their contacts with cultured spinal cord neurons (105).

Remodeling Target Innervation

After the initial process of synaptogenesis there is thought to be a period of activity-dependent competition for a limited supply of a growth factor, resulting in the elimination of some synapses and strengthening of others (12,21,106). For example, rat skeletal muscle fibers which are initially innervated by 5 to 6 axons are eventually innervated strongly by a single axon (107). In the absence of activity, or with excess bFGF or ciliary neurotrophic factor (CNTF) (108), polyinnervation persists much longer, indicating that synapse elimination is normally dependent both on a shortage of growth factor and electrical activity.

In sensory systems, axonal connections are topographically organized to produce a sensory map of the world. In the initial period during which axons contact and recognize their target, they sort themselves out to form a rough map. This pattern of connections is then refined in an activity-dependent fashion (see reviews in 3,109). There is no evidence that growth factors are involved in the initial formation of a coarse-grained representation; rather this function may be carried out by the Eph family of receptors and their recently cloned ligands (110,111). There are, however, hints that growth factors play a role in the transformation of the map into a highly-refined pattern of connections.

It is in the visual system that activity-dependent synaptic rearrangement has been best characterized, and here neurotrophins have been implicated as key players. The hypothesis is that axons of the lateral geniculate nucleus (LGN) compete in an activity-dependent fashion for a limited supply of a neurotrophin (112). Competition in the visual cortex of higher mammals is necessary to produce ocular-dominance columns (109), the anatomical basis of which is segregation of the LGN inputs from the two eyes into alternating patches. The importance of electrical activity for competition was demonstrated in kittens by covering one eye during a critical period of development and showing that most of the visual cortical neurons lost their functional LGN input from the deprived and electrically "quiet" eye (113). Subsequent experiments have tested the importance of growth factors, in particular neurotrophins, in activity-dependent competition. Intraventricular injection of NGF in rats that have had one eye covered inhibits the effects of monocular deprivation: both eyes effectively establish connections with visual cortical neurons (112). The authors suggest that competition is unnecessary in the presence of excess growth factor. While these results argue for NGF being involved in activity-dependent competition, the fact that neither visual cortical neurons nor LGN axons express trk A, the receptor for NGF, indicates that its effect is probably indirect.

It is more likely that some other neurotrophin is in limited supply. For instance, BDNF is expressed by the cellular targets of RGCs in chick and *Xenopus* during the period when axonal connections are being reorganized (114,115). In *Xenopus*, BDNF does not appear to be involved in the early formation of the topographic map, but is expressed when activity-dependent order is being established. While such temporal correlations are only suggestive, more substantive evidence has been provided recently by Shatz and colleagues (116), who tested whether activity-dependent synaptic competition in the visual cortex was mediated by BDNF. BDNF infused into the cat primary visual cortex inhibited formation of ocular dominance columns in the vicinity of the infusion site. Normally, competition for a limited amount of BDNF by LGN axons, which express the receptor for BDNF, trk B, may be important for segregation of the inputs from the LGN into eye-specific patches. Infusion of BDNF eliminates the competition and thus inhibits ocular dominance column formation. Similar results were observed with NT-4/5, indicating that other neurotrophins may play comparable roles.

If synaptic activity is an important determinant of synaptic viability, anything that regulates transmission at the synapse will be important in the pruning of developing connections. Interestingly, growth factors can directly influence the activity of developing synapses. At the immature NMJ, both applied BDNF and NT-3 increase the amplitude and frequency of spontaneous synaptic currents and also increase the size of evoked responses (117). These data suggest that neurotrophins increase the probability of quantal release and/or the number of available quanta. The ability of growth factors to modulate synaptic activity may be crucial in the further maturation and maintenance of synaptic contacts. Interestingly, recent evidence indicates that the neurotrophin and FGF families of growth factors can have long-lasting effects on synaptic transmission in the adult hippocampus both *in vivo* and *in vitro* (118-121). Apparently, growth factors may be employed at a synapse long after the appropriate connections have been formed.

CONCLUSIONS

Recent studies indicate that growth factors function at all stages of neural connectivity, from the initiation of the axon to the regulation of electrical transmission at the synapse. In addition to promoting outgrowth of axons, growth factors may steer these same axons towards their target. Growth cone steering could result from a direct chemoattractant action of target-secreted growth factors, or indirectly *via* modulation of the extracellular environment. One of the most intriguing recent observations is that growth cones are influenced by differences in the levels of growth factor expressed in the pathway and the target. This difference signal appears critical for triggering the events of target innervation. The field is clearly at the frontier of discovering how growth factors act in concert with other guidance and targeting molecules to establish the appropriate neuronal connections. Transgenic technology provides new opportunities to study growth factor signaling in axonal navigation and targeting. In transgenic mice, many knock-out studies are hampered ironically by the plethora of effects produced by eliminating growth factor signaling. To overcome this problem, such studies clearly require the use of cell-specific heterologous promoters. The developing *Xenopus* visual system shows great promise for investigating growth factor involvement in axonal pathfinding and target recognition. We are able currently to transfect individual RGCs *in vivo* with mutant growth factor receptor constructs to determine how RGC axons respond to this insult while navigating in an otherwise normal environment.

ACKNOWLEDGMENTS

The authors would like to thank R. Riehl, J. Lewis, T. Worley, B. Lom, A. Walz, L. McNeill and C.B. Chien for critical comments on the manuscript. S. McFarlane is supported by an MRC (of Canada) postdoctoral fellowship.

REFERENCES

1 Bixby, J.L. and Harris, W.A. (1991) Annu. Rev. Cell. Bio. 7, 117-159.
2 Dodd, J. and Jessell, T.M. (1988) Science 242, 692-699.
3 Goodman, C.S. and Shatz, C.J. (1993) Cell 72, 77-98.
4 Tessier-Lavigne, M. and Placzek, M. (1991) Trends Neurosci. 14, 303-310.
5 Kennedy, T.E. and Tessier-Lavigne, M. (1995) Curr. Opin. Neurobiol. 5, 83-90.
6 Pini, A. (1993) Science 261, 95-98.
7 Tessier-Lavigne, M. (1994) Curr. Opin. Genet. Dev. 4, 596-601.
8 Serafini, T., Kennedy, T.E., Galko, M.J., Mirzayan, C., Jessell, T.M. and Tessier-Lavigne, M. (1994) Cell 78, 409-424.
9 Kennedy, T.E., Serafini, T., de la Torre, J.R. and Tessier-Lavigne, M. (1994) Cell 78, 425-435.
10 Colamarino, S.A. and Tessier-Lavigne, M. (1995) Cell 81, 621-629.
11 Sensebrenner, M. (1993) Prog. Neurobiol. 41, 683-704.
12 Barde, Y.A. (1989) Neuron 2, 1525-1534.
13 Chao, M.V. (1992) Neuron 9, 583-593.
14 Johnson, D.E. and Williams, L.T. (1993) Adv. Cancer Res. 60, 1-41.
15 Mason, I.J. (1994) Cell 78, 547-552.
16 Burrus, L.W., Zuber, M.E., Lueddecke, B.A. and Olwin, B.B. (1992) Mol. Cell. Biol. 12, 5600-5609.
17 Givol, D. and Yayon, A. (1992) FASEB J. 6, 3362-3369.
18 Klagsbrun, M. (1990) Curr. Opin. Cell. Biol. 2, 857-863.
19 Neufeld, G., Gospodarowicz, D., Dodge, D. and Fujii, D.K. (1987) J. Cell. Physiol. 131, 131-140.
20 Nurcombe, V., Ford, M.D., Wildshut, J.A. and Bartlett, P.F. (1993) Science 260, 103-106.
21 Thoenen, H. (1991) Trends Neurosci. 14, 165-170.
22 Davies, A.M. (1987) Development 101, 185-208.
23 Barbacid, M. (1993) Oncogene 8, 2033-2042.
24 Hantzopoulos, P.A., Suri, C., Glass, D.J., Goldfarb, M.P. and Yancopoulos, G.D. (1994) Neuron 13, 187-201.
25 Lumsden, A.G. and Davies, A.M. (1983) Nature 306, 786-788.
26 Crowley, C., Spencer, S.D., Nishimura, M.C., Chen, K.S., Pitts-Meek, S., Armanini, M.P., Ling, L.H., McMahon, S.B., Shelton, D.L., Levinson, A.D. and Phillips, H.S. (1994) Cell 76, 1001-1011.
27 Ernfors, P., Lee, K.F., Kucera, J. and Jaenisch, R. (1994) Cell 77, 503-512.
28 Ernfors, P., Lee, K.F. and Jaenisch, R. (1994) Nature 368, 147-150.
29 Fariñas, I., Jones, K.R., Backus, C., Wang, X.Y. and Reichardt, L.F. (1994) Nature 369, 658-661.
30 Jones, K.R., Fariñas, I., Backus, C. and Reichardt, L.F. (1994) Cell 76, 989-999.
31 Klein, R., Silos-Santiago, I., Smeyne, R.J., Lira, S.A., Brambilla, R., Bryant, S., Zhang, L., Snider, W.D. and Barbacid, M. (1994) Nature 368, 249-251.
32 Klein, R., Smeyne, R.J., Wurst, W., Long, L.K., Auerbach, B.A., Joyner, A.L. and Barbacid, M. (1993) Cell 75, 113-122.
33 Smeyne, R.J., Klein, R., Schapp, A., Long, L.K., Bryant, S., Lewin, A., Lira, S.A. and Barbacid, M. (1994) Nature 368, 246-249.
34 Snider, W.D. (1994) Cell 77, 627-638.
35 Hoyle, G.W., Mercer, E.H., Palmiter, R.D. and Brinster, R.L. (1993) Neuron 10, 1019-1034.

36 Hendry, I.A. (1989) in IBRO Handbook Series: Nerve growth factors (Rush, R.A., ed.) Vol. 12, Wiley, New York, NY.
37 Menesini, C.M., Chen, J.S. and Levi, M.R. (1978) Arch. Ital. Biol. 116, 53-84.
38 Zhang, L., Schmidt, R.E., Yan, Q. and Snider, W.D. (1994) J. Neurosci. 14, 5187-5201.
39 Chien, C.B., Rosenthal, D.E., Harris, W.A. and Holt, C.E. (1993) Neuron 11, 237-251.
40 Chien, C.B. and Harris, W.A. (1994) Curr. Top. Dev. Biol. 29, 135-169.
41 Chien, C.B. and Harris, W.A. (1995) Persp. Dev. (in press).
42 McFarlane, S., McNeill, L. and Holt, C.E. (1995) Neuron (in press).
43 Lillienbaum, A., Reszka, A.A., Horwitz, A.F. and Holt, C.E. (1995) Mol. Cell. Neurosci. 6, 139-152.
44 Riehl, R.H., Lillienbaum, A. and Holt, C.E. (unpublished data).
45 McFarlane, S., Cornel, E., Amaya, E. and Holt, C.E. (1995) Soc. Neurosci. Abstr. 21, 1293.
46 Bovolenta, P. and Mason, C. (1987) J. Neurosci. 7, 1447-1460.
47 Holt, C.E. (1989) J. Neurosci. 9, 3123-3145.
48 Tosney, K.W. and Landmesser, L.T. (1985) J. Neurosci. 5, 2345-2358.
49 Snider, W.D. and Johnson, E.J. (1989) Ann. Neurol. 26, 489-506.
50 Hatten, M.E., Lynch, M., Rydel, R.E., Sanchez, J., Joseph-Silverstein, J., Moscatti, D. and Rifkin, D.B. (1988) Dev. Biol. 125, 280-289.
51 Lipton, S.A., Wagner, J.A., Madison, R.D. and D'Amore, P.A. (1988) Proc. Nat. Acad. Sci. U.S.A. 85, 2388-2392.
52 Rydel, R.E. and Greene, L.A. (1987) J. Neurosci. 7, 3639-3653.
53 Walicke, P., Cowan, W.M., Veno, N., Baird, A. and Gullemin, R. (1986) Proc. Nat. Acad. Sci. U.S.A. 83, 3012-3016.
54 Walicke, P.A. (1988) Exp. Neurol. 102, 144-148.
55 Cohen, A., Bray, G.M. and Aguayo, A.J. (1994) J. Neurobiol. 25, 953-959.
56 Davies, A.M., Thoenen, H. and Barde, Y.A. (1986) J. Neurosci. 6, 1897-1904.
57 Collins, F. and Dawson, A. (1983) Proc. Nat. Acad. Sci. U.S.A. 80, 2091-2094.
58 Lindsay, R.M. (1988) J. Neurosci. 8, 2394-2405.
59 Greene, L.A. and Shooter, E.M. (1980) Annu. Rev. Neurosci. 3, 353-402.
60 Sarder, M., Saito, H. and Abe, K. (1993) Brain Res. 625, 347-350.
61 Haugen, P.K. and Letourneau, P.C. (1990) J. Neurosci. Res. 25, 443-452.
62 Levi-Montalcini, R. and Hamburger, V. (1951) J. Exp. Zool. 116, 321-361.
63 Hagg, T., Vahlsing, H.L., Manthorpe, M. and Varon, S. (1990) J. Neurosci. 10, 3087-3092.
64 Kawaja, M.D., Rosenberg, M.B., Yoshida, K. and Gage, F.H. (1992) J. Neurosci. 12, 2849-2864.
65 Williams, E.J., Furness, J., Walsh, F.S. and Doherty, P. (1994) Neuron 13, 583-594.
66 Amaya, E., Musci, T.J. and Kirschner, M.W. (1991) Cell 66, 257-270.
67 Levi-Montalcini, R. and Angeletti, P.U. (1968) Physiol. Rev. 48, 534-569.
68 Gundersen, R.W. and Barrett, J.N. (1979) Science 206, 1079-1080.
69 Davies, A.M., Bandtlow, C., Heumann, R., Korsching, S., Rohrer, H. and Thoenen, H. (1987) Nature 326, 353-358.
70 O'Leary, D.D. and Daston, M.M. (1994) Soc. Neurosci. Abstr. 20, 1685.
71 O'Leary, D.D. and Terashima, T. (1988) Neuron 1, 901-910.
72 Heffner, C.D., Lumsden, A.G. and O'Leary, D.D. (1990) Science 247, 217-220.
73 Sato, M., Lopez-Mascaraque, L., Heffner, C.D. and O'Leary, D.D. (1994) Neuron 13, 791-803.

74 Schnell, L., Schneider, R., Kolbeck, R., Barde, Y.-A. and Schwab, M.E. (1994) Nature 367, 170-173.
75 Klein, R. (1994) FASEB J. 8, 738-744.
76 Bentley, D. and Caudy, M. (1983) Nature 304, 62-65.
77 Elkabes, S., Dreyfus, C.F., Schaar, D.G. and Black, I.B. (1994) J. Comp. Neurol. 341, 204-213.
78 Rosenberg, M.B., Friedmann, T., Robertson, R.C., Tuszynski, M., Wolff, J.A., Breakefield, X.O. and Gage, F.H. (1988) Science 242, 1575-1578.
79 Tuszynski, M.H., Peterson, D.A., Ray, J., Baird, A., Nakahara, Y. and Gage, F.H. (1994) Exp. Neurol. 126, 1-14.
80 Panni, M.K., Atkinson, J. and Lund, R.D. (1994) Soc. Neurosci. Abstr. 20, 679.
81 Meiners, S., Marone, M., Rittenhouse, J.L. and Geller, H.M. (1993) Dev. Biol. 160, 480-493.
82 Rettig, W.J., Erickson, H.P., Albino, A.P. and Garin, C.P. (1994) J. Cell. Sci., 487-497.
83 Perez, R.G. and Halfter, W. (1993) Dev. Biol. 156, 278-292.
84 Taylor, J., Pesheva, P. and Schachner, M. (1993) J. Neurosci. Res. 35, 347-362.
85 Tuttle, R. and Matthew, W.D. (1995) Development 121, 1301-1309.
86 Kinoshita, Y., Kinoshita, C., Heuer, J.G. and Bothwell, M. (1993) Development 119, 943-956.
87 Bixby, J.L., Lilien, J. and Reichardt, L.F. (1988) J. Cell. Biol. 107, 353-361.
88 Neugebauer, K.M., Tomaselli, K.J., Lilien, J. and Reichardt, L.F. (1988) J. Cell. Biol. 107, 1177-1187.
89 Reichardt, L.F. and Tomaselli, K.J. (1991) Annu. Rev. Neurosci. 14, 531-570.
90 Delannet, M. and Duband, J.L. (1992) Development 116, 275-287.
91 Goodman, C.S., Bastiani, M.J., Doe, C.Q., du Lac, S., Helfand, S.L., Kuwada, J.Y. and Thomas, J.B. (1984) Science 225, 1271-1279.
92 Ure, D.R., Campenot, R.B. and Acheson, A. (1992) Dev. Biol. 154, 388-395.
93 Segal, R.A., Pomeroy, S.L. and Stiles, C.D. (1995) J. Neurosci. 15, 4970-4981.
94 Harris, W.A., Holt, C.E. and Bonhoeffer, F. (1987) Development 101, 123-133.
95 Kaethner, R.J. and Stuermer, C.A. (1992) J. Neurosci. 12, 3257-3271.
96 Albers, K.M., Wright, D.E. and Davis, B.M. (1994) J. Neurosci. 14, 1422-1432.
97 Davis, B.M., Albers, K.M., Seroogy, K.B. and Katz, D.M. (1994) J. Comp. Neurol. 349, 464-474.
98 Guidry, G., Davis, B.M., Landis, C.S. and Albers, K.M. (1994) Soc. Neurosci. Abstr. 20, 1090.
99 Lemaster, A.M., Davis, B.M., Perronne, T.N., Davis, F., Pang, Z. and Albers, K.M. (1994) Soc. Neurosci. Abstr. 20, 1090.
100 Fitzgerald, M., Kwiat, G.C., Middleton, J. and Pini, A. (1993) Development 117, 1377-1384.
101 Davenport, R.W., Löschinger, J., Huf, J., Jung, J. and Bonhoeffer, F. (1994) Soc. Neurosci. Abstr. 20, 1065.
102 Olson, L. (1967) Z. Zellforsch. Microsk. Anat. 81, 155-173.
103 Miyagawa, T., Saito, H. and Nishiyama, N. (1993) Neurosci. Lett. 153, 29-31.
104 Peng, H.B., Baker, L.P. and Chen, Q. (1991) Neuron 6, 237-246.
105 Dai, Z. and Peng, H.B. (1995) J. Neurosci. 15, 5466-5475.
106 Oppenheim, R.W., Prevette, D., Qin-Wei, Y., Collins, F. and MacDonald, J. (1991) Science 251, 1616-1618.
107 Brown, M.C., Jansen, J.K. and Van, E.D. (1976) J. Physiol. London 261, 387-422.

108 English, A.W. and Schwartz, G. (1995) Dev. Biol. 169, 57-64.
109 Shatz, C.J. (1990) Neuron 5, 745-756.
110 Drescher, U., Kremoser, C., Handwerker, C., Löschinger, J., Noda, M. and Bonhoeffer, F. (1995) Cell 82, 359-370.
111 Cheng, H.J., Nakamoto, M., Bergemann, A.D. and Flanagan, J.G. (1995) Cell 82, 371-386.
112 Maffei, L., Berardi, N., Domenici, L., Parisi, V. and Pizzorusso, T. (1992) J. Neurosci. 12, 4651-4662.
113 Wiesel, T.N. and Hubel, D.H. (1963) J. Neurosphysiol. 26, 1003-1017.
114 Cohen Cory, S. and Fraser, S.E. (1994) Neuron 12, 747-761.
115 Herzog, K.H., Bailey, K. and Barde, Y.A. (1994) Development 120, 1643-1649.
116 Cabelli, R.J., Hohn, A. and Shatz, C.J. (1995) Science 267, 1662-1666.
117 Lohof, A.M., Ip, N.Y. and Poo, M.M. (1993) Nature 363, 350-353.
118 Kang, H. and Schuman, E.M. (1995) Science 267, 1658-1662.
119 Ishiyama, J., Saito, H. and Abe, K. (1991) Neurosci. Res. 12, 403-411.
120 Hisajima, H., Saito, H., Abe, K. and Nishiyama, N. (1992) J. Neurosci. Res. 31, 549-553.
121 Sasaki, K., Oomura, Y., Figurov, A. and Yagi, H. (1994) Brain Res. Bull. 33, 505-511.

GENE IDENTIFICATION BY 3′ TERMINAL EXON TRAPPING

David B. Krizman

National Institutes of Health
National Center for Human Genome Research
Bldg. 49, Room 4A15
9000 Rockville Pike
Bethesda, MD 20892

INTRODUCTION

The fact that most eukaryotic genes are split into exons and introns accounts for the difficulty in identifying those regions of the eukaryotic genome that code for protein. To compound the problem, introns in higher eukaryotes are generally quite long while the exons are usually very short. In order to translate correct peptides from split genes, the cell utilizes the mechanism of mRNA splicing to bring the coding regions together while removing the non-coding intervening sequences. The technology of exon trapping, sometimes called exon amplification, strives to exploit the phenomenon of mRNA splicing to discover genes directly from genomic DNA.

There are 3 distinct exon trapping methodologies that differ simply in the genomic target of interest. The original experimental design was to capture isolated 3′ splice sites residing within fragments of genomic DNA (1), while later approaches focused on either complete internal exons (2-5) or entire 3′ terminal exons (6). The requirement of complete, intact exons as targets has proven absolutely essential and only the trapping of either internal exons or 3′ terminal exons is practical.

Plasmid, phage, cosmid, P1, YAC clones or pooled clones of any type containing mammalian genomic DNA, can be used as a substrate for either exon trapping approach. Both internal exon and 3′ terminal exon trapping have successfully trapped exons from all of the above-mentioned substrates (6-9). Small numbers of exons are trapped from individual clones such as plasmid, phage or cosmid clones, while larger numbers can be trapped if collections of clones such as phage or cosmid pools are used. Trapping from YACs is desirable when attempts are made to discover exons from larger genomic regions that may be difficult to cover with the small-insert cloning vectors.

The strategy discussed here for the identification of 3′ terminal exons from vertebrate genomic DNA sources is a slight modification of the approach developed earlier (6). The vector strategy employed was developed on the basis of earlier work suggesting that the vertebrate exon is the unit of recognition of the RNA processing machinery (10). Further results suggested a very strict biochemical definition of 3′ terminal exons which makes

Genetic Engineering, Vol. 18, Edited by J.K. Setlow
Plenum Press, New York, 1996

them promising candidates for an efficient exon trapping approach (11,12).

3′ terminal exon trapping involves the use of the trapping vector pTAG4 which harbors an incomplete transcription unit. The basis of this exon trapping approach is to ask foreign DNA inserted downstream of vector-derived internal exons to donate a 3′ terminal exon and produce stable, mature, chimeric mRNA that can be amplified by the RT/PCR technique of 3′ rapid amplification of cDNA ends (3′ RACE) (13).

HOW 3′ TERMINAL EXON TRAPPING IS CARRIED OUT

General Principle

Figure 1 diagrams the general protocol. The genomic target of 3′ terminal exon trapping is the last exon of a gene and is accomplished by the ligation of restricted genomic fragments of DNA with the trapping vector pTAG4. Ligation products are subsequently transfected directly to Cos7 cells that are able to support transcription from this vector. This vector was engineered to contain a trapping cassette that is an incomplete transcription unit that lacks only a last exon. The foreign DNA fragments are expected to donate a 3′ terminal exon to the vector to complete the transcription unit and generate a stable vector-derived mRNA molecule upon transfection into Cos7 cells. Exons trapped in this manner can be amplified by the 3′ RACE procedure that was designed to amplify mRNA species specifically from the 3′ end of the molecule with the use of the poly(A) tail as an anchor. Primers used for the PCR reaction are specific for vector exons, thus imparting the specificity needed when mRNA preparations from transfected mammalian cells are used. PCR product is subcloned and sequenced for further analysis.

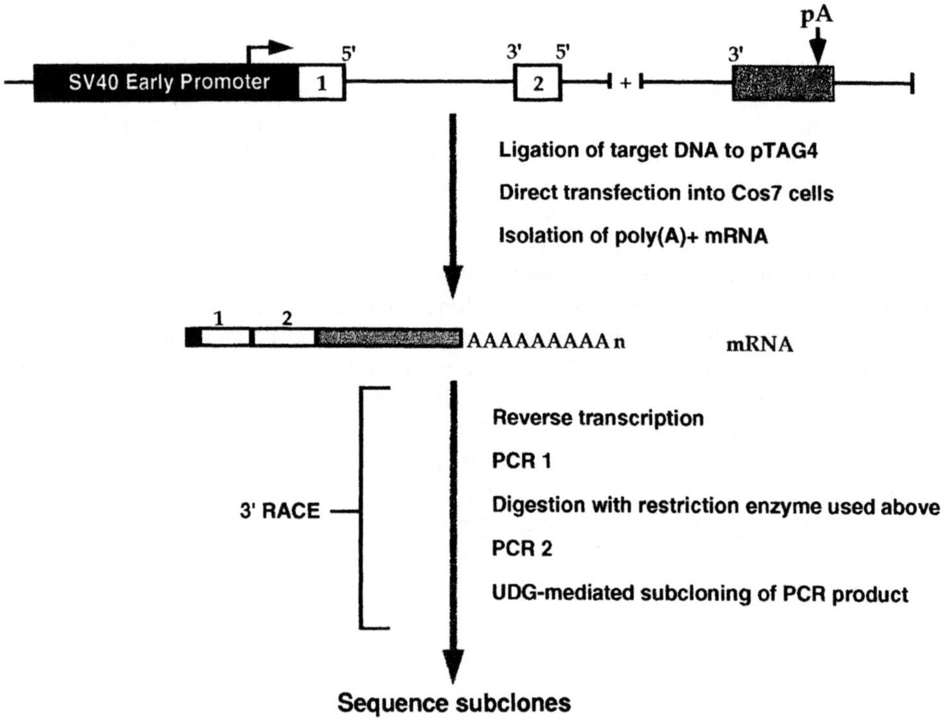

Figure 1. Schematic representation of the trapping procedure.

There is only one 3′ terminal exon trapping vector, pTAG4, and it is illustrated in Figure 2. The vector is 3980 bases in length with an ampicillin resistance gene and bacterial origin of replication for propagation in *E. coli*. The trapping cassette consists of the SV40 early region to drive transcription, exons 1 and 2, as well as intron 1, from the leader exons of the human adenovirus 2 genome, and a multiple cloning site for the ligation of genomic DNA. pTAG4 lacks a last exon containing a poly(A) signal thus no mature polyadenylated mRNA is produced from the vector when transfected into Cos7 cells. The multiple cloning site (MCS) downstream of the 5′ splice site of exon 2 contains the following unique restriction enzyme sites: *Eco*RI, *Bam*HI, *Bgl*II, *Bss*HII, *Sph*I, *Nhe*I, *Eag*I, *Not*I, *Pst*I, *Nar*I, *Mlu*I and *Spl*I.

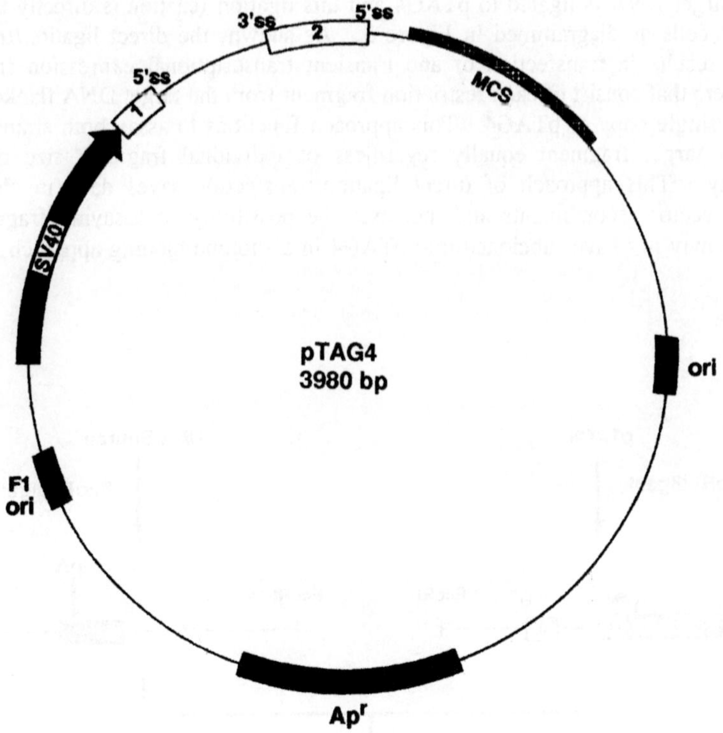

Figure 2. pTAG4 vector.

Methodology

The target DNA can reside in any cloning vector (plasmid, phage, cosmid, P1, BAC or YAC vector) used for propagating genomic DNA and is digested to completion with one of the restriction enzymes within the pTAG4 MCS to be used for the ligation reaction. The target DNA must be sufficiently purified from host DNA (*E. coli* and yeast) to keep nonspecific background to a minimum. It is not necessary to purify the DNA insert from the cloning vector since the vectors do not contain exons and introns and will not score positive. Many protocols exist that are designed to purify cloned DNA within the various cloning vectors and most of them work well. If the target DNA is from a YAC, this clone will need to be purified by 1.0% low-melting agarose pulse field gel electrophoresis followed by excision of an agarose gel slice containing the YAC DNA. This technology has been described elsewhere (9). Purification of YACs usually shears the DNA to roughly 50 kb in size; however, the DNA is still large enough for restriction digestion and suitable for exon trapping.

The trapping vector pTAG4 is double digested with *Aat*II and the same restriction enzyme used to digest the target DNA. The linear 1.5 kb vector fragment to be used for trapping needs to be gel purified from the other fragments generated from pTAG4 double digest. Digestion and purification of pTAG4 in this way functions to remove the ampicillin resistance gene, leave 1 kb of vector DNA upstream of the transcriptional start in the SV40 promoter, and make available a single cloning site 50 to 300 bases downstream of the second exon of pTAG4.

The target DNA is ligated to pTAG4 and this ligation reaction is directly transfected into Cos7 cells as diagrammed in Figure 3. As shown, the direct ligation/transfection approach results in transfection of and transient transcriptional expression from linear concatamers that consist of each restriction fragment from the target DNA flanked on each side by a single copy of pTAG4. This approach functions to assay both strands of each restriction target fragment equally regardless of individual fragment size or mixture complexity. This approach of direct ligation/transfection saves days in cloning and purifying vector recombinants and increases the possibility of assaying fragments that otherwise may not have subcloned into pTAG4 in a shotgun cloning approach.

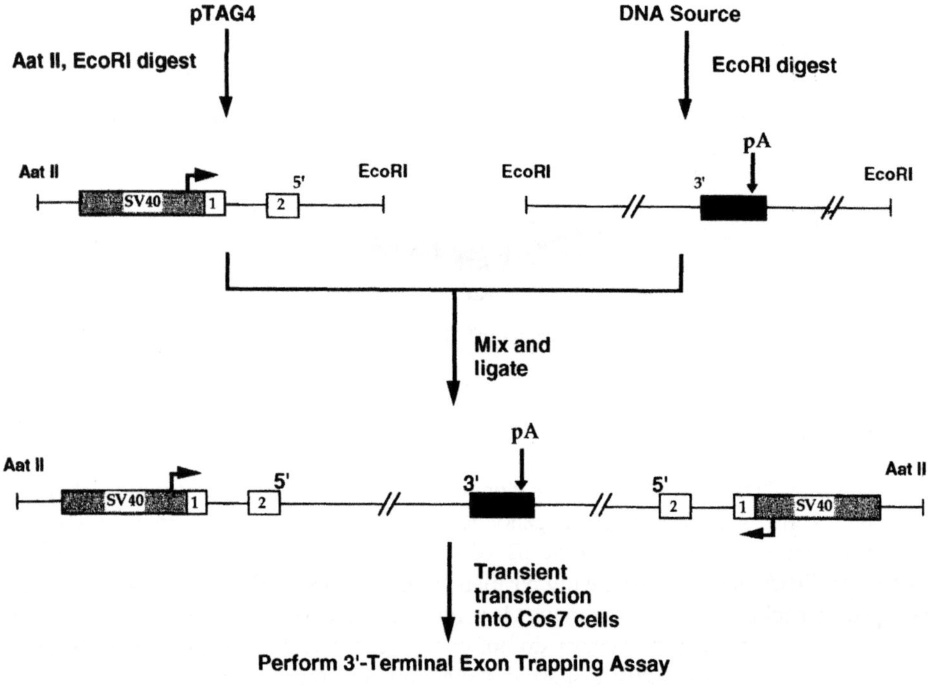

Figure 3. Schematic of the direct ligation/transfection approach.

Transiently-expressed mRNA is harvested 16 to 24 hours after transfection. The mRNA is reverse transcribed with the use of an oligo(dT)-based adapter primer (AP) for the 3′ RACE technique. The oligonucleotide AP is 45 nucleotides in length and consists of $(T)_{17}$ at the 3′ end that will prime cDNA off the poly(A) tail of all mRNA. The remaining 28 nucleotides of the AP primer consist of an engineered sequence that will not base pair with any endogenous sequences from COS-7 mRNA species. The resulting cDNA primed with the AP primer can be amplified with 5′ primers specific for vector sequences and a 3′ primer specific for the engineered tail sequence of the AP. This is the basis of 3′ RACE technology (13).

Exons are then amplified by a hemi-nested polymerase chain reaction (PCR) approach. The primary PCR reaction uses one primer specific to the tail sequence on the AP primer (UAP) while the other is specific to vector exonic sequence (SV40P). The oligonucleotide SV40P is designed for the portion of any transcript that is derived from the SV40 promoter. Primary PCR product is then digested with the same restriction enzyme used for preparative digestion of the target DNA and pTAG4. This step functions to remove any false positives resulting from reverse transcription of unspliced precursor RNA or residual contaminating DNA present in the mRNA preparation. The secondary PCR uses a new vector-derived primer (Ad2) with the UAP primer to amplify product from the digested primary PCR reaction. The oligonucleotide Ad2 is specific for the second exon from the adenovirus sequence within pTAG4. This PCR protocol requires the use of HotStart PCR by the addition of the Taq polymerase after the temperature has been raised to above the denaturing temperature. This functions to lower the amount of nonspecific priming and polymerization induced when the oligonucleotides anneal to the cDNA before the optimal annealing temperature has been reached. The primers Ad2 and UAP have been engineered to contain 12 base pair extensions to impart UDG-clonable ends on resulting PCR product. Thus the PCR products generated from 3′ terminal exon trapping can be efficiently cloned utilizing the UDG cloning technology (14).

RESULTS OF TRAPPING EXPERIMENT

Analysis of PCR Results

As far as has been determined, this protocol works equally well with cosmid, P1, bacteriophage, plasmid, BAC or YAC clones. However, the efficiency of trapping from these various clones has yet to be determined. It is impossible to predict how many genes are present in some clones *versus* others, and individual regions of the genome will no doubt be variable in gene density. Thus it is difficult to predict how many exons will be trapped from any one substrate. However, the larger and more complex clones such as BAC and YAC clones most certainly would contain many genes.

The complexity of the final PCR product from this protocol will be proportional to the amount and complexity of target DNA used for the original ligation reactions. For example, a small number (1 to 3) of ethidium bromide stained bands will be seen when secondary PCR product is analyzed by agarose gel electrophoresis in a trapping experiment with single cosmid or phage clones. If there is trapping from a YAC clone or a pool of smaller clones the numbers of bands seen on a gel will be proportionally greater. Thus subcloning of the products may involve excision of single bands from the gel (single clones) to a shotgun approach with total PCR product (YAC clone and pools of clones). Figure 4 shows the resulting PCR product generated from trapping a pool of 5 cosmid clones. Many individual bands can be visualized among a slightly complex mixture.

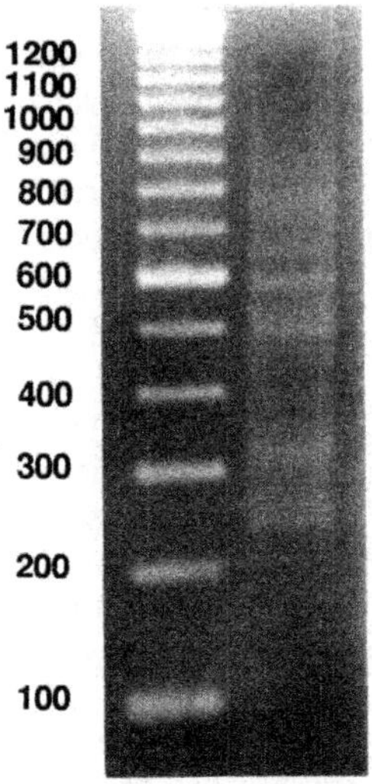

Figure 4. Final PCR product resulting from trapping 3′ terminal exons from a pool of 5 cosmid clones.

Product from the secondary PCR reactions should be subcloned for further analysis by either UDG-mediated cloning or TA cloning (14,15). The primers used for the secondary PCR reaction are designed to impart UDG cloning capabilities upon PCR product, and this approach is generally more efficient and yields greater numbers of useful subclones (14). For these reasons UDG cloning is recommended; however, TA cloning can be used and should yield sufficient numbers of subclones for sequencing (15).

Identification of Trapped Sequences

Subclones containing candidate exons should first be sequenced to determine exon validity. The published average size of 3′ terminal exons is 627 base pairs; however, in practice the majority appear to be from 300 to 500 bases in length (16). The sequence of an individual trapped last exon will show evidence of a splice event at the 5′ end of the sequence and a poly(A) signal/tail at the 3′ end of the sequence. A splicing event can be determined to have occurred by comparing vector sequences in the resulting PCR product with those in the unspliced vector as shown in Figure 5. If EcoRI restriction enzyme was used for ligation of the target DNA to pTAG4, then specific nucleotides that define the splice junction at the 5′ splice site of the vector-derived exon should be present at the 3′ end of vector exon 2 and at the 5′ end of the newly-trapped sequence (sequence shown above PCR product in Figure 5). If one of the other restriction sites was chosen for pTAG4 and the target DNA, then an alternative splice site within the multiple cloning site of the vector will be used and different nucleotides that define a different splice junction will be at the 3′ end of vector sequences (sequence shown below the PCR product in Figure 5). If a splicing event did not occur then much of the multiple cloning site derived from the vector will not have been spliced out and those sequences will be present downstream of the vector-derived splice junction and any new sequence. PCR products that have not been spliced and yet still contain new sequence most probably represent reverse transcribing off A-rich regions in precursor (non-spliced) RNA derived from vector-driven transcripts. This problem is a very rare event and can be overcome by more

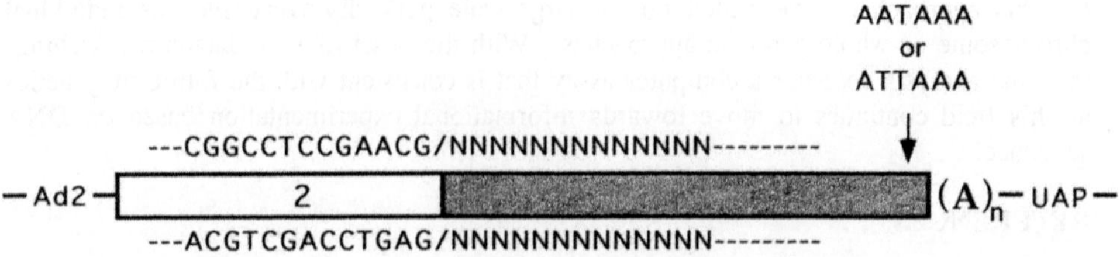

Figure 5. Schematic of the anticipated result of sequence analysis of a hypothetical trapped 3′ terminal exon.

efficient digestion of the primary PCR reaction, preparing higher quality mRNA from the transfected Cos7 cells, and/or DNAse I treatment of the mRNA preparation.

In approximately 90% of the cases, a consensus poly(A) signal consisting of one of the hexanucleotides AATAAA or ATTAAA will be present 12 to 30 bases upstream of the poly(A) tail. Multiple termination codons will also be present in all reading frames. The great majority of 3′ terminal exon sequence is 3′ untranslated region, thus very little coding sequence exists.

The most efficient use of the sequence information is to search the genome databases with the BLAST program from the National Center for Biotechnology Information (NCBI) (17). This program is accessed through computer-based avenues including the World Wide Web and e-mail. This program searches many genome databases simultaneously and assigns a score according to how close your sequence of interest matches any sequence that is currently in the databases. This program is especially useful when searching with trapped exons because of the amount of human expressed sequence tags (ESTs) that currently are in the databases. A number of the large human genome projects have sequenced random cDNAs and downloaded that information into the databases. Complete validation of exon sequence of human origin can be made if a trapped sequence hits one or more ESTs. If a hit is confirmed, further experiments to prove exon validity are not necessary and gene mapping and expression studies of the trapped exon should begin. If a trapped sequence gives either no match or shows weak homology to ESTs or other genes in the databases, further experimental exon validation needs to be done which includes RT/PCR, S1, or Northern blot expression analyses.

Both PCR-based and hybridization-based experiments to map trapped exons and perform gene expressions studies can be performed. Full-length cDNA clones can also be obtained in hybridization-based screens. Oligonucleotide primers can be designed to the unique sequence for use in a number of subsequent experiments including: 1) mapping the candidate exon to its genomic region of origin, 2) RT/PCR expression studies, and 3) generating a specific probe for cDNA screening and Southern/Northern blot mapping. However, due to the small size of many trapped exons it may be more informative to perform PCR-based *versus* hybridization-based experiments with DNA and RNA from interesting sources as templates.

SUMMARY

3′ terminal exon trapping offers a powerful and efficient technology for rendering fragments of transcribed genes from large stretches of unsequenced, vertebrate, genomic DNA. The products from this methodology are the starting point for further studies of gene discovery and analysis. Future considerations for this technology include answering questions about the degree of efficiency from various trapping substrates and whether or not this approach can be scaled up for large-scale gene discovery such as individual chromosome or whole genomic approaches. With the onset of EST database searching, the final analysis becomes a computer assay that is consistent with the future of genetics as this field continues to move towards informational experimentation based on DNA sequence.

REFERENCES

1 Duyk, G.M., Kim, S., Myers, R.M. and Cox, D.R. (1990) Proc. Nat. Acad. Sci. U.S.A. 87, 8995-8999.
2 Buckler, A.J., Chang, D.D., Graw, S.L., Brook, J.D., Haber, D.A., Sharp, P.A. and Housman, D.E. (1991) Proc. Nat. Acad. Sci. U.S.A. 88, 4005-4009.
3 Auch, D. and Reth, M. (1991) Nucl. Acids Res. 18, 6743-6744.
4 Hamaguchi, M., Sakamoto, H., Tsuruta, H., Sasaki, H., Muto, T., Sugimura, T. and Terada, M. (1992) Proc. Nat. Acad. Sci. U.S.A. 89, 9779-9783.
5 Nehls, M., Pfeifer, D. and Boehm, T. (1994) Oncogene 9, 2169-2175.
6 Krizman, D.B. and Berget, S.M. (1993) Nucl. Acids Res. 21, 5198-5202.
7 Huntington Disease Collaborative Research Group (1993) Cell 72, 971-983.
8 Church, D.M., Stotler, C.J., Rutter, J.L., Murrell, J.R., Trofatter,J.A. and Buckler, A.J. (1994) Nature Genet. 6, 98-105.
9 Krizman, D.B., Hofmann, T.A., DeSilva, U., Green, E.D., Meltzer, P.S. and Trent, J.M. (1994) PCR Methods and Appl. 4, 322-326.
10 Robberson, B.L., Cote, G.J. and Berget, S.M. (1990) Mol. Cell. Biol. 10, 84-94.
11 Niwa, M. and Berget, S.M. (1991) Genes Dev. 5, 2086-2095.
12 Niwa, M. and Berget, S.M. (1992) Nature 360, 277-280.
13 Frohman, M.A., Dush, M.A. and Martin, G.R. (1988) Proc. Nat. Acad. Sci. U.S.A. 85, 8998-9002.
14 Nisson, P.E., Rashtchian, A. and Watkins, P.C. (1991) PCR Methods and Appl. 1, 120-123.
15 Marchuk, D., Drumm, M., Saulino, A. and Collins, F.S. (1991) Nucl. Acids Res. 19, 1154.
16 Hawkins, J.D. (1988) Nucl. Acids Res. 16, 9893-9908.
17 Altschul, S.F., Gish, W., Miller, W., Myers, E.W. and Lipman, D.J. (1990) J. Mol. Biol. 215, 403-410.

ENGINEERING TRANSGENES FOR USE IN THE MAMMARY GLAND

Sinai Yarus, Darryl Hadsell and Jeffrey M. Rosen

Department of Cell Biology
Baylor College of Medicine
One Baylor Plaza
Houston, TX 77030

INTRODUCTION

Our original interest in the milk protein genes stems from studies of Yale Topper and his colleagues performed almost thirty years ago in which casein expression in mammary explant cultures was employed as a model system for understanding how peptide and steroid hormones regulate milk protein gene expression (1). Once the milk protein cDNAs and genes were cloned in the early 80s (2-4), we were confronted with the question, "What is the appropriate biological model in which to study milk protein gene regulation?" Most of the available breast tumor cell lines, even those that retain hormone receptors, do not express the milk protein genes, and transfection studies with these cells are not particularly informative. Primary mammary epithelial and explant cultures are also highly variable and difficult to transfect. Fortunately, the development of transgenic mice as a basic research tool provides an ideal model in which to study the hormonal and developmental regulation of milk protein gene expression. Transgenic mice are especially valuable for the analysis of complex systems such as mammary gland where no single cell line can mimic the complete spectrum of developmental, hormonal, cell-substratum, and epithelial-stromal interactions. In 1984 studies were begun to identify the *cis*-acting sequences required to elicit the appropriate tissue-specific and hormonal regulation of milk protein gene expression in transgenic mice (5). These studies focus on two rat milk protein genes encoding ß-casein and the most abundant rodent whey protein, whey acidic protein (3,4).

Our studies and those from a number of other laboratories have led to the identification of tissue-specific regulatory elements in several milk protein genes from different mammalian species. These are composite regulatory elements containing binding sites for several transcription factors (6). This basic research in the area of gene regulation provides the foundation for the development of the mammary gland as a bioreactor. Extension of these studies performed initially in transgenic mice has confirmed the feasibility of targeting heterologous gene products to the mammary gland in transgenic livestock (7).

The focus of this chapter is to review the general utility of the mammary gland as a bioreactor with emphasis on two areas:

Genetic Engineering, Vol. 18, Edited by J.K. Setlow
Plenum Press, New York, 1996

1. The effects of ectopic production of foreign proteins on mammary gland development--their importance as basic research tools, as well as their impact on the use of the mammary gland as a bioreactor.
2. Its ability to produce biologically active proteins contrasting heterologous proteins with those proteins normally expressed in mammary gland and secreted into milk.

TRANSGENIC TECHNOLOGIES AS TOOLS IN MAMMARY GLAND BIOLOGY

Mammary Gland Development

The development of the mammary gland into an organ capable of secreting milk is regulated by several types of signals: those generated by cell-cell interactions, those generated by extracellular matrix, paracrine signals produced within the mammary gland and endocrine signals. In the intact animal the proper orchestration of these signals results in a developmental program that can be divided into five stages based on age of the animal, pattern formation within the gland and functional properties of the gland. The purpose of this section is to describe these stages briefly as a basis for discussion of the effects of specific transgenes on mammary gland development and lactation.

Prenatal development. The development of the embryonic mammary gland occurs as a result of specific interactions between epidermal cells and their underlying mesenchyme (see Sakakura (8) for a detailed review). In the mouse this interaction is characterized by two distinct steps. The first occurs at embryonic day 12 to 13 and involves condensation of ventral epidermis into several discrete buds along the dorso-ventral axis. The second occurs at embryonic days 16 to 17 and consists of elongation and branching of these epithelial buds to form a ductile tree that begins to penetrate the developing mammary fat pad, a mesenchymal structure that develops from cells situated deeply within the subcutis. By the time of birth, this rudimentary anlagen consists of cords of partially canalized ducts made up of epithelial cells that, though not committed to forming mammary gland structures *per se*, are committed to differentiation into secretory mammary epithelial cells. Although this process has been thoroughly described, the specific endocrine and paracrine factors that regulate it are still poorly understood.

Ductile development. Postnatal development of the mammary gland is characterized by a stage of ductile morphogenesis (reviewed in detail by Daniel and Silberstein (9)) followed by a stage of lobulo-alveolar development (discussed below). Little or no ductile development occurs in the neonate until the age of 3 wks. From 3 wks of age until maturity at 12 wks, terminal end buds invade the surrounding fat pad forming a highly ordered arbor. This process requires estrogen and growth hormone and is driven by the local effects of several growth factors. A major factor in the pattern formation itself is transforming growth factor β (TGF-β). This growth factor is found within the extracellular matrix surrounding existing ducts and acts as a cell cycle inhibitor preventing terminal end buds from growing into existing ducts. Ductile development ceases when the terminal end buds reach the limit of the fat pad. The epithelium then becomes relatively quiescent until the animal becomes pregnant.

Lobulo-alveolar development. With pregnancy, the mammary epithelium is again stimulated to proliferate. This proliferation fills the remaining fat pad that exists between ducts with clusters of alveoli. These lobules originate from specialized side branches termed alveolar buds, and their proliferation is driven by progesterone, prolactin, adrenocorticotropic hormones and a variety of growth factors (see Vonderhaar and Plaut (10) for a detailed review). With this continued development comes gradual differentiation of the alveolar epithelial cells. This process is evident by the increased expression of milk protein genes and the accumulation of secretion within alveoli and ducts. Complete differentiation occurs with parturition, at which time cell proliferation slows and the epithelial cells display high-level expression of genes involved with the synthesis of lactose and triglycerides, as well as the major milk proteins.

Lactation. The subject of lactation is a very complex topic that can be subdivided into many areas. These include the regulation of milk protein synthesis (11), of lipid biosynthesis (12), of carbohydrate metabolism (13) and the physiological impact of milk secretion on the lactating female (14). In general, this process is characterized by the secretion of large quantities of milk containing various amounts of lipid, protein and carbohydrate depending on the species. The duration of lactation also varies with species, and depends upon the continued stimulation of the gland and the removal of secretion. If the secretion is not removed the gland involutes to a structural and physiological state very similar to that which is found prior to pregnancy.

Involution. The involution of the mammary gland normally results from the cessation of suckling and the failure to remove secretions. The physiological factors that regulate this process are not completely identified, but in part involve decreased stimulation of the gland by prolactin (15) and milk stasis. Detailed studies of the process in the mouse (16) have shown that epithelial apoptosis accompanied by dramatic changes in the expression of matrix metalloproteinases is responsible for the loss of secretory lobules following lactation that remodels the once lactating tissue into an organ ready for another round of proliferation and lactation.

Biological Limits to the Mammary Gland Bioreactor

As illustrated above, mammary development and lactation are complex processes whose regulation is not totally understood. As such, the use of this organ as a bioreactor in the production of pharmaceuticals presents potential problems associated uniquely with the effects of each expressed product on the bioreactor itself. These problems can be classified into three basic types: those that interfere with the physical process of milk secretion or ejection, those associated with the physiology of the lactating female and those associated with the development and differentiated function of the mammary gland. The following section will discuss and provide specific examples of each of these. Subsequent sections will then focus on what has been learned from these types of biological phenomena about mammary gland development and lactation.

Biophysical effects. The continued high-level production of milk by the mammary gland requires the efficient secretion of products from each epithelial cell as well as the regular removal of the such products from the organ in a timely fashion. To ensure that these processes occur, both the structure of the gland and the composition of milk have likely evolved together giving rise to compatible biophysical properties between the secretion and the gland (17). Consequently the overexpression of transgene products (18) or the use of homologous recombination to modify milk composition may have drastic effects on the ability of the animal to maintain lactation. For example, the targeted ablation of the α-lactalbumin (*α-lac*) gene in mice causes the production of lactose-free, abnormally viscous milk (19). These females fail to lactate because of the inability of the myoepithelium and the suckling pups to remove the milk. Likewise, expression of human surfactant protein C at high levels leads to a failure to lactate (18). This phenomenon may be linked to the insertion of the protein into the apical plasma membrane of the mammary epithelial cell, and consequent interference with the secretion of cellular products.

Systemic physiological effects. Alteration in the systemic physiology of the transgenic female by transgene products also represents a problem encountered in the production of hormones or cytokines. This phenomenon occurs when the transgene product is secreted inappropriately into the bloodstream by the mammary epithelium or when it is inappropriately expressed in organs other than the mammary gland. For example, with the use of whey acidic protein (WAP)-based constructs that are designed to target human growth hormone (hGH) expression to the mammary gland (20), both mechanisms may account for increased hGH in blood resulting in infertility. Previous studies demonstrate (personal communication, P. Neville) that mammary epithelial cells secrete hGH both basalaterally and apically. Furthermore, both rat and mouse WAP-promoter based hGH constructs express hGH in ectopic sites such as brain, liver, kidney and thymus. The resulting infertility is thought to be due to suppression of luteal function (21). Infertility is also a problem observed in some lines of mice carrying a ß-casein promoter-driven, modified bovine follicle-stimulating hormone (FSH) transgene (personal communication, N.M. Greenberg). This again occurs due to FSH in the bloodstream derived from ectopic transgene expression. While fertility problems occur, they are relatively minor as lactation can be induced artificially without the need for a pregnancy. However, other more life-threatening effects may result from abnormal secretion or expression of other transgene products.

Direct developmental effects. Direct autocrine effects of secreted transgene products on the proliferation and/or differentiation of the mammary epithelial cells, or paracrine effects on their supporting stroma, create a problem that is likely the most difficult to overcome. Striking examples are found with transgenic mice that express mWAP, des(1-3)hIGF-I and hGH. Failure to lactate, or the *milchlos* phenotype, frequently occurs in primiparous mWAP transgenic mice (22). This is thought to result from the inhibition of alveolar development resulting from the premature expression of this milk whey protein. With des(1-3)hIGF-I, lactational failure also occurs (23). In this case, decreased alveolar development also occurs, but much more slowly and to a more variable extent. In contrast, hGH transgenic mice exhibit precocious mammary gland development (20,24). In fact, strains of transgenic mice in which hGH expression is driven by the phosphoenolpyruvate carboxykinase (PEPCK)-promoter are capable of pregnancy-independent lactation (25). The potential impact of this precocious development on the productivity of these animals is not certain and may even be beneficial. Regardless, precocious development, as well as *milchlos*, represent the types of potential roadblocks to the efficient production of recombinant proteins in the milk of transgenic animals for commercial purposes.

The preceding sections illustrate some potential consequences encountered in the production of bioactive recombinant proteins in the mammary glands of transgenic animals. These consequences, while having negative implications for the biotechnology industry, represent a boon to biologists interested in understanding the mechanisms governing the development and differentiation of the various cells that constitute the mammary gland. Transgene constructs which target oncogenes, growth factors and cytokines have provided insights into mammary gland development, lactation and tumorigenesis for more than a decade, and continue to be useful in the production of experimental models. The remainder of this section will focus primarily on the use of such models to understand mammary gland development and lactation with only minor aspects of tumorigenesis being discussed where necessary. For a more complete review of transgenic models of tumorigenesis see Cardiff and Munn (26).

Current Transgene Targeting Strategies

Several types of transgene constructs have been used to study the impact of oncogenes, growth factors, cytokines and viral proteins on the mammary gland (Table 1), or to target heterologous proteins of pharmaceutical value to the mammary gland (Table 2). These constructs, though frequently made with some regard for the inclusion of introns and RNA processing signals, often still rely on integration into a transcriptionally active location of the genome for efficient expression. Regardless, useful lines can be obtained providing that the tissue specificity and temporal pattern of expression do not compromise the design of the experiment or the life of the animal. In general, constructs fall into several categories based on when and where they are expressed. These factors are critical in determining the experimental limitations of a particular line of mice and will each be addressed separately below.

Temporal regulation. The timing of transgene expression is a critical parameter which determines whether the expressed transgene will effect early mammary gland development or lactation, or both. There currently exists no promoter capable of targeting specifically to the embryonic mammary gland. This is a critical shortcoming because it precludes the use of transgenes to study lineage commitment, which has already occurred in the epithelium by this time (27,28). However, promoters do exist that are capable of allowing the use of transgenic technology in studies of ductile and lobulo-alveolar development, as well as lactation. In general these promoters show some degree of inducibility by such things as hormones of dietary manipulation and vary broadly in their temporal patterns of expression and tissue specificity.

The mouse mammary tumor virus LTR (MMTV) is most frequently used to target transgene expression to the developing mammary gland of the post-pubertal mouse. Expression of MMTV generally depends on glucocorticoids and prolactin, and increases with pregnancy and lactation (29). However, many MMTV-based transgenes show readily detectable expression in mature virgin females of 8 to 12 weeks (Table 1). Some studies suggest that MMTV-driven constructs may be expressed as early as one to three weeks of age (30-32), making MMTV well-suited for studying virgin mammary gland development from a temporal standpoint.

The WAP gene promoter also displays glucocorticoid dependence (33) and transcriptional activity in the virgin mammary gland (34,35), but this activity is much lower than that of MMTV, making WAP less suited to targeting the virgin mammary gland. However, during pregnancy and lactation WAP promoter activity increases by 1000-fold (36), making it well suited to the targeting studies focused on late pregnancy and lactation. This is also the case for the majority of the other milk protein-based transgenes. In fact all but a few (Table 1) (34,37) of the studies with the WAP promoter require at least a single lactation to observe phenotypic effects that often occur during pregnancy, lactation, or involution (23,37-40).

Other milk protein gene promoters of potential value for experiments requiring targeted expression to the pregnant or lactating mammary gland include those for the caseins and other whey proteins such as α-lactalbumin, and β-lactoglobulin. Transgenes derived from the rat β-casein gene (5,41) express only during pregnancy and lactation.

Table 1. Transgenes Used to Study Mammary Development, Lactation and Tumorigenesis

Promoter	CDS[a]	Earliest evidence for expression[b]	Tissue distribution[c]	Preneoplastic Developmental effects	Site of action	Ref.
MMTV	*myc*	NA[d]	MG,SG,Te,Lu,I	none	intracellular	(78)
	Ha-*ras*	4 weeks	MG,SG,HG,Sp,Th, SV	none	intracellular	(74)
	wnt-1	2 weeks	MG,SG,Te,SV,VE	alveolar hyperplasia	secreted, localized	(137)
	SV40t	14 weeks	MG,Lu,Ki,Sp	dysplasia	intracellular	(138)
	c-*neu*	NA	MG,SG,Ep	alveolar hyperplasia	intracellular	(139)
	neu-NT	8 weeks	MG,SG,Ep	alveolar hyperplasia	intracellular	(140)
	TGF-α	12 weeks	MG,SG,Te,SV	alveolar hyperplasia/lactational failure	secreted, localized	(79)
	int-2	lactation-dependent	MG,SG,Pr,SV,VD	alveolar hyperplasia	secreted, localized	(53)
	ret	lactation-dependent	MG,SG,Te,HG	some alveolar dysplasia	intracellular	(141)
	N-*ras*	NA	MG,SG,HG,Lu,Te, Ov,Pr	none	intracellular	(142)
	PV-mt	3 weeks	MG,SG,Ov	alveolar hyperplasia	intracellular	(31)
	int-3	7 weeks	MG,SG,Ep	ductile dysplasia	membrane bound	(84)
	TGF-β1	7 weeks	MG,SG	ductile dysplasia	secreted, localized	(59)
	cyclin D1	8 weeks	MG	alveolar hyperplasia	intracellular	(80)
	RARβ4-like	NA	MG,SG,Te,Lu	cystic ductal hypertrophy	intracellular	(143)
	stromelysin	NA	NA	ductile hyperplasia precocious alveolar development	secreted, localized	(67)
	mWAP	NA	MG,SG,Sk,Lu, Sp,Pr,Ep,SV	alveolar dysplasia	secreted, localized	(86)
mWAP	Ha-*ras*	lactation	MG,Br,SG	none	intracellular	(77)
	PS2	lactation	MG	none	secreted, localized	(144)
	GH	pregnancy	MG,Br	none	secreted, localized	(51)
	mWAP	pregnancy	NA	alveolar dysplasia	secreted, localized	(22)
	TGF-β1	5 weeks	MG	alveolar dysplasia	localized	(39)

	SV40t	pregnancy	NA	lactational failure	intracellular	(145)
	TGF-α	lactation	NA	alveolar hyperplasia	secreted, localized	(40)
	stromelysin	pregnancy	NA	ductile hyperplasia	secreted, localized	(37)
rWAP	rWAP	pregnancy	MG,Ki,To	none	secreted, localized	(36)
	p53	pregnancy	MG,Lu,Sp,Ov	alveolar dysplasia	intracellular	(52)
	des(1-3)hIGF-I	pregnancy	MG,SG	ductile hyperplasia and alveolar dysplasia	secreted, localized	(23)
bWAP	hGH	pregnancy	variable: MG,Li,Th,Br,SG	precocious alveolar development	secreted, systemic	(20)
MT	TGF-α		B,Li,Lu,St,Ki,Br, SM,Ut,Sk,C,Pa, MG,Te	alveolar hyperplasia	secreted, ubiquitous	(146)
	GH			none reported	secreted, systemic	(147)
αs1-casein	IGF-I	NA	NA	none	secreted, local	(65)
PEPCK	GH	1 day	Li,MG,A,Ki,I	NA	secreted, systemic	(45)
HMGCoA-reductase	GH	NA	Ubiquitous	precocious alveolar development	secreted, systemic	(24)

[a] Coding sequence
[b] Includes biochemical or phenotypic observations
[c] Tissues examined: B, bone; Br, brain; C, colon; Ep, epididymis; HG, harderian gland; I, intestine; Ki kidney; Li, liver; Lu, lung; MG, mammary gland; Ov, ovary; Pa, pancreas; Pr, prostate SG, salivary gland; Sk, skin; SM, skeletal muscle; Sp, spleen; St, stomach; SV, seminal vesticles; Te, testis; Th, thymus; Tc, tongue; Ut, uterus; VD, vas deferens; VE, vas dererens/epididymis
[d] NA, not applicable

Table 2. Use of the Mammary Gland as a Bioreactor

Transgene	Promoter	Host species	Processing	*In vitro* activity	Ref.
α and β FSH (bovine)	β casein (rat)	mouse	subunit assembly	yes	(105)
urokinase (human)	αs1-casein (bovine)	mouse	correct folding assumed from activity and Western blot data	yes	(106)
IGF-I (human)	αs1-casein (bovine)	rabbit	signal peptide cleavage	yes	(65)
IGF-I (human) des (1-3) IGF-I (human)	WAP (rat)	mouse	roughly 30% of transgene protein is cleaved to yield mature form	yes	(23)
lactoferrin (human)	αs1-casein (bovine)	mouse	indirect analyses: RIA Western blot	ND	(88)
growth hormone (human)	α-s1 casein (bovine) β casein (bovine) α-lactalbumin (bovine) WAP (murine)	rat rat rat rat	inferred from RIA inferred from RIA inferred from RIA inferred from RIA	ND ND ND ND	(148)
trophoblast interferon (ovine)	α-lactalbumin (bovine)	mouse	ND	yes	(133, 149)
anti-hemophilic Factor-IX (human)	ovine β-lactoglobulin	sheep	incomplete glycosylation inferred from Western blot data γ-carboxylation presumed to be correct due to activity; β-hydroxylation not assayed	yes	(100)
α_1-antitrypsin (human)*	ovine β-lactoglobulin	mouse	ND	yes	(103)
α_1-antitrypsin (human)*	ovine β-lactoglobulin	sheep	N-glycosylation	yes	(104)
γ-interferon (human)	ovine β-lactoglobulin	mouse	ND	yes	(121)
α_1-antitrypsin (human)	rabbit WAP	mouse	glycosylation (inferred from Western blot)	yes	(101)
tissue plasminogen activator (human)	mouse WAP	mouse	inferred from Western blot and ELISA	ND	(97)
LA tissue plasminogen activator (human)	mouse WAP	goat	ND	yes	(98, 99)
protein C (human)	mouse WAP	mouse	aberrant endoproteolysis partial γ-carboxylation	yes	(109)
protein C (human)	mouse WAP	pig	aberrant endoproteolysis partial γ-carboxylation	yes	(107, 108)
surfactant protein C (human)	mouse WAP	mouse	not processed	ND	(18)
superoxide dismutase (human)	mouse WAP	mouse	glycosylation (not 100% similar to native protein)	yes	(102)

*Same transgene construct

Likewise, expression of transgenes derived from the *α-lac* (42) and the β-lactoglobulin (43) genes occurs only during pregnancy and lactation. The temporal regulation of these genes makes them potentially valuable tools for studies on lactation in transgenic mice. To date they have been employed primarily for the production of pharmaceuticals in the mammary gland bioreactor, as will be discussed in more detail in subsequent sections.

Other promoters that may be useful tools in studying mammary gland development and lactation include those derived from the metallothionein 1 (MT1) (44), PEPCK (45) and hydroxylmethylglutaryl CoA(HMGCoA)-reductase (24) genes. Because these promoters show high expression in the mammary glands of prepubertal mice, they may facilitate studies during the early stages of mammary development without the need for screening different lines for appropriate temporal regulation. Furthermore the MT1 and PEPCK promoters have the advantage that they are inducible by dietary factors (44,46). However all of these promoters may have limitations stemming from the altered tissue-specificity which they exhibit.

Tissue distribution. One of the most valued attributes of transgenic models in developmental studies is their potential for allowing the localized exposure of a particular cell type within a tissue to the specific gene product of interest. Of the promoters discussed thus far, the WAP (36), β-casein (5,41), β-lactoglobulin (47,48) and *α-lac* (42,49) promoters all show a high degree of tissue-specificity. For WAP and β-lactoglobulin, the salivary gland is the only other organ where expression is observed. For β-casein, the thymus shows some expression, and for *α-lac*, expression has also been observed in sebaceous glands. In all of these cases the transgene is expressed to a much greater extent in the mammary gland than in other organs. Occasional constructs (50-52) also show expression in other organs, but this ectopic expression can be attributable to either sequences derived from the structural gene or integration into a transcriptionally active locus within the genome (50).

The MMTV promoter tends to display a more relaxed tissue-specificity than the milk protein gene promoters and is often expressed in salivary gland, Harderian gland, testis and spleen. Consequently, transgenic mice carrying MMTV-based constructs are often difficult to propagate due to infertility in males (53). Regardless, MMTV is still the most valuable promoter available for developmental studies as it shows a preference for mammary tissue over the other expressing tissues (54).

The other promoters discussed above display dramatically altered tissue-specificity relative to the milk protein genes and MMTV. The HMGCoA-reductase promoter is ubiquitously expressed (55), while PEPCK shows greater expression in liver and kidney than mammary tissue (45). Consequently with these promoters, specific attention must be paid to the possibility that biological effects of the transgene on organs other than the mammary gland may complicate the interpretation of results, as well as interfere with the propagation of the mice.

While numerous examples of the value of transgenic mice in the study of mammary gland biology exist, the temporal and spatial problems discussed above often translate into an inability to propagate a specific line either due to transgene lethality or infertility. One strategy that is potentially useful in dealing with the limitations discussed above is tissue transplantation. Transplanted mammary tissue, when placed into the cleared fat pad of a syngeneic host, will typically develop in a manner similar to that observed for the endogenous mammary epithelium (56). These types of studies are thus useful in tissue recombination experiments between transgenic and non-transgenic animals. For example, transplantation studies with *wnt*-1 mammary epithelial cells support the conclusions that *wnt*-1 activity does not require a transgenic stroma and is capable of driving ductile morphogenesis in the absence of ovarian steroids (30). For WAP-TGF-ß (34) and *int*-2 (57), transplantation studies support the conclusion that these factors act as autocrine effectors of mammary development. Thus, from the standpoint of gaining specific information about epithelial-stromal interactions, transplantation studies may be quite informative. More importantly, transplantation may be a useful strategy to gain information about mammary development from transgenic mice where the transgene product causes early lethality.

Modulation of Transgene Product Bioactivity

Once the issues of temporal and tissue specificity are addressed with any transgene construct, the final issue must be appropriate biological activity of the transgene product. For many of the transgenes listed in Tables 1 and 2 this issue is simply a matter of ensuring that the coding sequence is accurate and that the appropriate translational signals are present. However, in some instances, specific post-translational steps are required to activate or maintain activity of the transgene product. These steps include dissociation of the transgene product from inhibitory proteins, alterations in protein conformation as a result of proteolysis or ionic interactions and/or cooperative interactions with other gene products. In the case of TGF-ß1 and insulin-like growth factor I (IGF-I), specific protein-protein interactions inhibit biological activity. For TGF-β1, this inhibition results from a post-translational non-covalent association between the pro-region of TGF-ß1 and the mature region of the molecule causing the secretion of a latent form that can be activated *in vitro* by exposure to acidic pH (58). Mutation of Cys residues 223 and 225 to Ser within the pro-protein region inhibits its association with the mature region (58) in transfection studies, and also allows activated TGF-ß1 to be targeted to the mammary gland of transgenic mice (39,59). For IGF-I, inhibition of activity may result from the non-covalent association with any one of six IGF-binding proteins (60) that are normally found in extracellular fluids. By deletion of the first three residues in the mature ligand (61) or by modifying the sequence in the B-domain of the protein to more closely resemble that of insulin (62), the interaction with these binding proteins is inhibited and the potency of the resulting molecule is increased dramatically (61,63,64). Transgenic mice that express the N-terminal deleted form of IGF-I not only show increased IGFBP expression but also show alterations in mammary gland morphology (23). Transgenic rabbits (65) that overexpress the normal IGF-I molecule in the mammary gland have not been reported to show developmental abnormalities. For stromelysin, the coordination of Zn^{++} by a Cys residue within the "autoinhibitor" region causes inhibition that can be overcome by mutating an adjacent Val or Pro residue to Gly or Val, respectively (66). The use of cDNAs containing either of these mutations to target expression to transgenic mice with either MMTV- (67) or WAP-driven constructs (37) results in profound alterations in mammary gland development and tumorigenesis. All of the above examples support the idea that specific modifications may need to be made in the coding sequences of some transgenes to ensure biological activity in the context of the transgenic animal.

Other factors to consider when proteins are targeted to the mammary gland for specific studies include the genetic background of the mice in which the studies are to be conducted, and the possibility that cooperative interactions between the transgene protein and other gene products may be required to observe biological effects. Markedly different phenotypic effects have been observed because of this "genetic background" effect. The most recent example occurs in homozygous mice containing a targeted disruption of the epidermal growth factor (EGF) receptor gene (68,69). However, the effect has also been observed in transgenic mice (70,71) and may be due to strain-dependent differences in methylation (71). These genetic background effects, while often causing difficulty in interpretation of experimental results, point to the requirement for cooperative protein-protein or protein-DNA interactions for certain biological processes to occur.

The importance of cooperative interactions between the transgene and other gene products is best illustrated by studies performed with transgenic mice that overexpress polyoma virus middle t (pv-mt). These mice display a dramatic alveolar hyperplasia at a very young age in part as a consequence of the activation of c-*src* (31). The critical importance of c-*src* was established by crossing these mice with mice containing a targeted disruption of the c-*src* gene (72). The resulting offspring fail to exhibit pv-mt induced hyperplasia. Likewise, transgenic mice that express *wnt*-1 under the control of MMTV display accelerated tumorigenesis when crossed with p53-deficient mice (73). This effect is believed to be due to increased genomic instability observed in the absence of p53 in these bigenic mice. Other examples of specific cooperative effects can be found in studies involving bigenic mice expressing such gene products as *ras* and *myc* (74,75), or TGF-α and *myc* (32,40) or *wnt* -1 and *int* -2 (76). These examples illustrate the kinds of studies that can be performed with the use of transgenic mouse models.

Advances Due to Transgene Technology

Studies on mammary gland biology have been limited by the fact that no cell culture models exist that accurately reflect the *in vivo* biology of this organ. Transgenic mice provide an opportunity to ask critical questions concerning the regulation of mammary gland development and lactation. Such questions may serve to confirm predictions based on experiments conducted in cell culture models or to elucidate the function of novel gene products or even to stimulate reevaluation of the mechanisms by which well-characterized gene products function. For example, transgenic experiments with *ras* (74,77), *myc* (78), TGF-α (79) or cyclin D1 (80) have confirmed predictions based on cell culture studies that these gene products would be tumorigenic, and that transformation usually requires multiple "hits" as well as the cooperative effects of several oncogenes and the loss of tumor suppressor genes. The *wnt*-1 mice, on the other hand, have also proven extremely valuable in elucidating the function of novel gene products. Through the use of these mice in retroposon tagging studies, such factors as *int*-2/Fgf-3 (81), *hst*/Fgf-4 (81) and FGF-8 (82) have been identified as important cooperating gene products in the process of mammary tumorigenesis. Likewise, the use of *int*-2 mice has identified a novel member of the *wnt* family that may be involved in tumorigenesis (83). In another example, studies on transgenic mice that overexpress *int*-3 (84,85), a protein homologous to the *Drosophila* Notch protein, support the conclusion that Notch-related proteins may play a role in the determination of cell fate within the mammary gland. Finally, in terms of altering our understanding of the function of well-characterized proteins, recent studies with mice that overexpress WAP either under control of the WAP or MMTV promoters (22,86) now suggest a novel function for this milk protein in the regulation of mammary epithelial cell differentiation. Studies with stromelysin 1 support the suggestion that matrix metalloproteinases, through their effects on extracellular matrix proteins, control both mammary cell differentiation (37) and apoptosis (67). Other examples of this type can be found in mice that overexpress TGF-ß (34,39), p53 (52) and des(1-3)hIGF-I (23). All of these studies emphasize that transgenic mice, despite some current limitations, serve as a valuable model for studies involving mammary gland development, lactation and tumorigenesis.

The Mammary Gland as a Bioreactor

A mammary gland transgene construct should ideally meet several criteria if production of a foreign protein in milk for pharmaceutical or other use is intended. First, the construct should target expression primarily to the mammary gland. Second, the construct should be regulated appropriately during development so that expression occurs primarily during late pregnancy and lactation. Third, the construct should provide a high level of expression (mg/ml or g/l in this context) of the correctly-processed protein to facilitate large-scale production. Much of the previous research in this field has been directed at achieving these objectives (6,41,43,50,87,88). Thus regulatory elements from the major milk protein genes have been used with varying degrees of success to direct expression of heterologous genes to the mammary gland in transgenic mice (5,36,43,48,50,89,90). A number of excellent reviews on milk gene regulatory sequences and their ability to target transgene expression to the mammary gland have been published previously (7,91-94). In addition, a searchable database of mammary gland-directed transgenes is available through the World Wide Web (WWW) (95).

Despite these successes there are still several important scientific issues that need to be addressed before this technology can be applied in transgenic livestock. First, it is important to demonstrate the fidelity of the post-translational modifications of the heterologous proteins expressed in milk, and that they exhibit the appropriate biological activities. Second, because of the high cost of generating transgenic livestock, methods need to be developed to improve the frequency of high-level transgene expression in these animals. While these goals have not yet been accomplished, some recent progress has been made in these areas. The importance of the regulatory status of products likely to arise from this research is discussed in a separate section, although it is not a scientific issue.

Post-Translational Modifications

Having demonstrated the feasibility of expressing high levels of heterologous proteins in milk in transgenic animals, more recent research has been directed towards understanding the ability of the mammary gland to perform various post-translational modifications and the generation of biologically active molecules. Although, these analyses are still incomplete in many cases, the following section summarizes some of the successes and failures of the mammary gland in performing post-translational modifications of heterologous proteins.

One of the principal rationales for using a mammary gland bioreactor is that the mammary gland is capable of complex post-translational processing which cannot be achieved in bacterial or yeast expression systems (96). This processing capability includes cleavage of signal peptides from the nascent peptide chain, both N- and O- glycosylation, γ-carboxylation and O-phosphorylation (96). The evaluation of post-translational processing events such as cleavage, subunit assembly and side chain addition in some cases involves direct biochemical analyses, while in other cases *in vitro* bioactivity serves to provide an overall picture of the mammary gland's potential for faithfully producing non-milk proteins. A summary of some of the published studies is included in Table 2.

The ability of the mammary gland to perform post-translational modifications of a non-milk protein is presumed to depend on several factors. These include: 1) the presence of the correct processing enzyme in the gland that performs a similar function on a milk (or cellular) protein, 2) the ability of such an enzyme to recognize a site in the transgene protein, 3) the presence of enzyme and substrate in the same subcellular compartment and

4) a sufficient quantity or activity of enzyme to process the large amounts of the transgene protein being produced. The interaction between these factors is complicated and at present the empirical evaluation of the success of post-translational processing is the only available approach.

Glycosylation. The addition of carbohydrate side chains to a transgene protein is reported for human tissue plasminogen activator (tPA) (97-99), human anti-hemophilic Factor-IX (FIX) (100), human α_1-antitrypsin (α_1AT) (101), human lactoferrin (hLF) (88) and human extracellular superoxide dismutase (EC-SOD) (102).

Correct glycosylation of tPA produced in the mammary gland is inferred from ELISA and Western blot data, as well as the fact that the milk has biological activity in a fibrin clot assay (97). Subsequent studies on a tPA variant in transgenic sheep include a more detailed analysis of the glycosylation pattern of the transgenic tPA isolated from milk in comparison to tPA isolated from tissue culture cells (98). The milk-derived tPA contains significantly lower levels of galactose, N-acetylglucosamine and sialic acid. In contrast, it contains N-acetylgalactosamine which is not present in the cell culture-derived tPA. Fucose and mannose levels are similar in both preparations. Since there is no evidence of O-glycosylation, the N-acetylgalactosamine residue is probably part of an N-linked oligosaccharide as is the case for α-lactalbumin (98). In addition to these stoichiometric analyses, the milk-derived tPA is variably glycosylated at asparagine 184 as observed in native tPA. Both milk-derived and native tPA are invariantly glycosylated at asparagine 448, while the glycosylation of asparagine 117 reported for native tPA is not reported for the milk-derived material (98).

Analyses of clotting factor IX (FIX) in sheep milk also indicate partial success in side chain addition by the mammary gland (100). Samples of FIX purified from transgenic sheep milk with the use of an anti-FIX monoclonal antibody do not contain native FIX, based on electrophoretic mobility and Western blot data (100). This may indicate incomplete glycosylation. However, since the FIX glycosylation sites are primarily on the activation peptide, this did not interfere with FIX activity. Conversely, the presence of activity in a one-stage clotting assay implies γ-carboxylation, which is known to be necessary for function. The issue of β-hydroxylation is not addressed by either the Western blot experiments or the clotting assay (100). For human lactoferrin, N-glycosylation is also inferred from radioimmunoassay data (88).

Human α_1-antitrypsin has been produced in both mice (101,103) and sheep (104). The α_1AT from milk of transgenic mice displays the same electrophoretic mobility as purified human α_1AT which is known to be glycosylated. Recombinant α_1AT from *E. coli*, which is not glycosylated, has enhanced electrophoretic mobility. This suggests glycosylation of the milk-derived α_1AT (101). Glycosylation of recombinant transgenic α_1AT is confirmed by analyses of sheep milk by both electrophoresis and N-glycosidase analyses (104). Analyses of whole milk suggest that all of the transgenic α_1AT is glycosylated completely (104). While the exact composition of the carbohydrate side chains is not reported, there is agreement on the activity of the recombinant transgenic α_1AT in an *in vitro* assay (101,103,104). This implies that the glycosylation pattern of native and milk-derived α_1AT is either similar, or not critical in the protein's function.

The glycosylation pattern of human extracellular superoxide dismutase (hEC-SOD) in the milk of transgenic mice is reported based upon lectin binding studies (102). The native protein contains a terminal galactose which is apparently absent or blocked in both milk-derived and CHO cell-derived recombinant EC-SOD. The binding of milk-derived EC-SOD to *Sambucus nigra* agglutinin (SNA) lectin suggests blocking of the terminal galactose with a sialic acid residue in the α2,6 or α2,3 position (102). Despite this difference, the purified milk-derived EC-SOD is active in an *in vitro* assay (102), again implying that the exact glycosylation pattern is not critical for enzyme function.

Cleavage, folding, and subunit assembly. Post-translational cleavage, subunit assembly and folding of transgenic proteins has also been reported for heterologous proteins expressed in the mammary gland (65,105-107). Correct assembly of a heterodimeric FSH molecule in milk of transgenic mice has been reported in bigenic mice carrying both α and β FSH transgenes. The milk-derived hormone is active in both radioreceptor and granulosa assays (105). SOD also exhibits subunit assembly, since the functional enzyme is a homotetramer (102).

Correct folding of urokinase produced in milk of transgenic mice is inferred from zymograph and Western blot data (106). Milk-derived urokinase has the correct molecular weight and blocks plasminogen activator activity. Native urokinase contains nine disulfide bonds, presumed to be involved in its activity (106). Correct folding suggests that the secondary structure of the protein is dictated by its primary structure, or that the necessary chaperone molecules are normally present in the mammary gland.

The protein C molecule is normally cleaved by an endoproteolytic enzyme at Lys^{-2}-Arg^{-1} and Lys^{156}-Arg^{157}. Recombinant human protein C (hPC) produced in the milk of transgenic swine contains a higher percentage of single-chain material than native hPC (107). This indicates that the Lys^{156}-Arg^{157} dipeptide normally cleaved in the processing of this molecule is not cleaved with 100% efficiency in the mammary gland. More detailed analysis of hPC from swine milk reveals that 10 to 20% of the molecules retained their pro-peptide while 30 to 40% were in the single-chain form (108). In addition, cleavage at an additional site (Lys^{151}-Arg^{152}) apparently occurs in the mammary gland (108). These observations are similar to those made for hPC in murine milk (109). The heavy- and light-chain material produced in the mammary gland are slightly smaller than those of native hPC (107), suggesting differences in glycosylation between the recombinant and native proteins. The milk-derived hPC is apparently only partially γ-carboxylated based upon its low specific activity and amino acid sequence analysis (107,109). Purification of hPC from the milk of transgenic swine with enzymatic and anticoagulant activities similar to those of hPC from plasma has been reported despite these processing problems (110).

Recombinant human IGF-I has been purified from the milk of transgenic rabbits and is biologically active (65). Cleavage of a bovine αs1-casein signal peptide from the transgene protein is at the correct site and occurs in all of the transgenic material (65). IGF-I and des (1-3)hIGF-I are also produced in the milk of transgenic mice with the use of a rat WAP-based construct (23). In this case about 30% of the transgene product is correctly cleaved as determined by Western blot analysis. The remainder is secreted as an uncleaved precursor.

Secretion of an uncleaved pro-protein is also reported for human pulmonary surfactant protein C (18) produced with the use of the same rat WAP vector employed in the IGF-I and des(1-3)IGF-I studies (23). Human pulmonary surfactant protein B is also secreted as an unprocessed pro-protein (S. Yarus, unpublished data) with the use of the same vector as surfactant protein C (18).

Summary. In summary, the mammary gland seems able to perform a wide range of post-translational modifications on non-milk proteins with enzymes already in place for processing of milk proteins. Successful processing is dependent upon the ability of the enzyme in question to recognize a target site in the heterologous protein. In addition, the mammary cells may attach a different side chain than the cells normally producing the transgene product (98). There is also the possibility that the amount of transgene protein might exceed the capacity of an endogenous enzyme to process it, as reported for endogenous casein phosphorylation (96). This may be the case for protein C (107) and des (1-3)IGF-I (23), in which only some of the transgene protein is correctly cleaved. Even assuming relatively low levels of transgene protein, and the existence of the correct enzyme for processing in the mammary gland, cleavage is not necessarily assured. For example, pulmonary surfactant protein B is known to be cleaved by cathepsin D (111), an enzyme known to be present in the lactating mammary gland (112,113), yet transgenic pulmonary surfactant protein B is secreted as an uncleaved pro-protein in milk (S. Yarus, unpublished data). This suggests that there is an issue of compartmentalization and trafficking in the secretory cell as well as the other factors discussed above. Thus, it cannot be assumed that the mammary gland will carry out all the necessary post-translational modifications of a heterologous protein with complete fidelity. It may, therefore, be necessary to engineer these modifications by expressing the correct processing enzyme in bigenic mice. For example, expression of the furin protease in bigenic mice expressing protein C has been suggested to improve the efficiency of post-translational processing of the pro-protein (H. Lubon, personal communication). Another novel solution to this problem used to express IGF-I (65) has been to fuse the processed protein sequence directly to the casein signal peptide, thus obviating the need for additional processing once the signal peptide has been removed. Alternatively, it should be possible to perform some of these post-translational modifications, such as modification of the terminal sugars, *in vitro* on purified pro-proteins. This may be critical to ensure the appropriate stability for these proteins *in vivo*. The issue of stability is unresolved for most of the heterologous transgene products. A unique combination of the above approaches may be required for each transgene protein.

Alternative Approaches for Improving Transgene Expression

Even with the advances in the identification of important regulatory sequences in the milk protein genes needed to target heterologous gene expression selectively to the mammary gland, problems in gene delivery and expression are still of paramount importance in this technology. Microinjection, the method most commonly employed for the introduction of DNA into the germline of transgenic animals, results in the relatively low efficiency, nonhomologous integration of DNA into the genome, often resulting in position-dependent expression. The design of the transgene construct can, therefore, markedly influence its frequency, and in some cases, its level of expression. The importance of introns for efficient transgene expression in mice has been reported previously and genomic constructs are usually expressed more efficiently than cDNA constructs (114). While the use of genomic sequences for transgene constructs is desirable, this has not always been feasible. In cases where only cDNA sequences are available, or very large genomic sequences are required, several alternative strategies have been employed with varying degrees of success. These consist of inclusion of one or more introns from the proposed transgene (115,116), exon replacement (23), co-integration of genomic and cDNA constructs (117,118), and inclusion of matrix attachment regions (MAR) in the transgene locus (119,120).

Intron inclusion. As expected, a genomic β-lactoglobulin transgene expresses at higher levels than a similar cDNA-containing construct (116). Addition of the first intron of β-lactoglobulin to the cDNA construct improves expression only marginally (116). Addition of the same βLG intron to an α_1AT cDNA construct did not improve α_1AT expression in the mammary gland although a genomic α_1AT construct was expressed at levels up to 10 mg/ml (116). Analysis of human serum albumin (HSA) transgenes shows greater effects of HSA intron 2 than intron 1 on HSA transgene expression (115). Synergy among HSA introns is observed (115). Interestingly, a construct containing introns 2 and introns 7-14 performs better than a genomic construct containing all 14 introns (115). A good correlation with tissue culture results with the use of the same constructs is observed for HSA (115), but not for βLG/α_1AT (116). A human pulmonary surfactant protein B cDNA transgene is not expressed efficiently (J. Rosen, unpublished data) but an SP-B genomic sequence driven by the same WAP promoter is expressed in the mammary gland (S. Yarus, unpublished data). In general, the importance of introns for efficient transgene expression in the mammary gland is confirmed, but their number and location must be assessed empirically.

Exon replacement. IGF-I and des(1-3)IGF-I are expressed at high levels from a construct containing rat WAP 5′ and 3′ sequences and the rat WAP introns (23). Exon-sized fragments of DNA, flanked by splice-site consensus sequences, are substituted for the endogenous exons in a transgene by sequential splice overlap extension PCR. This approach permits the construction of an approximately 3.0 kb transgene from a >80 kb gene. While this method appears promising for large genes with complex intron/exon architecture, its general applicability remains to be established primarily because of the possibility of mutations due to multiple rounds of PCR, and the difficulty in avoiding the activation of cryptic splice sites that may exist in the engineered exons.

Genomic and cDNA co-integration. An alternative approach that has been employed for mammary gland-targeted transgenes is to induce expression of a cDNA construct by co-integation with a genomic construct (117,118). Initial reports appeared promising and indicated that co-injection of a cDNA construct with a genomic construct could facilitate expression of the cDNA transgene (118). Unfortunately, expression levels of the cDNAs are variable. This variability is presumably due to variation in the orientation of the two transgenes within each specific integration site, and supports the suggestion that a "double-construct" might express more efficiently. More recent studies with co-integrated mouse WAP genomic and cDNA transgenes indicate that a cDNA transgene co-integrated with a genomic transgene is activated, but at only about 1% of the level of the genomic transgene (117). Production of human γ-interferon with the "double-construct" strategy resulted in low levels of interferon being produced from the cDNA portion of the construct, although some biologically active material was isolated (121). Thus, the use of co-integrated genomic transgenes, whether *via* co-injection or as part of a "double-construct", to rescue expression of cDNA transgenes does not seem very promising.

Matrix attachment regions. The use of matrix attachment regions has been suggested to improve the integration site independence of transgene expression both in transfected cells (119) and transgenic mice (120). Co-injection of a complex chicken lysozyme MAR element increases the proportion of transgenic lines expressing a WAP transgene. In addition, the appropriate hormonal regulation and expression of the mouse WAP transgene during mammary gland development was reported, suggesting that MARs can help establish independent genetic domains (120). This ability of MARs to reduce integration site effects makes them potentially useful for mammary gland bioreactor applications in livestock where the cost of generating a transgenic line is high. However, their general utility for other transgene constructs expressed in the mammary gland remains to be established. The mechanism by which the chicken lysozyme MAR elicits this effect is also unclear at present. It has not been determined if the reported effects are due to a true insulator or enhancer activity within the MAR. The same chicken lysozyme MAR element used in transgenic mice is being employed to generate transgenic cattle carrying a casein/human erythropoietin transgene (122) although analysis of transgene expression in lactating females has not yet been reported.

Prenatal identification of founders. Since the frequency of generating positive transgenic embryos by microinjection is low, the ability to screen for stably-integrated DNA prior to reimplantation is an important economic consideration. While the technology is available for performing single cell PCR and fluorescence *in situ* hybridization (FISH) on cells surgically removed from the blastula, the application of this technology to embryos of livestock species is still far from routine. In one report, a single casein/human erythropoietin calf (122) was produced by *in vitro* fertilization coupled with methylation-sensitive DNA analysis prior to uterine implantation. Sex analysis was conducted concurrently so that only female embryos were implanted. The presence of the transgene was confirmed both during pregnancy and after the birth of the calf. Milk production data are still not available. Widespread application and success of these techniques may significantly reduce the cost of generating transgenic livestock.

Direct introduction of transgenes into the lactating gland. Alternatives to microinjection for the introduction of DNA stably into the mammary gland have been explored. The use of replication-defective retrovirus to carry an hGH transgene directly into mammary cells is reported as a possible way to achieve this goal (123). Virions carrying the transgene are infused directly into the teat canal of a hormonally-induced female. Lactation begins 14 days into the regimen. The transgenic protein is found in the milk at 60 ng/ml on day one of lactation and at 12 ng/ml on days 9 to 15 of lactation (123). While these levels are too low to make this method commercially feasible, improvement of the viral vectors employed may make this method the preferred one in the future. Advances in gene delivery that result from the increased emphasis of research in gene therapy may provide the necessary technological breakthrough to make this approach feasible, but its approval by regulatory agencies may be problematic.

Heritability of transgenes. One potential drawback of the direct gene delivery approach is the inability to propagate the transgene to a herd of offspring, to allow for the expansion of production. In contrast, the stability of transgene expression over generations has been documented in transgenic sheep (124). A male founder ram generated seven F_1 females which expressed the transgene in milk during two lactations. Two F_2 females also expressed the transgene in their milk at comparable levels (124). This result is not surprising, since there is no reason to expect that transgenes would not behave in a Mendelian fashion, yet experimental confirmation is important for the development of a transgenic livestock industry.

Improved temporal and spatial regulation. As discussed above, one principal limitation of transgenic models may be the inability to propagate a specific line either due to transgene lethality or infertility. Methods to overcome inappropriate temporal regulation and/or relaxed tissue specificity would clearly advance the utility of transgenic animals both as disease models and for their application as a bioreactor. One approach to overcome the problems of relaxed temporal regulation and tissue specificity of transgene expression would be the identification, or construction, of new promoters with more restricted tissue specificity and more flexible temporal regulation. Early attempts at modifying transgene expression relied on the use of binary systems in which a single responder line of mice contained a silent target gene encoding the gene product of interest. The target gene could be activated by crossing this line with a separate line carrying a transgene encoding for a transactivator or recombinase (125,126). While these approaches have had success in allowing the propagation of lines which suffer from infertility (125), they still rely on the temporal and spatial expression patterns of the promoter driving the transactivator. To gain greater control of temporal regulation of transgene expression, operator sequences from *E. coli* operons may also be useful (127). One of these that has shown particular promise is the tetR-VP16 system (128). Through the fusion of the Herpes simplex VP16 activation domain to the DNA-binding domain of the tetracycline repressor, an inducible transcriptional activator has been produced (128). Studies in transgenic mice with the cytomegalovirus (CMV) (129) and the MMTV promoters (130) support the conclusion that this approach can allow temporal regulation of transgene expression to be controlled by the researcher rather than by the inherent properties of the transgene. However, the issue of tissue specificity still remains a problem that will only be addressed by the engineering or discovery of novel promoters and the elimination of position effects.

Proposed Food Technology Applications

Altering milk composition by homologous recombination. Production of modified milk proteins is subject to less obstacles post-translationally than production of completely heterologous proteins in milk. Transgenes encoding milk proteins are subject to the same limitations as those encoding pharmaceutical proteins: low expression levels and strong integration site effects. In addition, any protocol which requires replacement of an endogenous milk gene with a modified milk gene would be difficult to implement in the absence of embryonic stem (ES) cell lines which are not currently available for livestock species. Until the development of pluripotent bovine ES cells, the production of calves by nuclear transfer from cultured inner cell mass cells (131) may provide an alternative approach.

The potential for use of homologous recombination in mouse ES cells for manipulating the composition of milk has been demonstrated recently (132-134). For example, the targeted deletion of the mouse ß-casein gene has been reported (132), resulting in mice capable of lactation but containing a reduced total protein concentration in milk. This is the first step in targeted replacement to modify the amino acid content of casein. Such strategic modification of the amino acid content of casein has been suggested as a mechanism to alter production properties of cheese and yogurt (135). Another example of targeted recombination of milk protein genes in ES cells includes elimination of α-lactalbumin as a means of reducing lactose content in milk (19,134). Two independent groups have eliminated the α-lactalbumin gene and both observe disruption of lactation by increased milk viscosity (19,134). Replacement of the mouse gene by the human α-lactalbumin gene restored normal protein concentrations, milk viscosity and lactational competence (134). Another application of targeted recombination is the removal or modification of β-lactoglobulin from milk, which has the potential to reduce gelation of milk and make it less of an allergen (7). This application is still theoretical.

Altering milk composition by overexpression of milk genes. It is also possible to manipulate the composition of milk by overexpression of intrinsic milk proteins (48). For example, increasing the content of αs1-casein has been suggested as a method to improve cheese yield (136). The production of human lactoferrin in cow's milk (88) may provide a "designer" milk to improve infant formula, as well as serving as a potential source of lactoferrin as a pharmaceutical. Notwithstanding the development of new technologies which would allow implementation of any proposed changes in milk composition, there are three additional obstacles to development of modified milk products - price, regulatory acceptance, and consumer acceptance. Addressing these non-scientific issues is perhaps more critical to the ultimate success of this technology than solving the scientific problems listed above.

CONCLUSIONS AND OUTLOOK

Our understanding of the genetic, biochemical and physiological factors governing transgene expression has increased dramatically in the last ten years. Along with this increase has come dramatic advancement in our understanding of the biology of the mammary gland. While the mammary gland bioreactor has many limitations, it still offers significant advantages over other expression systems and will likely continue to serve as a source of novel insight into biology. The production of biologically-active transgene products in milk is clearly a reality in livestock species. This production will hopefully lead to establishment of reasonable regulatory requirements for recombinant milk products and increased public acceptance of transgenic livestock. Development of food-technology applications of transgenic livestock will likely follow.

REFERENCES

1 Juergens, W.G., Stockdale, F.E., Topper, Y.J. and Ellias, J.J. (1965) Proc. Nat. Acad. Sci. U.S.A. 54, 629-633.

2 Richards, D.A., Rodgers, J.R., Supowit, S. and Rosen, J.M. (1981) J. Biol. Chem. 256, 526-532.

3 Blackburn, D.E., Hobbs, A.A. and Rosen, J.M. (1982) Nucl. Acids Res. 10, 2295-2306.

4 Hennighausen, L.G., Sippel, A.E., Hobbs, A.A. and Rosen, J.M. (1982) Nucl. Acids Res. 10, 3733-3743.

5 Lee, K.F., DeMayo, F.J., Atiee, S.H. and Rosen, J.M. (1988) Nucl. Acids Res 16, 1027-1041.

6 Rosen, J.M., Li, S., Raught, B. and Hadsell, D. (1996) Amer. J. Clin. Nutrition, Suppl. (in press).

7 Clark, A.J. (1992) J. Cell. Biochem. 49, 121-127.

8 Sakakura, T. (1987) in The Mammary Gland: Development, Regulation, and Function (Neville, M.C. and Daniel, C.W., eds.) pp. 37-63, Plenum Press, New York, NY.

9 Daniel, C.W. and Silberstein, G.B. (1987) in The Mammary Gland: Development, Regulation, and Function (Neville, M.C. and Daniel, C.W., eds.) pp. 3-31, Plenum Press, New York, NY.

10 Vonderhaar, B.K. and Plaut, K. (1992) in Breast Cancer: Biological and Clinical Progress (Dogliotti, L., Sapino, A. and Bussolati, G., eds.) vol. 69, pp. 59-80, Kluwer Academic Publishers, Norwell, MA.

11 Rosen, J.M. (1987) in The Mammary Gland: Development, Regulation, and Function (Neville, M.C. and Daniel, C.W., eds.) pp. 301-322, Plenum Press, New York, NY.

12 Munday, M.R. and Hardie, D.G. (1987) in The Mammary Gland: Development, Regulation, and Function. (Neville, M.C. and Daniel, C.W., eds.) pp. 563-603, Plenum Press, New York, NY.

13 Faulkner, A. and Peaker, M. (1987) in The Mammary Gland: Development, Regulation, and Function (Neville, M.C. and Daniel, C.W., eds.) pp. 536-558, Plenum Press, New York, NY.
14 Vernon, R.G. (1989) Proc. Nutr. Soc. 48, 23-32.
15 Ossowski, L., Biegel, D. and Reich, E. (1979) Cell 16, 929-940.
16 Strange, R., Li, F., Saurer, S., Burkhardt, A. and Friis, R.R. (1992) Development 115, 49-58.
17 Blackburn, D.G. (1992) J. Dairy Sci. 76, 3195-3212.
18 Wei, Y., Yarus, S., Greenberg, N.M., Whitsett, J. and Rosen, J.M. (1995) Transgenic Res. 4, 232-240.
19 Stinnakre, M.-G., Vilotte, J.-L., Soulier, S. and Mercier, J.-C. (1994) Proc. Nat. Acad. Sci. U.S.A. 91, 6544-6548.
20 Devinoy, E., Thépot, D., Stinnakre, M.-G., Fontaine, M.L., Grabowski, H., Puissant, C., Pavirani, A. and Houdebine, L.M. (1994) Transgenic Res. 3, 79-89.
21 Pomp, D., Geisert, R.D., Durham, C.M. and Murray, J.D. (1995) Biol. Reprod. 51, 170-178.
22 Burdon, T., Wall, R.J., Shamay, A., Smith, G.H. and Hennighausen, L. (1991) Mech. Dev. 36, 67-74.
23 Hadsell, D.L., Greenberg, N.M., Fligger, J.M., Baumrucker, C.R. and Rosen, J.M. (1996) Endocrinology (in press).
24 Bichini, O., Andres, A.C., Schubaur, B., Mehtali, M., LeMeur, M., Lathe, R. and Gerlinger, P. (1991) Endocrinology 128, 539-546.
25 Milton, S., Cecim, M., Li, Y.S., Yun, J.S., Wagner, T.E. and Bartke, A. (1992) Endocrinology 131, 536-538.
26 Cardiff, R.D. and Munn, R.J. (1995) Cancer Lett. 90, 13-19.
27 Ceriani, R.L. (1970) Dev. Biol. 21, 506-529.
28 Ceriani, R.L. (1970) Dev. Biol. 21, 530-546.
29 Ross, S.R. and Solter, D. (1985) Proc. Nat. Acad. Sci. U.S.A. 82, 5880-5884.
30 Lin, T.P., Guzman, R.C., Osborn, R.C., Thordarson, G. and Nandi, S. (1992) Cancer Res. 52, 4413-4419.
31 Guy, C.T., Cardiff, R.D. and Muller, W.J. (1992) Mol. Cell. Biol. 12, 954-961.
32 Amundadottir, L.T., Johnson, M.D., Merlino, G., Smith, G.H. and Dickson, R.B. (1995) Cell Growth Diff. 6, 737-748.
33 Li, S. and Rosen, J.M. (1994) Mol. Endocrinol. 8, 1328-1335.
34 Kordon, E.C., McKnight, R.A., Jhappan, C., Hennighausen, L., Merlino, G. and Smith, G.H. (1995) Dev. Biol. 168, 47-61.
35 Robinson, G.W., McKnight, R.A., Smith, G.H. and Hennighausen, L. (1995) Development 121, 2079-2090.
36 Bayna, E.M. and Rosen, J.M. (1990) Nucl. Acids Res. 18, 2977-2985.
37 Sympson, C.J., Talhouk, R.S., Alexander, C.M., Chin, J.R., Clift, S.M., Bissell, M.J. and Werb, Z. (1994) J. Cell Biol. 125, 681-693.
38 Mercier, J.C. and Vilotte, J.L. (1993) J. Dairy Sci. 76, 3079-3098.
39 Jhappan, C., Geiser, A.G., Kordon, E.C., Bagheri, D., Hennighausen, L., Roberts, A.B., Smith, G.H. and Merlino, G. (1993) EMBO J. 12, 1835-1845.
40 Sandgren, E.P., Schroeder, J.A., Qui, T.H., Palmiter, R.D., Brinster, R.L. and Lee, D.C. (1995) Cancer Res. 55, 3915-3927.
41 Lee, K.F., Atiee, S.H. and Rosen, J.M. (1989) Mol. Cell. Biol. 9, 560-565.
42 Vilotte, J.-L., Soulier, S., Stinnarkre, M.-G., Massoud, M. and Mercier, J.-C. (1989) Eur. J. Biochem. 186, 43-48.

43 Harris, S., McClenaghan, M., Simons, J.P., Ali, S. and Clark, A.J. (1991) Dev. Genet. 12, 299-307.

44 Jhappan, C., Stahle, C., Harkins, R.N., Fausto, N., Smith, G.H. and Merlino, G.T. (1990) Cell 61, 1137-1146.

45 McGrane, M.M., Yun, J.S., Moorman, A.F., Lamers, W.H., Hendrick, G.K., Arafah, B.M., Park, E.A., Wagner, T.E. and Hanson, R.W. (1990) J. Biol. Chem. 265, 22371-22379.

46 McGrane, M.M., de Vente, J., Yun, J., Bloom, J., Park, E., Wynshaw-Boris, A., Wagner, T., Rottman, F.M. and Hanson, R.W. (1988) J. Biol. Chem. 263, 11443-11451.

47 Whitelaw, C.B., Harris, S., McClenaghan, M., Simons, J.P. and Clark, A.J. (1992) Biochem. J. 286, 31-39.

48 Simons, J.P., McClenaghan, M. and Clark, A.J. (1987) Nature 328, 530-532.

49 Maschio, A., Brickell, P.M., Kioussis, D., Mellor, A.L., Katz, D. and Craig, R.K. (1991) Biochem. J. 275, 459-467.

50 Farini, E. and Whitelaw, C.B. (1995) Mol. Gen. Genet. 246, 734-738.

51 Gunzburg, W.H., Salmons, B., Zimmermann, B., Muller, M., Erfle, V. and Brem, G. (1991) Mol. Endocrinol. 5, 123-133.

52 Li, B., Greenberg, N., Stephens, L.C., Meyn, R., Medina, D. and Rosen, J.M. (1994) Cell Growth Diff. 5, 711-721.

53 Muller, W.J., Lee, F.S., Dickson, C., Peters, G., Pattengale, P. and Leder, P. (1990) EMBO J. 9, 907-913.

54 Henrard, D. and Ross, S.R. (1988) J. Virol. 62, 3046-3049.

55 Gautier, C., Mehtali, M. and Lathe, R. (1989) Nucl. Acids Res. 17, 8389.

56 DeOme, K.B., Faulklin, L.J., Bern, H.A. and Blair, P.B. (1959) Cancer Res. 19, 515-520.

57 Ornitz, D.M., Cardiff, R.D., Kuo, A. and Leder, P. (1992) J. Nat. Cancer Inst. 84, 887-892.

58 Brunner, A.M., Marquardt, H., Malacko, A.R., Lioubin, M.N. and Purchio, A.F. (1989) J. Biol. Chem. 264, 13660-13664.

59 Pierce, D.J., Johnson, M.D., Matsui, Y., Robinson, S.D., Gold, L.I., Purchio, A.F., Daniel, C.W., Hogan, B.L. and Moses, H.L. (1993) Genes Dev. 7, 2308-2317.

60 Clemmons, D.R. (1993) Mol. Rep. Dev. 35, 368-375.

61 Francis, G.L., Upton, F.M., Ballard, F.J., McNeil, K.A. and Wallace, J.C. (1988) Biochem. J. 251, 95-103.

62 Bayne, M.L., Applebaum, J., Chicchi, G.G., Hayes, N.S., Green, B.G. and Cascieri, M.A. (1988) J. Biol. Chem. 263, 6233-9623.

63 Gillespie, C., Read, L.C., Bagley, C.J. and Ballard, F.J. (1990) J. Endocrinol. 127, 401-405.

64 Clemmons, D.R., Cascieri, M.A., Camacho, H.C., McCusker, R.H. and Bayne, M.L. (1990) J. Biol. Chem. 265, 12210-12216.

65 Brem, G., Hartl, P., Besenfelder, U., Wolf, E., Zinovieva, N. and Pfaller, R. (1994) Gene 149, 351-355.

66 Park, A.J., Matrisian, L.M., Kelly, A.F., Pearson, R., Yuan, Z. and Navre, M. (1991) J. Biol. Chem. 266, 1584-1590.

67 Witty, J.P., Lempka, T., Coffey, R.J. and Matrisian, L.M. (1995) Cancer Res. 55, 1401-406.

68 Threadgill, D.W., Dlugosz, A.A., Hansen, L.A., Tennenbaum, T., Lichti, U., Yee, D., LaMantia, C., Mourton, T., Herrup, K., Harris, R.C., Barnard, J.A.,

Yuspa, S.H., Coffey, R.J. and Magnuson, T. (1995) Science 269, 230-234.
69 Sibilia, M. and Wagner, E.F. (1995) Science 269, 234-238.
70 Keller, S., Jones, J., Boyle, A., Barrow, L., Killen, D., Kapousta, N., Hitchcock, P., Swank, R.T. and Meisler, M. (1994) Genomics 23, 309-320.
71 Allen, N.D., Norris, M.L. and Surani, M.A. (1990) Cell 61, 835-861.
72 Guy, C.T., Muthuswamy, S.K., Cardiff, R.D., Soriano, P. and Muller, W.J. (1994) Genes Dev. 8, 23-32.
73 Donehower, L.A., Godley, L.A., Aldaz, C.M., Pyle, R., Shi, Y.P., Pinkel, D., Gray, J., Bradley, A., Medina, D. and Varmus, H.E. (1995) Genes Dev. 9, 882-895.
74 Sinn, E., Muller, W., Pattengale, P., Tepler, I., Wallace, R. and Leder, P. (1987) Cell 49, 465-475.
75 Andres, A.C., van der Valk, M.A., Schonenberger, C.A., Fluckiger, F., LeMeur, M., Gerlinger, P. and Groner, B. (1988) Genes Dev. 2, 1486-1495.
76 Kwan, H., Pecenka, V., Tsukamoto, A., Parslow, T.G., Guzman, R., Lin, T.P., Muller, W.J., Lee, F.S., Leder, P. and Varmus, H.E. (1992) Mol. Cell. Biol. 12, 147-154.
77 Andres, A.C., Schonenberger, C.A., Groner, B., Henninghausen, L., LeMeur, M. and Gerlinger, P. (1987) Proc. Nat. Acad. Sci. U.S.A. 84, 1299-1303.
78 Stewart, T.A., Pattengale, P.K. and Leder, P. (1984) Cell 38, 627-637.
79 Matsui, Y., Halter, S.A., Holt, J.T., Hogan, B.L. and Coffey, R.J. (1990) Cell 61, 1147-1155.
80 Wang, T.C., Cardiff, R.D., Zukerberg, L., Lees, E., Arnold, A. and Schmidt, E.V. (1994) Nature 369, 669-671.
81 Shackleford, G.M., MacArthur, C.A., Kwan, H.C. and Varmus, H.E. (1993) Proc. Nat. Acad. Sci. U.S.A. 90, 740-744.
82 MacArthur, C.A., Shankar, D.B. and Shackleford, G.M. (1995) J. Virol. 69, 2501-2507.
83 Lee, F.S., Lane, T.F., Kuo, A., Shackleford, G.M. and Leder, P. (1995) Proc. Nat. Acad. Sci. U.S.A. 92, 2268-2272.
84 Jhappan, C., Gallahan, D., Stahle, C., Chu, E., Smith, G.H., Merlino, G. and Callahan, R. (1992) Genes Dev. 6, 345-355.
85 Smith, G.H., Gallahan, D., Diella, F., Jhappan, C., Merlino, G. and Callahan, R. (1995) Cell Growth Diff. 6, 563-577.
86 Hennighausen, L., McKnight, R., Burdon, T., Baik, M., Wall, R.J. and Smith, G.H. (1994) Cell Growth Diff. 5, 607-613.
87 Lee, K.F., Atiee, S.H., Henning, S.J. and Rosen, J.M. (1989) Mol. Endocrinol. 3, 447-453.
88 Platenburg, G.J., Kootwijk, E.P., Kooiman, P.M., Woloshuk, S.L., Nuijens, J.H., Krimpenfort, P.J., Pieper, F.R., de Boer, H.A. and Strijker, R. (1994) Transgenic Res. 3, 99-108.
89 Dale, T.C., Krnacik, M.J., Schmidhauser, C., Yang, C.L., Bissell, M.J. and Rosen, J.M. (1992) Mol. Cell. Biol. 12, 905-914.
90 Rijnkels, M., Kooiman, P.M., Krimpenfort, P.J.A., de Boer, H.A. and Pieper, F.R. (1995) Biochem. J. 311 (in press).
91 Bawden, W.S., Passey, R.J. and MacKinlay, A.G. (1994) in Biotechnology and Genetic Engineering Reviews (Tombs, M.P., ed.) vol. 12, pp. 89-137, Intercept Ltd., Andover, Hampshire, U.K.
92 Clark, A.J., Ali, S., Archibald, A.L., Bessos, H., Brown, P., Harris, S., McClenaghan, M., Prowse, C., Simons, J.P., Whitelaw, C.B. and Wilmut, I. (1989) Genome 31, 950-955.

93 Hennighausen, L. (1990) Protein Expr. Purif. 1, 3-8.

94 Yom, H.C. and Bremel, R.D. (1993) Amer. J. Clin. Nutr. 58 (Suppl.), 299s-306s.

95 Yarus, S. (ed.) and Steffen, D. (Systems Analyst) (1995) http://condor.mbcr.bcm.tmc. edu/BEP/ERMB. Molecular Biology Computational Resource, Baylor College of Medicine, Houston, TX.

96 Mercier, J.-C. and Gaye, P. (1983) in Biochemistry of Lactation (Mepham, T.B., ed.) pp. 177-227, Elsevier, Amsterdam-New York, NY.

97 Pittius, C.W., Hennighausen, L., Lee, E., Westphal, H., Nicols, E., Vitale, J. and Gordon, K. (1988) Proc. Nat. Acad. Sci. U.S.A. 85, 5874-5878.

98 Denman, J., Hayes, M., O'Day, C., Edmunds, T., Bartlett, C., Hirani, S., Ebert, K.M., Gordon, K. and McPherson, J.M. (1991) Bio/Technology 9, 839-843.

99 Ebert, K.M., Selgrath, J.P., DiTullio, P., Denman, J., Smith, T.E., Memon, M.A., Schindler, J.E., Monastersky, G.M., Vitale, J.A. and Gordon, K. (1991) Bio/Technology 9, 835-838.

100 Clark, A.J., Bessos, H., Bishop, J.O., Brown, P., Harris, S., Lathe, R., McClenaghan, M., Prowse, C., Simons, J.P., Whitelaw, C.B.A. and Wilmut, I. (1989) Bio/Technology 7, 487-492.

101 Bischoff, R., Degryse, E., Perraud, F., Dalemans, W., Ali-Hadji, D., Thépot, D., Devinoy, E., Houdebine, L.-M. and Pavirani, A. (1992) FEBS Lett. 305, 265-268.

102 Hansson, L., Edlund, M., Edlund, A., Johansson, T., Marklund, S.L., Fromm, S., Strömqvist, M. and Törnell, J. (1994) J. Biol. Chem. 269, 5358-5363.

103 Archibald, A.L., McClenaghan, M., Hornsey, V., Simons, J.P. and Clark, A.J. (1990) Proc. Nat. Acad. Sci. U.S.A. 87, 5178-5182.

104 Wright, G., Carver, A., Cottom, D., Reeves, D., Scott, A., Simons, P., Wilmut, I., Garner, I. and Colman, A. (1991) Bio/Technology 9, 830-834.

105 Greenberg, N.M., Anderson, J.W., Hsueh, A.J.W., Nishimori, K., Reeves, J.J., deAvila, D.M., Ward, D.N. and Rosen, J.M. (1991) Proc. Nat. Acad. Sci. U.S.A. 88, 8327-8331.

106 Meade, H., Gates, L., Lacy, E. and Lonberg, N. (1990) Bio/Technology 8, 443-446.

107 Velander, W.H., Johnson, J.L., Page, R.L., Russell, C.G., Subramanian, A., Wilkins, T.D., Gwazdauskas, F.C., Pittius, C. and Drohan, W.N. (1992) Proc. Nat. Acad. Sci. U.S.A. 89, 12003-12007.

108 Lee, T.K., Drohan, W.N. and Lubon, H. (1995) J. Biochem. 118, 81-87.

109 Drohan, W.N., Zhang, D.W., Paleyanda, R.K., Chang, R., Wroble, M., Velander, W. and Lubon, H. (1994) Transgenic Res. 3, 355-364.

110 Drohan, W.N., Wilkins, T.D., Latimer, E., Zhou, D., Velander, W., Lee, T.K. and Lubon, H. (1994) in Advances in Bioprocess Engineering (Galindo, E. and Ramirez, O.T., eds.), pp. 501-507, Kluwer Academic Publishers, Amsterdam.

111 Weaver, T.E., Lin, S., Bogucki, B. and Dey, C. (1992) Amer. J. Physiol. 263, L95-L103.

112 Wong, V.L.Y., Compton, M.M. and Witorsch, R.J. (1986) Biochim. Biophys. Acta 881, 167-174.

113 Baldocchi, R.A., Tan, L., King, D.S. and Nicoll, C.S. (1993) Endocrinology 133, 935-938.

114 Brinster, R.L., Allen, J.M., Behringer, R.R., Gelinas, R.E. and Palmiter, R.D. (1988) Proc. Nat. Acad. Sci. U.S.A. 85, 836-840.

115 Hurwitz, D.R., Nathan, M., Barash, I., Ilan, N. and Shani, M. (1994) Transgenic Res. 3, 365-375.

116 Whitelaw, C.B., Archibald, A.L., Harris, S., McClenaghan, M., Simons, J.P. and Clark, A.J. (1991) Transgenic Res. 1, 3-13.

117 McKnight, R.A., Wall, R.J. and Hennighausen, L. (1995) Transgenic Res. 4, 39-43.

118 Clark, A.J., Archibald, A.L., McClenaghan, M., Simons, J.P., Wallace, R. and Whitelaw, C.B. (1993) Philos. Trans. Roy. Soc. Lond. B. Biol. Sci. 339, 225-232.

119 Kalos, M. and Fournier, R.E.K. (1995) Mol. Cell. Biol. 15, 198-207.

120 McKnight, R.A., Shamay, A., Sankaran, L., Wall, R.J. and Hennighausen, L. (1992) Proc. Nat. Acad. Sci. U.S.A. 89, 6943-6947.

121 Dobrovolsky, V.N., Lagutin, O.V., Vinogradova, T.V., Frolova, I.S., Kuznetsov, V.P. and Larionov, O.A. (1993) FEBS Lett. 319, 181-184.

122 Hyttinen, J.-M., Peura, T., Tolvanen, M., Aalto, J., Alhonen, L., Sinervirta, R., Halmekyto, M., Myohanen, S. and Janne, J. (1994) Bio/Technology 12, 606-608.

123 Archer, J.S., Kennan, W.S., Gould, M.N. and Bremel, R.D. (1994) Proc. Nat. Acad. Sci. U.S.A. 91, 6840-6844.

124 Carver, A.S., Dalrymple, M.A., Wright, G., Cottom, D.S., Reeves, D.B., Gibson, Y.H., Keenan, J.L., Barrass, J.D., Scott, A.R., Colman, A. and Garner, I. (1993) Bio/Technology 11, 1263-1270.

125 Ornitz, D.M., Moreadith, R.W. and Leder, P. (1991) Proc. Nat. Acad. Sci. U.S.A. 88, 698-702.

126 Lakso, M., Sauer, B., Mosinger, B.J., Lee, E.J., Manning, R.W., Yu, S.-H., Mulder, K.L. and Westphal, H. (1992) Proc. Nat. Acad. Sci. U.S.A. 89, 6232-6236.

127 Gossen, M., Bonin, A.L. and Bujard, H. (1993) Trends Biochem. Sci. 18, 471-475.

128 Gossen, M. and Bujard, H. (1992) Proc. Nat. Acad. Sci. U.S.A. 89, 5547-5551.

129 Furth, P.A., St. Onge, L., Boger, H., Gruss, P., Gossen, M., Kistner, A., Bujard, H. and Hennighausen, L. (1994) Proc. Nat. Acad. Sci. U.S.A. 91, 9302-9306.

130 Hennighausen, L., Wall, R.J., Tillmann, U., Li, M. and Furth, P.A. (1995) J. Cell Biochem. (in press).

131 Sims, M. and First, N.L. (1994) Proc. Nat. Acad. Sci. U.S.A. 91, 6143-6147.

132 Kumar, S., Clarke, A.R., Hooper, M.L., Horne, D.S., Law, A.J., Leaver, J., Springbett, A., Stevenson, E. and Simons, J.P. (1994) Proc. Nat. Acad. Sci. U.S.A. 91, 6138-6142.

133 Stinnakre, M.G., Vilotte, J.L., Soulier, S., L'Haridon, R., Charlier, M., Gaye, P. and Mercier, J.C. (1991) FEBS Lett. 284, 19-22.

134 Stacey, A., Schnieke, A., Kerr, M., Scott, A., McKee, C., Cottingham, I., Binas, B., Wilde, C. and Colman, A. (1995) Proc. Nat. Acad. Sci. U.S.A. 92, 2835-2839.

135 Jiminez-Flores, R. and Richardson, T. (1988) J. Dairy Sci. 71, 2640-2654.

136 Hennighausen, L., Ruiz, L. and Wall, R. (1990) Curr. Opin. Biotech. 1, 74-78.

137 Tsukamoto, A.S., Grosschedl, R., Guzman, R.C., Parslow, T. and Varmus, H.E. (1988) Cell 55, 619-625.

138 Choi, Y.W., Henrard, D., Lee, I. and Ross, S.R. (1987) J. Virol. 61, 3013-3019.

139 Guy, C.T., Webster, M.A., Schaller, M., Parsons, T.J., Cardiff, R.D. and Muller, W.J. (1992) Proc. Nat. Acad. Sci. U.S.A. 89, 10578-10582.

140 Muller, W.J., Sinn, E., Pattengale, P.K., Wallace, R. and Leder, P. (1988) Cell 54, 105-115.

141 Iwamoto, T., Takahashi, M., Ito, M., Hamaguchi, M., Isobe, K., Misawa, N., Asai, J., Yoshida, T. and Nakashima, I. (1990) Oncogene 5, 535-542.

142 Mangues, R., Seidman, I., Pellicer, A. and Gordon, J.W. (1990) Oncogene 5, 1491-1497.
143 B'erard, J., Gaboury, L., Landers, M., De, R.Y., Houle, B., Kothary, R. and Bradley, W.E. (1994) EMBO J. 13, 5570-5580.
144 Tomasetto, C., Wolf, C., Rio, M.C., Mehtali, M., LeMeur, M., Gerlinger, P., Chambon, P. and Lathe, R. (1989) Mol. Endocrinol. 3, 1579-1584.
145 Tzeng, Y.J., Guhl, E., Graessmann, M. and Graessmann, A. (1993) Oncogene 8, 1965-1971.
146 Sandgren, E.P., Luetteke, N.C., Palmiter, R.D., Brinster, R.L. and Lee, D.C. (1990) Cell 61, 1121-1135.
147 Tornell, J., Rymo, L. and Isaksson, O.G. (1991) Int. J. Cancer 49, 114-117.
148 Ninomiya, T., Hirabayashi, M., Sagara, J. and Yuki, A. (1994) Mol. Reprod. Dev. 37, 276-283.
149 Stinnakre, M.G., Vilotte, J.-L., Soulier, S., L'Haridon, R., Charlier, M., Gaye, P. and Mercier, J.C. (1991) FEBS Lett. 288, 247.

PROBLEMS THAT CAN LIMIT THE EXPRESSION OF FOREIGN GENES IN PLANTS: LESSONS TO BE LEARNED FROM *B.t.* TOXIN GENES

Scott H. Diehn[a,c], E. Jay De Rocher[a], and Pamela J. Green[a,b]

MSU-DOE Plant Research Laboratory[a]
Department of Biochemistry[b]
Department of Botany and Plant Pathology[c]
Michigan State University
East Lansing, Michigan 48824-1312

INTRODUCTION

It has been more than ten years since the first transgenic plants expressing foreign genes were regenerated following *Agrobacterium*-mediated transformation. During this time, significant progress has been made toward crop improvement by genetically engineered plants with a variety of desirable traits. New developments in the area of plant transformation and the isolation of agronomically important genes have contributed to this advancement. In some cases, endogenous plant genes have been re-engineered to produce advantageous characteristics, whereas in others, genes from non-plant sources have been used. Prominent examples include the antisense and sense configurations of several genes that have been used to manipulate ripening and senescence processes (1). In addition, genes from bacteria, fungi, plants and viruses have been expressed in plants to achieve resistance to various pathogens (2) or herbicides (3), to alter male fertility (4,5) or lipid composition (6), or to produce new compounds in plants such as biodegradable plastics (7,8). As a result, many exciting transgenic plants and their products should soon reach the market place.

Despite the many success stories, there are instances where genetically-engineered plants fail to adequately produce the desired gene product, particularly when the gene is introduced from a foreign source. This can occur even when the problematic gene is introduced into plants under the control of strong plant promoters, implying that the coding region or protein itself is responsible for the poor expression. Although most of these difficulties have not been published, the problems associated with one prominent group of genes encoding insecticidal proteins, the *B.t.* toxins, have been well documented.

B.t. toxins, also known as insecticidal crystal proteins (ICPs), *B.t.k.* insect control proteins, δ-endotoxins, crystal proteins, and *cry* gene products, are a family of insecticidal proteins produced by the gram-positive soil bacterium *Bacillus thuringiensis*. They have been divided into six classes based on their insecticidal spectra and structural homologies: CryI (effective against Lepidoptera), CryII (effective against Lepidoptera and Diptera), CryIII (effective against Coleoptera), CryIV (effective against Diptera), and CryV and

CryVI (both effective against nematodes), although the CryV designation has also been used to classify *B.t.* toxins effective against both lepidopteran and coleopteran larvae (9) (see references 10 and 11 for a more detailed description of these classes and additional subclasses). Upon sporulation, *Bacillus thuringiensis* generates crystalline parasporal inclusions which can be composed of one or more protoxin species. When ingested by the larvae of a susceptible insect, the crystalline inclusions are solubilized in the midgut releasing the protoxins. Proteases in the larval midgut cleave the protoxins into their active form. These toxic polypeptides act by binding to specific receptors in the membrane of epithelial cells in the midgut of susceptible insect larvae. There the toxin forms a cation-selective channel that disrupts the K^+ balance in the cell. As a result, water enters the epithelial cell causing it to swell and lyse (10,12).

Spore-crystal mixtures of these toxins have been used for many years as an alternative to conventional chemical pesticides because they are very potent and highly specific. They are not known to be toxic to vertebrates or non-targeted insects (13). However, widespread commercial use of these toxins as insecticidal sprays has been limited, in part due to high production costs and poor persistence in the field. An attractive approach that avoids these problems is the genetic engineering of plants with *B.t.* toxin genes. Plants that express *B.t.* toxin genes should be resistant to insect feeding damage throughout the growing season, reducing the need for pesticide applications. Considerable effort has been devoted to achieving this goal, particularly in the private sector, because of its high potential for commercial and ecological benefit.

Unfortunately, the path to success was much more challenging than it was first thought. *B.t.* toxin genes have proven to be the most difficult foreign genes reported to date to express at high levels in plants. As such, the work on *B.t.* toxin genes provides an excellent case study of problematic gene expression in higher plants. An informal poll of several biotechnology companies indicated that this problem is not unique to *B.t.* toxin genes; nearly all reported that a proportion of the foreign genes they had examined were not expressed to expected levels in transgenic plants. At present, the most common approach to increase the expression of a foreign gene is to resynthesize the gene to make it more plant-like, a rather time-consuming and expensive operation. This approach can also be somewhat risky if the mechanisms posing the original limitation are unknown.

In this review, we highlight the research done on *B.t.* toxin gene expression in plants and discuss the specific strategies that have been used successfully and unsuccessfully to circumvent the problems that were encountered. Emphasis will be placed on those studies that address the specific mechanisms that limit *B.t.* toxin gene expression in an effort to glean the greatest insight towards the effective engineering of new *B.t.* toxin genes and other problematic foreign genes for future crop improvement efforts. We will not address other aspects associated with *B.t.* toxin such as classification, structure-function relationships, or the evolution of insecticidal resistance, as these topics have been reviewed elsewhere (10-12,14).

B.t. TOXIN GENE EXPRESSION AND INSECTICIDAL ACTIVITY IN TRANSGENIC PLANTS

Initial Studies with *B.t.* Protoxins in Plants

The first *B.t.* toxin genes to be transformed into plants were members of the *cryIA* class of genes. Initially, the entire protoxin coding region was introduced into tobacco and tomato by *Agrobacterium*-mediated transformation. Unfortunately, regenerated plants displayed little or no insecticidal activity in insect feeding assays. A determination of the

B.t. toxin levels within these plants showed that very low levels of the protein (2 ng/mg protein) were synthesized (15-17). Higher levels of protoxin (10 to 50 ng/mg protein) were reportedly produced in tobacco calli, but plants could not be successfully regenerated from these calli. This prompted the suggestion that the protoxin is deleterious to plant cells (18).

Truncated *B.t.* Toxin Genes

In an attempt to circumvent the problems encountered with the intact protoxin genes, truncated *B.t.* toxin genes were engineered for expression in plants. Extensive deletion analyses performed in bacteria (reviewed in 10) and plants (16) have demonstrated that the N-terminal portion of the protoxin is effective in killing susceptible insect larvae. The function of the C-terminal portion is unknown, although in the case of the *cryI* genes, it is thought to be involved in the co-assembly of protoxins into the crystal (10,12).

The truncated genes, in comparison to full-length *B.t.* toxin genes, were expressed at significantly higher levels and protected plants to a much greater degree. Tobacco hornworm (*Manduca sexta*) mortality rates in isolated leaf bioassays generally ranged from 75 to 100% within 6 days, with most transformants above 90% (15,16,18). Up to 14 ng of *B.t.* toxin per mg total protein typically accumulated in these plants (15,18). The degree of truncation also appears to have an impact on the level of expression. Tobacco plants transformed with a gene fusion corresponding to the first 683 amino acids of a *cryIA(b)* gene linked to the kanamycin resistance coding region, *nptII*, exhibited a higher level of insecticidal activity than did plants expressing a similar gene fusion containing the first 724 amino acids (15). These results indicate that *B.t.* toxin genes containing only those sequences necessary for toxicity may provide the highest levels of *B.t.* toxin protein in plants.

Although plants expressing a truncated *B.t.* toxin gene are able to control the tobacco hornworm, they are often not able to manifest this effect against less susceptible insects. The tobacco hornworm is among the insects that are the most sensitive to *B.t.* toxin, having an LC_{50} of approximately 0.04 mg/ml for both CryIA(b) and CryIA(c) proteins (19). Other lepidopteran insect pests such as the tobacco budworm (*Heliothis virescens*), *Helicoverpa zea* (corn earworm, cotton bollworm, and tomato fruitworm, formerly known as *Heliothis zea*), and beet armyworm (*Spodoptera exigua*) are increasingly difficult to kill with both CryIA(b) and CryIA(c) proteins in this order, although the sensitivities between the two toxins do vary for some species (19). Thus, transgenic plants exhibiting high levels of resistance to the tobacco hornworm were only partially toxic to the tobacco budworm and even less resistant to *Helicoverpa zea* in laboratory and field tests (16,20). As a result of these studies, it was apparent that higher levels of *B.t.* toxin expression were still required to protect plants from a great majority of the agronomically important insect pests that are susceptible to *B.t.* toxin. Nonetheless, it was realized that the C-terminal portion was dispensable and most subsequent plant transformation experiments exploited this fact. All *B.t.* toxin genes referred to in the remainder of this review should be assumed to be truncated unless otherwise indicated.

<u>Attempts to enhance or control expression with 5′ and 3′ flanking sequences.</u> In many cases, *B.t.* toxin genes have been expressed under the control of the strong and relatively constitutive 35S promoter of the Cauliflower Mosaic Virus (35S) (16,18,21-27). Yet, these genes usually gave rise to very low levels of *B.t.* toxin mRNA in transgenic plants. One exception was a 35S-driven *cryIA(b)* gene that, for unknown reasons, produced readily detectable levels of mRNA and correspondingly high quantities of protein (~100 ng *B.t.* toxin/mg protein) in tobacco plants (28). However, efforts to achieve improved levels of

insect control continued, in part through the construction of *B.t.* toxin genes with alternative 5′ flanking sequences. For example, in one study phaseolin, 19S, 35S, soybean ribulose-1,5-bisphosphate carboxylase/oxygenase (rubisco) small subunit, and mannopine synthase promoters were each used to control the transcription of either a protoxin or truncated *cryIA(c)* gene in transgenic tobacco (21). Transcripts were detected only in young plants containing the mannopine synthase promoter-driven genes. Interestingly, expression of the genes disappeared with plant maturation (21).

The 35S promoter has been modified for greater transcriptional activity by duplicating the enhancer region (29). The activity of this enhanced 35S promoter is approximately ten-fold higher than the unmodified version and about 100 times as active as a NOS promoter. An enhanced 35S promoter has been used to direct the expression of a *cryIA(c)* gene in tobacco cells (30). *B.t.* toxin transcripts from polyA-enriched fractions were observed on Northern blots, but they were not observed in total RNA fractions, indicating that this promoter would not suffice as a remedy for the low accumulation of *B.t.* toxin in plants. Thus far, no promoters have been demonstrated to be more effective than an enhanced 35S promoter for achieving high levels of *B.t.* toxin gene expression, although most studies lack comparative analyses.

In other attempts to optimize the expression of *B.t.* toxin genes in plants, the 5′ untranslated region (5′ UTR) from alfalfa mosaic virus (AMV) RNA 4 was inserted between the promoter and the coding regions of several *cryI* genes (18,25,26,28). AMV RNA 4 is a well-translated transcript encoding the viral coat protein. The 5′ UTR of this message has been found to increase the translational efficiency of chimeric mRNAs *in vitro* and *in vivo* (reviewed in 31). Whether these sequences actually increase the accumulation of *B.t.* toxin in plants is unclear. In only one study were the levels of *B.t.* toxin produced with and without the translational enhancer directly compared. Inclusion of the AMV 5′ UTR in a *cryIA(b)* gene construction did not appear to have an effect; however, only one line expressing the AMV-BT chimeric gene was generated, making it difficult to distinguish between position effects and the effect of the leader sequence (28). Otherwise, the AMV 5′ UTR has been included in gene constructions under the assumption that it would benefit expression. Incorporation of the AMV leader into *cryIA(b)* *B.t.* toxin constructs gave rise to *B.t.* toxin levels of 2 to 12 ng/mg protein in experiments carried out in transgenic tobacco (18) and soybean (25) plants. Two to forty nanograms of *B.t.* toxin protein per milligram protein were detected in transgenic tomato plants (32); however, only 2 to 10 ng/mg protein were present in leaf tissues (25), sites where many insect larvae feed. Because none of these accumulation levels is of economic relevance, AMV 5′ UTR sequences in combination with strong promoters are not sufficient to overcome the expression problems associated with *B.t.* toxin in plants.

Translational fusions consisting of the *nptII* gene coupled to the 3′ end of a truncated *B.t.* toxin gene have been utilized in an attempt to select for transformants expressing high levels of *B.t.* toxin (15). These fusions do not increase the expression of the *B.t.* toxin gene, but increase the possibility of finding transformants with high levels of expression. Plants able to survive on growth media containing high concentrations of kanamycin displayed elevated levels of insect mortality. However, still higher levels of expression were needed to fully protect the plants from less sensitive insects (15).

The 5′ and 3′ flanking sequences from a construct known to direct the expression of a luciferase reporter gene in the interveinal regions of a leaf upon wounding (33) have also been fused to a *cryIA(c)-nptII* coding region in an attempt to increase *B.t.* toxin levels. Several species of plants have been engineered with this chimeric gene, including some woody perennials (22-24,27). However, very few plants expressed the fusion protein to high enough levels to kill larvae (23,24,27). In only one case were insects effectively controlled, but it is not clear whether or not wound induction of the *B.t.* toxin gene occurs (22).

Some genera of Coleoptera and Diptera feed on the roots and root nodules of economically important plants. These tissues are difficult to protect with conventional methods; however, an obvious approach would be to express an appropriate *B.t.* toxin gene under the control of a root-specific promoter. To the best of our knowledge this has not been pursued, possibly due in part to the fact that some of these host plants belong to the families Leguminosae and Gramineae, historically difficult plants to transform and regenerate. To overcome this obstacle, root-colonizing bacteria have been engineered to express a variety of *B.t.* toxin genes (34-37). An advantage to this approach is that the problems of low expression in plants can be avoided. The accumulation of *B.t.* toxin in these bacteria was significant, but only modest levels of protection were observed in inoculated plants (35-37). Thus far, the reasons for this poor protection have not been elucidated nor overcome. Expression of *B.t.* toxin in these plants may be more practical in the future owing to recent advances in plant transformation technology.

High expression levels achieved with modified *B.t.* toxin genes. C-terminal truncations, modification of 5′ and 3′ UTR's, protein fusions, and the use of strong plant promoters were all insufficient to overcome the low-level expression of otherwise "unmodified" *B.t.* toxin genes in plants. Much higher levels of expression have been achieved with the additional step of constructing "modified" *B.t.* toxin genes in which the coding sequences of wild-type *B.t.* toxin genes have been extensively altered. Fifteen modified *B.t.* toxin genes and their expression properties have been reported to date and are listed in Table 1. The expression of modified *B.t.* toxin genes has been examined in ten monocot and dicot species and has been shown to protect against more than ten types of insects including relatively insensitive species such as beet armyworm and European corn borer (*Ostrinia nubilalis*). Increased insecticidal activity and increased mRNA abundance were observed in all cases where these parameters were compared between modified and unmodified *B.t.* toxin genes. Similar N-terminal regions of the protoxin genes have been used in most instances but the degree of modification has covered a wide range. Sequence changes have altered from as few as 3% to as many as 85% of the codons, and the GC content has been raised from between 34% and 38% for unmodified genes to between 41% and 65% for modified genes.

The rationale for altering the coding sequence is based on the high AT content of *B.t.* toxin genes relative to plant gene coding sequences. Whereas *B.t.* toxin genes typically have an AT content of approximately 65% (Table 1), dicot coding regions are approximately 55% AT and monocot coding regions are approximately 45% AT (38). One consequence of the AT-richness is that *B.t.* toxin coding regions are similar to plant 3′ UTRs and introns. These properties increase the probability that *B.t.* toxin transcripts may by chance contain sequences recognized in plants as signals for polyadenylation, mRNA decay, splicing, or other processes that can affect the structure and accumulation of the mRNA. Another consequence of the AT-richness of *B.t.* toxin genes is the frequent occurrence of codons that are rare in plant genes. It has been proposed that these rare codons could slow translation, thereby inhibiting *B.t.* toxin protein synthesis (21).

One approach used for sequence modification has been the specific targeting of sequences suspected to inhibit *B.t.* toxin gene expression in plants. Modifications were made in AT-rich regions, especially those containing potential plant polyadenylation signals (39) or the ATTTA instability motif (40,41). Sequence changes that eliminated these possible deleterious elements were designed so as to convert the codons involved to synonymous codons preferred in plant genes (26,42-45). Thorough application of the criteria for sequence modification resulted in extensive changes, with more than half of the

Table 1. Characteristics of Modified *B.t.* Toxin Genes Expressed in Plant Cells

Gene	Length[a]	Change In GC (%)	Codons Changed	Plant Species	Expression: Increased mRNA[b]	Expression: Maximum Protein[c]	Expression: Increased Activity[d]	Active Against[e]	Reference
cryIA(a)	1-645	nr	141	poplar	nr	nr	nr	forest tent caterpillar	(68)
							nr	gypsy moth	
				cranberry	nr	nr	(-)		(69)
cryIA(b)	1-612	nr	nr	cotton	nr	0.1%	+	cabbage looper	(42)
							+	beet armyworm	
							+	cottom bollworm	
"	1-615	37 to 49	356	tomato	+	0.2-0.3%	+	tobacco hornworm	(43)
				tobacco	nr	0.2-0.3%	+	tobacco hornworm	
"	29-607	37 to 41	55	tomato	+	0.02%	+	tobacco hornworm	(43)
				tobacco	nr	0.02%	+	tobacco hornworm	
"	1-648	38 to 65	549	maize	nr	0.4%	nr	European corn borer	(46)
"	1-724	38 to 59	482	rice	nr	0.05%	nr	striped stemborer	(48)
							nr	leaffolder	
"	1-646	nr	21	tomato	+	<<	+	tobacco hornworm	(26)
				tobacco	+	<<	+	tobacco budworm	
							+	tobacco hornworm	
cryIA(c)	1-640	nr	nr	cotton	nr	0.1%	+	cotton bollworm	(42)
"	1-615	nr	306	cotton	nr	0.1%	+	cotton bollworm	(42)
				tomato	nr	0.2-0.3%	nr	tobacco hornworm	(43)
"	29-615	nr	nr	tobacco	nr	0.2-0.3%	nr	tobacco hornworm	(43)
"	10-613	34 to 64	496	maize	+	nr	nr	nr	(49)
				tobacco	+	nr	+	tobacco hornworm	
				Arabidopsis	+	nr	+	tobacco hornworm	
cryIC	1-654	nr	42	tobacco	+	<<	+	beet armyworm	(26)
								tobacco hornworm	
cryIIIA	48-644	36 to 49	nr	tobacco	nr	0.6%	nr	colorado potato beetle	(44)
"	48-645	37 to 49	347	potato	nr	0.3%	+	colorado potato beetle	(45)
"	1-610	36 to 45	190	potato	nr	0.025%	nr	colorado potato beetle	(47)
				carrot	+	0.005%	nr	nr	
				maize	+	0.005%	nr	nr	

a Amino acid residues from protoxin genes included in modified *B.t.* toxin genes.
b +indicates increased mRNA accumulation for the modified *B.t.* toxin gene relative to thge unmodified gene.
c Highest observed level of *B.t.* toxin protein as a percent of total protein.
d +indicated increased insecticidal activity resulting from expression of the modified *B.t.* toxin gene relative to the unmodified gene.
e Insecticidal activity observed against the indicated insect.
nr Not reported.
<< Below limit detection.
(-) No insecticidal activity; tested against blackheaded fireworm.

codons altered (see Table 1). This approach was first used successfully to generate modified *cryIA(b)* and *cryIA(c)* genes expressed in cotton, conferring resistance to cabbage looper (*Trichoplusia ni*), beet armyworm and *Helicoverpa zea* (42). The extensive modifications to these genes were estimated to increase expression at least 100-fold based on insect bioassays (42) and Western analysis (1 mg/mg total protein) (43). A *cryIIIA* gene that was modified in the same way and transformed into potato yielded expression levels up to 300-fold higher than unmodified *cryIIIA* as assayed by Western analysis (3 mg/mg total protein)(45).

Targeting AT-rich regions for modification on a limited scale has been attempted as an alternative to comprehensive modification in an effort to achieve useful levels of expression with a smaller investment of time and resources. Potential instability and polyadenylation signals were removed from a *cryIC* gene and a *cryIA(b)* gene with only 42 and 21 altered codons, respectively (26). These modifications increased mRNA levels from undetectable to detectable, but protein levels remained below the limit of detection by Western blot analysis. These genes did, however, provide insect protection to tobacco and tomato plants. A second modified *cryIA(b)* gene was constructed with limited alterations affecting only 55 codons in 9 AT-rich regions (referred to as A through I), compared to 356 codons changed in a fully modified *cryIA(b)* gene (43). These limited modifications provided an increase in *B.t.* toxin protein accumulation in tobacco and tomato that was ten-fold greater than the wild-type gene but ten-fold less than the fully modified gene. A series of modified genes in which subsets of the nine AT-rich regions were targeted demonstrated that the changes in one region (region B) provided a large share of the ten-fold increase in expression over wild type. From these results, it can be concluded that specific sequences limiting expression should be identifiable and that the modifications made to the nine targeted regions missed sequences responsible for the further ten-fold improvement in expression found in the fully-modified gene. This latter conclusion was supported by the finding that substitution of the 5′ one-third of the fully-modified gene (containing region B) into the wild-type gene gave only a ten-fold increase in expression. Substitution of the 3′ two-thirds of the fully modified gene into the wild-type gene gave nearly a ten-fold increase in expression, indicating that other deleterious sequences must exist in this region. These studies with partially modified genes demonstrate that a degree of improved expression can be achieved without resorting to complete re-engineering of the coding sequence. The fact that improved expression was obtained after the specific targeting of potential polyadenylation signals and instability sequences suggests that these processes may be important factors limiting *B.t.* toxin gene expression in plants. It remains to be shown, however, that the partial modifications described actually affect polyadenylation or mRNA stability.

Another approach used successfully in the modification of *B.t.* toxin sequences has been to alter the codon usage throughout the coding region to conform to that found in plant genes without changing the encoded amino acid sequence (46-49). Since plant codon usage is biased toward codons with G or C in the third position (50), the change from bacterial to plant codon bias automatically raises the GC content of the gene. The resulting increase in GC content eliminates by default most or all AT-rich sequences, thereby removing possible splicing, polyadenylation, and instability signals. The conversion to plant preferred codons has the added potential benefit of improved translational efficiency. In most cases, the codon bias was altered to conform to dicot codon usage. For three genes, modifications were designed to give a monocot codon bias, specifically like that found in maize (46,49) and rice genes (48). However, a monocot- or dicot-like codon bias does not appear to limit the expression of modified *B.t.* toxin genes to one or the other class of plants. A modified *cryIIIA* gene with a dicot codon bias was expressed well in maize cells

(47) and a modified *cryIA(c)* gene with a monocot codon bias was expressed well in *Arabidopsis* and tobacco cells (49). Taken together, these studies show that extensive modification of AT-rich sequences and codon usage can be an effective way to engineer high *B.t.* toxin gene expression in a variety of plant species. The exact molecular basis for the improved expression has been unclear due to the breadth of the sequence changes that were made in the modified genes. However, when studies on the modified *B.t.* toxin genes are interpreted together with expression analyses of their unmodified counterparts, important mechanistic insights are beginning to emerge, as discussed in the next section.

Further improvements with promoters and other 5′ flanking sequences. Many insect larvae feed in locations that are difficult to treat with conventional insecticides. Such is the case in controlling the European corn borer. Part of this insect's life cycle is spent consuming pollen in the leaf axils of maize plants and tunneling through the stalk, locations that provide protection from most chemical pesticides. Through the use of tissue-specific and pollen-specific promoters, greater control of this pest has been achieved. Transgenic maize expressing a modified *cryIA(b)* gene under the control of a maize pollen-specific promoter and a phosphoenolpyruvate carboxylase promoter, a promoter which is active in mesophyll cells, displayed high levels of *B.t.* toxin production and minimal damage to the stalk as a result of tunneling. Insect mortality on leaf pieces was 85 to 100%, which was significantly better than most control plants transformed with a 35S-driven gene (46).

It has been observed that a combination of the 5′ UTR and transit peptide from the *Arabidopsis thaliana* rubisco small subunit *ats1A* gene can increase the expression of a modified *cryIA(c)* gene 10- to 20-fold in tobacco. The increase is not a result of targeting the protein to the chloroplast and appears to be independent of the promoter, because the same gene driven by either the *ats1A* promoter in tobacco plants or an enhanced CMV 35S in tobacco protoplasts showed a similar increase in expression. However, the effect the *ats1A* 5′ UTR and transit peptide have on enhancing expression may depend on the coding region to which these sequences are attached since they increased the expression of a β-glucuronidase (GUS) reporter gene by no more than 6-fold (51).

ACCUMULATION OF *B.t.* TOXIN mRNA IN TRANSGENIC PLANTS AND FACTORS THAT AFFECT ITS ABUNDANCE

In all cases where mRNA levels have been examined, poor expression of unmodified *B.t.* toxin genes is always associated with low mRNA accumulation (15,18,23,24). Conversely, high expression of modified *B.t.* toxin genes always results in higher mRNA levels (see Table 1). In additional experiments on the induction of *B.t.* toxin that occurred upon anthesis in a series of transgenic tobacco plants, the relative differences in the accumulation of *B.t.* toxin mRNA among the lines reflected the differences in protein levels (28). Each of these observations underscores the strong correlation between *B.t.* toxin protein and the level of the corresponding mRNA that is generally observed in transgenic plants. They also emphasize the importance of the factors that control mRNA abundance in establishing the efficacy of a given *B.t.* toxin gene in transgenic plants.

What mechanisms, then, limit the accumulation of unmodified *B.t.* toxin transcripts in plants? The initial observation that poor transcript accumulation was characteristic of unmodified genes, regardless of the promoter used, argued that the problem was post-transcriptional. However, it was still possible that unmodified *B.t.* toxin-coding regions contained a repressor sequence capable of inhibiting transcription initiation or elongation from all promoters. Recent nuclear-run-on transcription experiments demonstrate that there is no significant difference in the transcriptional activities of a poorly-expressed, unmodified *cryIA(b)* gene and a highly-expressed control (52) or an unmodified *cryIA(c)* and a highly-expressed, fully-modified *cryIA(c)* gene (49). These results were obtained in experiments carried out with probes from either the 5′ (52) or the 3′ (49) ends of the genes, indicating that the expression problem is post-transcriptional. This prompts an examination of other aspects of *B.t.* toxin mRNA metabolism such as splicing, mRNA stability and polyadenylation for their role in limiting gene expression. These topics, as well as the role of codon usage and translation, are discussed in the sections below.

Splicing

Northern blot analyses of plants expressing different *B.t.* toxin genes often reveal the presence of *B.t.* toxin transcripts of a shorter length than expected. Tobacco plants transformed with a *cry IA(c)* protoxin gene produced a truncated, polyadenylated transcript of 1.7 kb (17), whereas plants expressing a *cryIA(b)* protoxin gene accumulated two short polyadenylated transcripts of approximately 1.6 and 0.9 kb (21). Two transcripts also of approximately 1.6 and 0.9 kb were observed in the polyA-enriched fractions of tobacco plants expressing a truncated *cryIA(b)* gene (21). Tobacco plants expressing another *cryIA(b)* gene generated short transcripts, although these transcripts were also generated in an *in vitro* transcription system (18). Two short polyadenylated transcripts of 0.9 kb and 0.6 kb were also identified in tobacco cells expressing a *cryIA(c)* gene (30). Some of these short transcripts have been shown to arise from polyadenylation in the *B.t.* toxin coding region, as discussed in the next section. However, others may be generated through the aberrant splicing of *B.t.* toxin messages.

Pre-mRNA splicing in plants is known to be facilitated by the presence of AU-rich elements within the intron (reviewed in (53) and (54)). Unmodified *B.t.* toxin transcripts have a high AU content; thus, it is possible that these sequences trigger aberrant splicing thereby limiting the accumulation of intact *B.t.* toxin transcripts in plants. This possibility was recently investigated in tobacco with the use of a reverse transcriptase-polymerase chain reaction (RT-PCR) assay (52). Plants expressing a truncated *cryIA(b)* gene were found to excise three regions of the *B.t.* toxin transcript. The sequences surrounding the boundaries of the excised regions conform to consensus splice junctions. Mutations were made to one or two of the 5′ splice junctions to assess the contribution splicing makes to the low accumulation of *B.t.* toxin mRNA in plants. The inhibition of splicing was found to increase expression on the order of 4- to 20-fold depending on the experiment. Unfortunately, these experiments do not address the magnitude of the contribution aberrant splicing makes to the overall poor accumulation of unmodified *B.t.* toxin transcripts because the mutations were evaluated within the context of a partially modified *B.t.* toxin gene. However, they do argue that splicing occurs and has a negative influence.

Polyadenylation

Another explanation for the existence of short, polyadenylated *B.t.* toxin messages in transformed plants is that they are generated through the aberrant polyadenylation of newly-synthesized transcripts (17,30). This has been most clearly demonstrated in the case of two short transcripts produced by an unmodified *cryIA(c) B.t.* toxin gene in stably-transformed tobacco cells. RT-PCR analysis identified two poly(A) addition sites within the *B.t.* toxin coding region that have positions corresponding to the sites where polyadenylation was predicted to occur based on the size (~600 and ~900 bases) of these short transcripts (30). Oligonucleotide-directed RNase H cleavage and Northern blot hybridization with probes from different regions of the *cryIA(c)* gene provided further confirmation that the two short transcripts resulted from polyA addition at the mapped sites. The sequences upstream of both these sites are typical of plant polyadenylation signals (30,55).

Due to their low abundance under normal growth conditions, the two short transcripts can be detected in polyA+ but not in total RNA of stably-transformed tobacco cells (30). However, if cells are treated with cycloheximide, a protein synthesis inhibitor known to stabilize many unstable transcripts (56) (e.g., by blocking the production of a labile mRNA degradation factor), the abundance of the two short transcripts increases significantly. Under these conditions, the transcripts are routinely observed in total RNA from cells expressing the unmodified *cryIA(c)* gene (30). Therefore, it seems likely that a significant proportion of the transcripts produced by unmodified *cryIA(c)* genes are polyadenlyated at the mapped sites but fail to accumulate because they are unstable. These polyadenylation sites were eliminated in a highly-expressed, modified version of the same *cryIA(c)* gene, as verified by the absence of short polyadenylated transcripts in both untreated and CHX-treated tobacco cells (49). Thus, the data indicate that limited accumulation of intact *B.t.* toxin transcripts in plants is in part due to aberrant polyadenylation in the coding region.

mRNA Stability

The failure of unmodified *B.t.* toxin transcripts to accumulate is often attributed to mRNA instability because little or no *B.t.* toxin mRNA can be detected, even when transcription is driven by a strong plant promoter. Further support for the idea that the transcripts are unstable derives from the observation that unmodified *B.t.* toxin genes support normal levels of transcriptional activity in nuclear run-on assays, as mentioned earlier (49,52). Moreover, the AT richness of *B.t.* toxin genes resembles well-known mRNA instability sequences present in the 3′ UTRs of labile mammalian proto-oncogene and cytokine mRNAs such as those encoding c-*fos* and GM-CSF (see 57 for a review). Multiple ATTTA motifs are a common feature of these 3′ UTRs and can be essential for them to function as signals for rapid decay (58). Unmodified *B.t.* toxin genes also contain multiple ATTTA sequences and these have been major targets of mutagenesis during the construction of modified genes (26,42-45). ATTTA repeats have recently been shown to trigger rapid decay of reporter transcripts in tobacco (41) so it is likely that some ATTTA sequences in *B.t.* toxin function in a similar manner. However, it should be noted that not all ATTTA sequences act to decrease mRNA stability (57,59), in part because the minimal domain appears to be longer in some cases (60,61).

In principle, it should be possible to test whether unmodified *B.t.* toxin transcripts are unstable and to examine the role of ATTTA sequences by measuring mRNA half-lives by established methods (e.g., blocking transcription with an inhibitor such as actinomycin D and then monitoring the disappearance of the mRNA over time (41,62,63)). Unfortunately, the scarcity of unmodified *B.t.* toxin mRNA in plant cells has presented a technical barrier to this type of analysis. To overcome this obstacle and evaluate the stability of unmodified *B.t.* toxin transcripts, three different approaches have been taken, as outlined below.

In the first study, accumulation of *B.t.* toxin mRNA was investigated in carrot protoplasts after introduction of plasmids containing unmodified *cryIA(b)* or *cryIIIA* genes under the control of the 35S promoter (21). Abundance of *B.t.* toxin mRNAs was monitored by Northern analysis of polyA+ RNA isolated from protoplasts harvested at 1, 2, 4, 8 and 18 hr after electroporation. A second plasmid carrying an octopine synthase (OCS) gene, also driven by a 35S promoter, was co-electroporated to serve as a control. A basic assumption of this experiment was that transcription of the plasmid-borne genes would stop at some time after electroporation, allowing decay of the transcripts to be observed during the time course. The disappearance pattern of both *B.t.* toxin transcripts differed from that of the OCS transcripts. Intact *cry* mRNA present at 8 hr was replaced by a more intense smear of lower molecular weight transcripts at 18 hr, whereas OCS mRNA remained intact throughout the time course. That the intact *B.t.* toxin transcripts disappeared at 18 hr compared to the OCS transcripts, indicated that the former were relatively unstable. In addition, the smear of smaller *cry* transcripts may have been degradation intermediates. These were proposed to arise from a 5′ to 3′ exonucleolytic degradation mechanism because they were present in the polyA+ fraction and multiple 5′ ends were mapped by nuclease protection (21). A series of constructs in which increasingly large portions of the *cryIA(b)* coding region were deleted from the 3′ end were also electroporated into carrot protoplasts. None of the deletions prevented the differential transcript disappearance with respect to OCS mRNA (21). Similar results were also reported for a *cryIA(c)* deletion series (unpublished data in 21). This suggests that sequences sufficient to trigger rapid decay of unmodified *B.t.* toxin mRNA in protoplasts are present in the 5′ portion of the coding region.

A second investigation addressing the stability of unmodified *B.t.* toxin transcripts was carried out in tobacco protoplasts with a *cryIA(b)* gene (52) similar to that in the first study. However, in the second study, different methodology was used and different results were obtained. *In vitro* synthesized *cryIA(b)* and *bar* (bialaphos resistance) RNAs that were 5′ capped and polyadenylated were co-electroporated into tobacco protoplasts. The decay of the introduced transcripts was then followed over a 5-hr time course. It was assumed that decay of *in vitro* synthesized transcripts introduced into protoplasts by electroporation in these experiments accurately reflected the decay of endogenously generated transcripts. The half-lives of the *cryIA(b)* and *bar* transcripts after electroporation were determined to be 7.8 ± 3.0 and 4.9 ± 1.3 hr, respectively. Since the *in vitro* synthesized *cryIA(b)* and *bar* transcripts decayed in protoplasts with half-lives that were not statistically different, it was concluded that the low accumulation of *cryIA(b)* mRNA relative to *bar* mRNA in transgenic plants was not due to the instability of *cryIA(b)* transcripts. It is interesting to note that a *bar* mRNA half-life of 2.3 hr was calculated from a time course experiment following actinomycin D treatment of protoplasts from a plant transgenic for *bar* (64). It appears that when similar experiments were carried out with protoplasts from a *cryIA(b)* transgenic plant, the *B.t.* toxin mRNA half-life was more reflective of the electroporation results (5.3 hr; unpublished results in 52) than in the case of *bar*. Therefore, there was no evidence for rapid decay of unmodified *B.t.* toxin transcripts in this study. In addition, differences in transcriptional activity, measured in run-on assays, could not account for the poor accumulation of unmodified *cryIA(b)* mRNA (52). Because the run-on data were obtained with a probe against the 5′ region of *cryIA(c)*, the simplest explanation remaining is that B.t. toxin mRNA accumulation is limited at some step between transcription initiation and mRNA degradation.

The third study compared the stabilities of transcripts from unmodified and extensively modified *cryIA(c)* genes in stably transformed tobacco cells (49). The approach was to measure the half-lives of the transcripts to determine whether highly-modified *B.t.* toxin mRNA levels were due to an increase in mRNA stability caused by the sequence modifications. A transient treatment of tobacco cells expressing the unmodified gene with the translational inhibitor cycloheximide was used to induce *B.t.* toxin mRNA to levels that would allow half-life measurements to be made. The ability to induce mRNA accumulation with cycloheximide was itself an indication that the unmodified *B.t.* toxin transcripts might be unstable, since cycloheximide induction is characteristic of other unstable mRNAs (56). Accumulation of the modified *B.t.* toxin mRNA was not increased by cycloheximide. After removal of cycloheximide from the tobacco cells, transcription was blocked with actinomycin D and transcript decay monitored over a 2-hr time course. The half-life of unmodified *B.t.* toxin mRNA was determined to be several-fold less than for the modified *B.t.* toxin mRNA. The actual difference may be even greater if the presence or effect of the cycloheximide was not completely reversed before the time course began. In any event, this study showed that unmodified *cryIA(c)* mRNA is unstable in plants and that sequence modifications made to generate the modified *B.t.* toxin gene improved mRNA accumulation, at least in part, by increasing mRNA stability. The specific sequence modifications responsible for the increased stability remain to be identified.

At present it is unclear why the results of the second study are inconsistent with those of the other two which found the *B.t.* toxin mRNA to be unstable. It is tantalizing to propose that unmodified *B.t.* toxin transcripts are unstable only if they are transported to the cytoplasm *via* normal pathways from the nucleus as in the first and third approaches, rather than by electroporation as in the second study. However, an experiment cited in the latter, in which the unmodified *B.t.* toxin mRNA appears to be stable in protoplasts from transgenic leaves (unpublished results in 52), seems to argue against this possibility. Therefore, alternative explanations, such as differences among the transcripts examined or the plant systems used, would appear to be more viable. In any event, the data indicate that some unmodified *B.t.* toxin transcripts are rapidly degraded in plant cells, which undoubtedly contributes to their poor accumulation.

Codon Usage

The difference in codon usage between plants and unmodified *B.t.* toxin genes is frequently pointed to as an explanation for poor *B.t.* toxin expression in plants. This is because the codon bias of *B.t.* toxin genes also introduces codons that are rare in plant genes (21). In actuality, codon usage covers several possible mechanisms one or more of which could limit expression. The chance of introducing sequences that cause splicing, polyadenylation or mRNA decay is heightened by the AT-rich sequences that correspond to rare codons, as discussed in previous sections. If pool sizes for the tRNAs specified by rare codons are small in plants, the presence of these codons could cause ribosomes to stall during translation of *B.t.* toxin transcripts. Stalling of ribosomes at rare codons could simply reduce the rate of protein synthesis. Alternatively, stalled ribosomes could cause mRNA destabilization by leaving *B.t.* toxin transcripts exposed to components of the RNA turnover machinery. Little is known about tRNA pool sizes in plants but an especially restrictive codon bias is found in some highly-expressed plant genes (50).

To date there have been no studies directly addressing whether ribosome pausing at rare codons in unmodified *B.t.* toxin transcripts reduces protein synthesis. Some insight could be gained from a careful comparison of the ratios of *B.t.* toxin mRNA and protein produced by unmodified and modified *B.t.* toxin genes. However, the products of unmodified genes are generally extremely low and hard to quantify. In one case, increases in *B.t.* toxin protein and mRNA in tomato plants expressing modified *cryIA(b)* genes were compared with the *B.t.* toxin protein and mRNA levels in a small number of plants expressing the unmodified gene (unpublished data in 43). Partial modification of the *cryIA(b)* gene was reported to give a 2.5-fold increase in mRNA but a 10-fold increase in protein over wild-type levels and full modification a 5-fold increase in mRNA but a 50-fold increase in protein over wild type. The method used for RNA quantification (nuclease protection with a probe covering a 5′ segment of the *B.t.* toxin genes) cannot differentiate between intact and aberrantly processed transcripts, although the latter were not evident on Northern blots of total RNA. Assuming they are absent, the data may suggest that translational problems figure more prominently than mRNA accumulation problems for limiting the expression of the unmodified *cryIA(b)* gene (43).

The effect, if any, of rare codons in the *B.t.* toxin transcript on mRNA stability is unknown. It is difficult to evaluate the effect of rare codons on *B.t.* toxin transcript stability separately from the effect of the RNA sequences that encode them since conversion of rare codons to common codons unavoidably alters the mRNA sequence. One approach used to address this problem was to determine the effect on mRNA half-life of introducing a *B.t.* toxin sequence containing rare codons into a gene encoding a stable transcript. A 26-codon segment of a *cryIA(c)* gene, rich in rare codons, was inserted in frame into a phytohemagglutinin (PHA) gene (65). This *B.t.* toxin segment was chosen in part because it could be easily frameshifted to common codons to control for RNA sequence effects that were independent of the rare codons. However, this control proved to be unnecessary because the PHA transcripts were not destabilized by either a single or double insertion of the *cryIA(c)* sequence. While these results suggest that rare codons do not have a strong effect on mRNA stability, they do not rule out the possibility that an even greater number or a different set of rare codons would accelerate mRNA decay.

CHLOROPLAST TRANSFORMATION

Poor expression of wild-type *B.t.* toxin genes in plants appears to result from incompatibility of AT-rich *B.t.* toxin sequences with mechanisms that can affect mRNA accumulation and possibly translational efficiency. An alternative to modifying the sequences to be compatible with nuclear and cytoplasmic mechanisms is to express the gene from the chloroplast genome instead of the nuclear genome. Taking advantage of the bacterial ancestry of the chloroplast circumvents the problem of expressing an AT-rich bacterial gene in the context of GC-rich nuclear genes. Recently, an unmodified *cryIA(c)* protoxin gene was introduced into tobacco chloroplasts by particle bombardment followed by homologous recombination (66). Expression of the protoxin in regenerated plants with transformed chloroplasts was at least three times greater on a percent total protein basis (>30 μg/mg protein) than that reported for the best modified *B.t.* toxin gene transcribed in the nucleus. Interestingly, no obvious effects on plant growth or fertility were observed which suggests the chloroplast is insensitive to the ill effects the protoxin has been suggested to have in the cytoplasm (18). A high level of *B.t.* toxin mRNA was observed in total RNA preparations from these plants, most of which appears to be in a dicistronic form. The significant accumulation of *B.t.* toxin transcripts in plants with transformed chloroplasts indicates that the mechanisms that limit RNA accumulation in this organelle are likely to differ from those acting on nuclear encoded transcripts. Therefore, this approach could be a powerful solution to the problem of expressing *B.t.* toxin genes in plants if chloroplast transformation can be effectively applied to a wide range of species. It may also prove useful for expression of other problematic genes that can fulfill their intended function when expressed in the chloroplast. The maternal inheritance of transgenes located in the chloroplast would be beneficial in that dispersal of transgenes by pollen would be prevented. However, hybrid lines would have to be generated from female plants containing the transgene. This would be a disadvantage for those breeding programs which utilize paternal inbred gene donors as well as maternal inbred gene donors to efficiently introduce transgenes into multiple hybrid lines.

CONCLUSIONS AND FUTURE PROSPECTS

Now that the hurdles limiting the expression of *B.t.* toxin genes in plants have been largely overcome, it is clear that a great deal has been learned in the process. It can no longer be assumed that attaching a strong promoter to a gene will insure high expression in plants. As illustrated by the work on *B.t.* toxins, expression can be hindered at the mRNA level through mechanisms such as aberrant splicing, aberrant polyadenylation and rapid mRNA degradation. Avoiding these problems as well as those affecting translation and post-translational processes has become a popular consideration in plant biotechnology, in large part because of work on *B.t.* toxin genes.

The favored way to achieve high expression of *B.t.* toxin genes has been to create fully-modified genes removing all seemingly deleterious sequences. But is this strategy the best means by which to increase the expression of new *B.t.* toxin genes and other problematic genes based on current knowledge? There is no universal answer to this question because different genes have different characteristics and constraints. In some cases it may be possible to circumvent the complications presented by nuclear and cytoplasmic processes through chloroplast transformation. The initial results with *B.t.* toxin expression in chloroplasts havé been very exciting and the utility of this approach will become even greater if the technique can be adapted to a variety of crop plants. For gene products that cannot achieve their desired function in chloroplasts or in plants not amenable to chloroplast transformation, gene modification may be the best option. Although resynthesis of *B.t.* toxin genes to be more plant-like without prior knowledge of the underlying problems has been fairly successful, this process is still rather labor-intensive and costly. In addition, there is no guarantee that a fully modified gene will work as expected in transgenic plants. Therefore, placing some emphasis on elucidating the molecular basis of the problem would be advantageous. Determining whether the poor expression of a given foreign gene is a result of low mRNA levels and/or low protein levels is relatively simple and might indicate whether the situation is similar to that encountered with unmodified *B.t.* toxin genes.

Considerable progress has been made toward delineating the sequences responsible for poor expression of unmodified *B.t.* toxin genes, including the identification of anomalous polyadenylation and splicing sites. As this type of analysis progresses, it should be possible to address whether elimination of a limited number of sequences *known* to be deleterious will be sufficient to elevate gene expression to near maximal levels. The significant improvements that have been observed with partially-modified *B.t.* toxin genes are very encouraging in this regard.

The knowledge gained from investigating *B.t.* toxin genes in plants has basic as well as applied significance. The mechanisms that limit the expression of unmodified *B.t.* toxin genes are vital plant mechanisms that naturally affect endogenous RNA and protein levels. There is considerable potential to further our understanding of these basic processes through the study of foreign genes in plants, particularly for those processes where knowledge is most limited. For example, only a few sequences that trigger rapid mRNA decay have been identified in plants so far (67). If those in *B.t.* toxin mRNAs can be identified, our basic knowledge of *cis*-acting sequences controlling mRNA stability in plants will increase appreciably. This information may also help identify the natural targets of the corresponding decay pathways. Moreover, with respect to biotechnology, the identification of problematic sequences in *B.t.* toxin genes will make it easier to identify and remove similar sequences from other genes engineered for high expression in plants. Overall, it is hoped that the efforts devoted to *B.t.* toxin genes in plants will result in maximum benefits for crop improvement and basic knowledge, and a minimum of litigation.

ACKNOWLEDGMENTS

We are grateful to Dr. Michael Koziel for comments on the manuscript, to Ms. Karen Bird for editorial assistance, and to several colleagues who provided preprints of their work. *B.t.* toxin research in the authors' laboratory was supported by grants from the DOE, MSU Research Excellence Funds and the Consortium for Plant Biotechnology Research and matching funds to P.J.G. S.H.D. was supported in part by an NIH predoctoral traineeship.

REFERENCES

1 Fray, R.G. and Grierson, D. (1993) Trends Genet. 9, 438-443.
2 Beachy, R.N., Loesch-Fries, S. and Tumer, N.E. (1990) Annu. Rev. Phytopathol. 28, 451-474.
3 Botterman, J. and Leemans, J. (1988) Trends Genet. 4, 219-222.
4 Mariani, C., De Beuckeleer, M., Truettner, J., Leemans, J. and Goldberg, R.B. (1990) Nature 347, 737-741.
5 Mariani, C., Gossele, V., De Beuckeleer, M., De Block, M., Goldberg, R.B., De Greef, W. and Leemans, J. (1992) Nature 357, 384-387.
6 Voelker, T.A., Worrell, A.C., Anderson, L., Bleibaum, J., Fan, C., Hawkins, D.J., Radke, S.E. and Davies, H.M. (1992) Science 257, 72-74.
7 Poirier, Y., Dennis, D.E., Klomparens, K. and Somerville, C. (1992) Science 256, 520-523.
8 Nawrath, C., Poirier, Y. and Somerville, C. (1994) Proc. Nat. Acad. Sci. U.S.A. 91, 12760-12764.
9 Tailor, R., Tippett, J., Gibb, G., Pells, S., Pike, D., Jordan, L. and Ely, S. (1992) Mol. Microbiol. 6, 1211-1217.
10 Höfte, H. and Whiteley, H.R. (1989) Microbiological Reviews 53, 242-255.
11 Feitelson, J.S., Payne, J. and Kim, L. (1992) Bio/Technology 10, 271-275.
12 Aronson, A.I. (1993) Mol. Microbiol. 7, 489-496.
13 Burges, H.D. (1982) Parasitology 84, 79-117.
14 McGaughey, W.H. and Whalon, M.E. (1992) Science 258, 1451-1455.
15 Vaeck, M., Reynaerts, A., Höfte, H., Jansens, S., De Beuckeleer, M., Dean, C., Zabeau, M., Van Montagu, M. and Leemans, J. (1987) Nature 328, 33-37.
16 Fischhoff, D.A., Bowdish, K.S., Perlak, F.J., Marrone, P.G., McCormick, S.M., Niedermeyer, J.G., Dean, D.A., Kusano-Kretzmer, K., Mayer, E.J., Rochester, D.E., Rogers, S.G. and Fraley, R.T. (1987) Bio/Technology 5, 807-813.
17 Adang, M.J., Firoozabady, E., Klein, J., DeBoer, D., Sekar, V., Kemp, J.D., Murray, E., Rocheleau, T. A., Rashka, K., Staffeld, G., Stock, C., Sutton, D. and Merlo, D.J. (1987) in Expression of a *Bacillus thuringiensis* Insecticidal Crystal Protein Gene in Tobacco Plants. (Arntzen, C.J. and Ryan, C., eds.) 46. pp. 345-353, Alan R. Liss, Inc., New York, NY.
18 Barton, K.A., Whiteley, H.R. and Yang, N.S. (1987) Plant Physiol. 85, 1103-1109.
19 MacIntosh, S.C., Stone, T.B., Sims, S.R., Hunst, P.L., Greenplate, J.T., Marrone, P.G., Perlak, F.J., Fischhoff, D.A. and Fuchs, R.L. (1990) Journal of Invertebrate Pathology 56, 258-266.
20 Delannay, X., LaVallee, B.J., Proksch, R.K., Fuchs, R.L., Sims, S.R., Greenplate, J.T., Marrone, P.G., Dodson, R.B., Augustine, J.J., Layton, J.G. and Fischhoff, D.A. (1989) Bio/Technology 7, 1265-1269.
21 Murray, E.E., Rocheleau, T., Eberle, M., Stock, C., Sekar, V. and Adang, M. (1991) Plant Mol. Biol. 16, 1035-1050.

22 Hoffmann, M.P., Zalom, F.G., Wilson, L.T., Smilanick, J.M., Malyj, L.D., Kiser, J., Hilder, V.A. and Barnes, W.M. (1992) J. Econ. Entomol. 85, 2516-2522.
23 Cheng, J., Bolyard, M.G., Saxena, R.C. and Sticklen, M.B. (1992) Plant Sci. 81, 83-91.
24 Dandekar, A.M., McGranahan, G.H., Vail, P.V., Uratsu, S.L., Leslie, C. and Tebbets, J.S. (1994) Plant Sci. 96, 151-162.
25 Parrott, W.A., All, J.N., Adang, M.J., Bailey, M.A., Boerma, H.R. and Stewart Jr., C.N. (1994) In Vitro Cell. Dev. Biol. 30, 144-149.
26 van der Salm, T., Bosch, D., Honee, G., Feng, L., Munsterman, E., Bakker, P., Stiekema, W.J. and Visser, B. (1994) Plant Mol. Biol. 26, 51-59.
27 Shin, D.I., Podila, G.K., Huang, Y. and Karnosky, D.F. (1994) Can. J. For. Res. 24, 2059-2067.
28 Carozzi, N.B., Warren, G.W., Desai, N., Jayne, S.M., Lotstein, R., Rice, D.A., Evola, S. and Koziel, M.G. (1992) Plant Mol. Biol. 20, 539-548.
29 Kay, R., Chan, A., Daly, M. and McPherson, J. (1987) Science 236, 1299-1302.
30 Diehn, S., De Rocher, E.J., Chiu, W. and Green, P. (1996) (unpublished data).
31 Gallie, D.R. (1993) Annu. Rev. Plant Physiol. Plant Mol. Biol. 44, 77-105.
32 Adang, M., DeBoer, D., Endres, J., Firoozabady, E., Klein, J., Merlo, A., Merlo, D., Murray, E., Rashka, K. and Stock, C. (1988) in Manipulation of *Bacillus thuringiensis* Genes for Pest Insect Control. (Roberts, D.W. and Granados, R.R., eds.) pp. 31-37, Cornell University.
33 Barnes, W.M. (1990) Proc. Nat. Acad. Sci. U.S.A. 87, 9183-9187.
34 Obukowicz, M.G., Perlak, F.J., Kusano-Kretzmer, K., Mayer, E.J. and Watrud, L.S. (1986) Gene 45, 327-331.
35 Nambiar, P.T.C., Ma, S.W. and Iyer, V.N. (1990) Appl. Environ. Microbiol. 56, 2866-2869.
36 Skøt, L., Timms, E. and Mytton, L.R. (1994) Plant and Soil 163, 141-150.
37 Bezdicek, D.F., Quinn, M.A., Forse, L., Heron, D. and Kahn, M.L. (1994) Soil-biol-biochem. 26, 1637-1646.
38 Sinibaldi, R.M. and Mettler, I.J. (1992) Progr. Nucl. Acid Res. Mol. Biol. 42, 229-257.
39 Dean, C., Tamaki, S., Dunsmuir, P., Favreau, M., Katayama, C., Dooner, H. and Bedbrook, J. (1986) Nucl. Acids Res. 14, 2229-2240.
40 Shaw, G. and Kamen, R. (1986) Cell 46, 659-667.
41 Ohme-Takagi, M., Taylor, C.B., Newman, T.C. and Green, P.J. (1993) Proc. Nat. Acad. Sci. U.S.A. 90, 11811-11815.
42 Perlak, F.J., Deaton, R.W., Armstrong, T.A., Fuchs, R.L., Sims, S.R., Greenplate, J.T. and Fischhoff, D.A. (1990) Bio/Technology 8, 939-943.
43 Perlak, F.J., Fuchs, R.L., Dean, D.A., McPherson, S.L. and Fischhoff, D.A. (1991) Proc. Nat. Acad. Sci. U.S.A. 88, 3324-3328.
44 Sutton, D.W., Havstad, P.K. and Kemp, J.D. (1992) Transgenic Research 1, 228-236.
45 Perlak, F.J., Stone, T.B., Muskopf, Y.M., Petersen, L.J., Parker, G.B., McPherson, S.A., Wyman, J., Love, S., Reed, G., Biever, D. and Fischhoff, D.A. (1993) Plant Mol. Biol. 22, 313-321.
46 Koziel, M.G., Beland, G.L., Bowman, C., Carozzi, N.B., Crenshaw, R., Crossland, L., Dawson, J., Desai, N., Hill, M., Kadwell, S., Launis, K., Lewis, K., Maddox, D., McPherson, K., Meghji, M.R., Merlin, E., Rhodes, R., Warren, G.W., Wright, M. and Evola, S.V. (1993) Bio/Technology 11, 194-200.

47 Adang, M.J., Brody, M.S., Cardineau, G., Eagan, N., Roush, R.T., Shewmaker, C.K., Jones, A., Oakes, J.V. and McBride, K.E. (1993) Plant Mol. Biol. 21, 1131-1145.
48 Fujimoto, H., Itoh, K., Yamamoto, M., Kyozuka, J. and Shimamoto, K. (1993) Bio/Technology 11, 1151-1155.
49 De Rocher, E.J., Diehn, S. and Green, P. (1996) (unpublished data).
50 Murray, E.E., Lotzer, J. and Eberle, M. (1989) Nucl. Acids Res. 17, 477-498.
51 Wong, E.Y., Hironaka, C.M. and Fischhoff, D.A. (1992) Plant Mol. Biol. 20, 81-93.
52 Van Aarssen, R., Soetaert, P., Stam, M., Dockx, J., Gosselé, V., Seurinck, J., Reynaerts, A. and Cornelissen, M. (1995) Plant Mol. Biol. 28, 513-524.
53 Luehrsen, K.R., Taha, S. and Walbot, V. (1994) Progr. Nucl. Acid Res. Mol. Biol. 47, 149-193.
54 Filipowicz, W., Gniadkowski, M., Klahre, U. and Liu, H. (1995) in Pre-mRNA Splicing in Plants. (Lamond, A.I., ed.) pp. 65-77, R.G. Landes Company, Georgetown, TX.
55 Hunt, A.G. (1994) Annu. Rev. Plant Physiol. Plant Mol. Biol. 45, 47-60.
56 Ross, J. (1995) in mRNA Stability in Mammalian Cells (Joklic, W.K., ed.) American Society of Microbiology, Washington, DC.
57 Greenberg, M.E. and Belasco, J.G. (1993) in Control of the Decay of Labile Protooncogene and Cytokine mRNAs. (Belasco, J.G. and Brawerman, G., eds.) pp. 199-218, Academic Press, New York, NY.
58 Chen, C.Y.A. and Shyu, A.B. (1994) Mol. Cell. Biol. 14, 8471-8482.
59 Walker, E.L., Weeden, N.F., Taylor, C.B., Green, P.J. and Coruzzi, G.M. (1995) Plant Mol. Biol. (in press).
60 Lagnado, C.A., Brown, C.Y. and Goodall, G.J. (1994) Mol. Cell. Biol. 14, 7984-7995.
61 Zubiaga, A.M., Belasco, J.G. and Greenberg, M.E. (1995) Mol. Cell. Biol. 15, 2219-2230.
62 Newman, T.C., Ohme-Takagi, M., Taylor, C.B. and Green, P.J. (1993) Plant Cell 5, 701-714.
63 Taylor, C.B. and Green, P.J. (1995) Plant Mol. Biol. 28, 27-38.
64 Cornelissen, M. (1989) Nucl. Acids Res. 17, 7203-7209.
65 van Hoof, A. and Green, P.J. (1996) (unpublished data).
66 McBride, K.E., Svab, Z., Schaaf, D.J., Hogan, P.S., Stalker, D.M. and Maliga, P. (1995) Bio/Technology 13, 362-365.
67 Sullivan, M.L. and Green, P.J. (1993) Plant Mol. Biol. 23, 1091-1104.
68 McCown, B.H., McCabe, D.E., Russell, D.R., Robison, D.J., Barton, K.A. and Raffa, K.F. (1991) Plant Cell Reports 9, 590-594.
69 Serres, R., Stang, E., McCabe, D., Russell, D., Mahr, D. and McCown, B. (1992) J. Amer. Soc. Hortic. Sci. 117, 174-180.

RENATURATION AND RECONSTITUTION OF FUNCTIONAL HOLOENZYME FROM RECOMBINANT SUBUNITS OF CASEIN KINASE II EXPRESSED AS INCLUSION BODIES IN *E. COLI*

Wey-Jinq Lin, Rolf Jakobi and Jolinda A. Traugh

Department of Biochemistry
University of California
Riverside, California 92521

INTRODUCTION

Casein kinase II (protein kinase CKII) is a heterotetramer consisting of two α and two β subunits (1,2). The α subunit can be expressed as two different gene products which are present as α and/or α', depending on the tissue and species of origin. The β subunit is a single gene product in higher eukaryotes. A comparison of the sequences from a number of different species shows the α and β subunits are highly conserved (1,3). The heterotetramer is stable and the structure is dissociable only under denaturing conditions. Thus, complete physical, chemical and functional analysis of the enzyme at the molecular level is dependent on obtaining subunits capable of reconstituting into active CKII. This reconstituted CKII must have the capacity to interact with and react to the surrounding environment in an appropriate manner.

The α and α' subunits contain a catalytic domain similar to that of all protein kinases (4). Unlike most other protein kinases, CKII utilizes GTP (K_m of ~40 μM) to almost the same extent as ATP (K_m of ~10 μM) (1,2). The α subunit binds the protein substrate, and is sufficient for optimal activity with some substrates [e.g., glycogen synthase, eukaryotic initiation factor 3 (eIF-3) p120] (5). The catalytic activity is directly inhibited by acidic compounds (e.g., heparin, 2,3-bisphosphoglycerate, polyglutamic acid), and heparin has been shown to be competitive with the protein substrate (5,6). Formation of the holoenzyme enhances the catalytic activity about 5-fold by stimulating V_{max} (5,7,8). The β subunit is required for binding/phosphorylation of some substrates (eIF-2β, eIF-3 p67), and for stimulation by basic compounds or proteins (polyamines, polylysine, histone) (5,9).

To accomplish the goal of functional recombination, the α and β subunits of CKII from *Drosophila* and human are expressed in *E. coli* with the pET3a and pT7-7 expression vectors (9,10). Levels of expressed protein reaching as high as 30% of the total protein are obtained two hours after induction of expression. As a consequence, a significant portion of the expressed protein from human and *Drosophila* (60 to 100%) is present in inclusion bodies. A modified denaturation/renaturation system (9) based on the work of

Genetic Engineering, Vol. 18, Edited by J.K. Setlow
Plenum Press, New York, 1996

Buchner and Rudolph (11) has been used to obtain active subunits of CKII from inclusion bodies, and to prepare active holoenzyme from the purified subunits or from a mixture of inclusion bodies. The solubilized, renatured subunits and reconstituted holoenzyme can be rapidly purified from other proteins in the inclusion bodies in a single chromatography step. The procedures described below are for the preparation of wild-type and mutant subunits and holoenzymes of casein kinase II and the characterization thereof. It is also adaptable to other proteins, as described in **CONCLUDING REMARKS.**

REAGENTS

[γ-^{32}P]ATP (30 Ci/mmol) was obtained from Amersham. Urea, glutathione (oxidized), dithiothreitol, IPTG (isopropyl-β-D-thiogalactopyranoside), PMSF (phenylmethylsulfonyl fluoride), spermine, heparin and polylysine-HCl (36,500 kD) were from Sigma. Casein from Matheson, Coleman and Bell was dephosphorylated prior to use (12). Mono S HR 5/5, Mono Q HR 5/5 and Superose 12 HR 10/30 columns were from Pharmacia. The Heparin TSK column was from Tosohaas. CKII from rabbit reticulocytes was purified to apparent homogeneity as described previously (13). One enzyme unit is defined as the amount of enzyme that incorporates one pmol of phosphate/min into casein at 30°C.

PROTOCOLS

Subcloning and Expression of the α and β Subunits of CKII in *E. coli*

cDNAs encoding the α and β subunits of CKII from *Drosophila* were provided by C.V.C. Glover, Department of Biochemistry, University of Georgia, Athens, GA, in the λgtll vector. The subunits were excised from *Eco*RI sites in the vector and subcloned into M13 phage. The nucleotide sequences immediately adjacent to the 5′ end of the translation initiation codon were changed by site-directed mutagenesis to create a *Nde*I site (CATATG) (5). The cDNAs encoding the full-length polypeptide were then subcloned into *Nde*I and *Bam*HI sites in the pET3a vector.

cDNA clones for the α and β subunits of CKII from human were isolated from a Hela cell DNA library (14; R. Jakobi and W. Pyerin, unpublished data) and subcloned (15) into pET3a and pT7-7 vectors (16,17). The vectors containing recombinant cDNA for the α and β subunits from *Drosophila* and human were used to individually transform *E. coli* BL21(DE3).

Due to the high levels of protein produced, significant amounts of each subunit (1 to 2 mg) can be produced in a relatively small culture volume (25 ml). Larger amounts are readily obtained under the same conditions by scaling up the preparations. Overnight cultures of transformed *E. coli* are added to 500 ml of LB or M9ZB medium which contain 50 μg/ml ampicillin. The bacteria are incubated with high aeration at 37° C until the $O.D_{600}$ reaches 0.4 to 0.6. At that time, 0.5 mM IPTG is added to induce expression of the recombinant protein. After incubation at 37°C for 2 hr, the bacteria are harvested, suspended in 10 ml of extract buffer [50 mM Tris-HCl, pH 7.4, 1 mM EGTA, 1 mM EDTA, 10 mM 2-mercaptoethanol, 1 mM PMSF (added just prior to use) and 0.02% NaN_3]. The cells are disrupted by sonication with five 15-second cycles and a 1 min pause on ice between pulses. The sonicate is centrifuged at 10,000 x g for 20 min at 4°C. The supernate is collected and bacterial debris is removed by washing the inclusion body pellet twice, by resuspending the debris carefully with a pipette in 1.0 ml of extraction buffer without disturbing the inclusion bodies. The pellet is resuspended in 5.0 ml of extraction buffer, centrifuged as before, and resuspended in 5.0 ml of the same buffer. The concentration of resuspended subunits is 2 to 4 μg/μl. The inclusion body proteins, including the expressed α and β subunits of CKII, are analyzed by electrophoresis in sodium dodecyl sulphate on 11% polyacrylamide gels by electrophoresis, followed by staining with Coomassie blue (12).

Renaturation and Reconstitution of the α and β Subunits from Inclusion Bodies

To recover soluble active recombinant subunits from inclusion bodies, the resuspended subunits (2 to 4 mg) are denatured directly in 8 M urea and 50 to 300 mM dithiothreitol in a final volume of 1.0 ml at room temperature (25°C) for 2 hr. The final concentration of the reducing agent is dependent on the protein (e.g., human β subunit can be denatured at concentrations of 50 to 300 mM dithiothreitol, whereas the *Drosophila* β subunit requires the higher level of reducing agent) (18). With some proteins, we have found that addition of Tris-HCl, pH 7.5 and NaCl to final concentrations of 50 and 100 mM, respectively, in the denaturation buffer assists in the solublization and final recovery of active enzyme.

Renaturation of the subunits is initiated by diluting the sample 65-fold with cold renaturation buffer [100 mM Tris-HCl pH 8.5, 2 mM EDTA, 0.02% NaN_3, 0.2 mM PMSF, 0.4 M arginine, 1 mM glutathione (oxidized) and 5 μg/ml casein], which would result in final concentrations of approximately 3.0 to 4.5 mM dithiothreitol. Incubation is overnight at 10°C, at which time renaturation is complete. The 65-fold dilution reduces the urea to a concentration where accurate activity measurements can be obtained. Alternatively, a 10-fold dilution with renaturation buffer is sufficient for renaturation to occur, although the renatured protein needs to be diluted 4- to 5-fold prior to activity measurements and for chromatography to reduce the conductivity resulting from the high level of arginine in the buffer.

Equimolar amounts of α and β subunits in inclusion bodies are denatured and renatured together under the conditions described above, with holoenzyme as the final product. As a positive control for activity measurements, a similar amount of native CKII is denatured, diluted with renaturation buffer and incubated in conjunction with the experimental. For a negative control, denatured protein is diluted with buffer A (100 mM Tris-HCl, pH 8.5, 2 mM EDTA, 0.02% NaN_3 and 0.2 mM PMSF) and incubated under the same conditions. Activity assays for CKII are used to monitor the α subunit and reconstituted holoenzyme. The β subunit is detected by Coomassie blue staining following polyacrylamide gel electrophoresis.

For mammalian CKII, the concentration of dithiothreitol in the renaturation step is 3 mM, with oxidized glutathione increased to 4 mM; incubation is at either 25°C or 10°C. The differences in conditions of denaturation and renaturation between mammals and *Drosophila* have been shown to be due to differences in the aminoterminus of the β subunits (18).

Purification of the α and β Subunits and Reconstitution of CKII

For purification and concentration of reconstituted CKII, the sample is diluted 4- to 5-fold with buffer B (50 mM Tris-HCl, pH 7.4, 0.3 M NaCl, 1 mM EDTA, 1 mM EGTA, 10 mM 2-mercaptoethanol, 10% glycerol and 0.02% NaN_3) and loaded onto a heparin-TSK column previously equilibrated with buffer B. Protein is eluted with a 10 ml linear gradient from 0.3 to 1.0 M NaCl in buffer B, and 0.3 ml fractions are collected. Aliquots (0.010 ml) of each fraction are assayed for casein kinase activity. The reconstituted holoenzyme is present in fractions eluting at 750 mM NaCl.

The renatured α subunit is diluted 4-fold with buffer C (50 mM phosphate buffer, pH 6.8, 1 mM EGTA, 1 mM EDTA, 0.02% NaN_3, 10 mM 2-mercaptoethanol and 10% glycerol) and purified by chromatography on a Mono S column previously equilibrated with buffer C. A linear gradient of 15 ml of buffer C from 0 to 1.0 M NaCl is used to elute the protein; 0.5 ml fractions are collected.

The renatured β subunit is diluted 5-fold with buffer B containing 200 mM NaCl and loaded onto a Mono Q column equilibrated in the same buffer. After washing with 10 ml of the same buffer, the protein is step-eluted with 10 ml of buffer B containing 500 mM NaCl.

Assay for CKII

The activity of CKII (10 μl) is monitored in a final volume of 70 μl containing 50 mM Tris-HCl, pH 7.4, 100 mM KCl, 5 mM $MgCl_2$, 5 mg/ml dephosphorylated casein and 140 μM (γ-^{32}P)ATP (specific activity 500 dpm/pmol). Incubation is for 15 min at 30°C. The data provide values for the rate of phosphorylation, since both substrates, nucleotide and protein (casein), are in excess. Incorporation of radiolabel is measured by trichloroacetic acid precipitation of protein on filter paper (12).

Assays to examine sensitivity to acidic compounds also contain 1 μg/ml heparin. Assays examining stimulation by polyamines contain 1 mM spermine.

Analysis of Molecular Weight of CKII by Gel Filtration

Fractions from the heparin-TSK chromatography step containing high levels of CKII activity are pooled. A 0.5 ml sample is loaded onto a Superose 12 column equilibrated with buffer B containing 0.6 M NaCl, and 0.5 ml fractions are collected. Casein kinase activity is assayed as described above. The molecular weight is calibrated with the use of immunoglobulin, 160 kD, bovine serum albumin, 67 kD, and β-lactoglobulin, 35 kD, as standards.

EXPRESSION, DENATURATION AND RENATURATION OF CKII FROM INCLUSION BODIES

Expression of the α and β Subunits of CKII in *E. coli*

The cDNAs encoding the α and β subunits of CKII from *Drosophila* and human are expressed with the pET3a or pT7-7 vector in *E. coli* following induction with IPTG. The protein for both subunits is expressed at significant levels with 1 to 2 mg obtained from a 25 ml culture (Table 1). The subunit proteins are identifiable in the *E. coli* extracts by polyacrylamide gel electrophoresis, as shown for the *Drosophila* protein (Figure 1, left panel). Following low-speed centrifugation of the cell extract, the α subunit is observed both in the supernate and pellet fractions, with the majority of the subunit (85%) present in inclusion bodies. The β subunit from *Drosophila* is present only in the pellet fraction (Figure 1, right panel). The majority of the human α and β subunits, 60 and 75%, respectively, are also present as inclusion bodies (Table 1). Since the α and β subunits are major constituents of the inclusion bodies from both human and *Drosophila*, the majority of the CKII protein is highly purified from the vast majority of the bacterial proteins by the initial low-speed centrifugation step. The soluble subunits of CKII can be purified from the *E. coli* supernate in several steps by conventional ion-exchange chromatography and FPLC (Fast Protein Liquid Chromatography) (5,10).

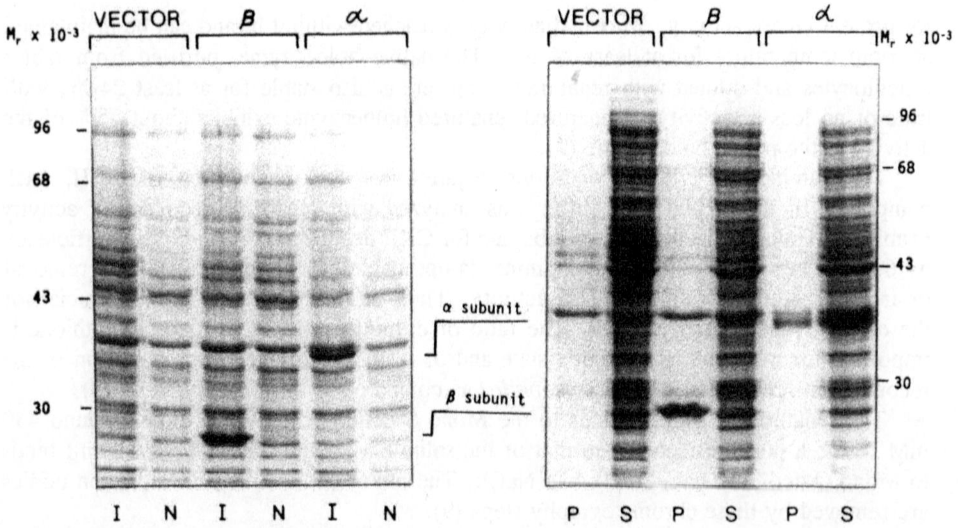

Figure 1. Expression of the α and β subunits of CKII from *Drosophila* in *E. coli*. *E. coli* BL21(DE3) were transformed with pET3a plasmids without a cDNA insert (Vector) or containing cDNA inserts for the β subunit or α subunit. Expression of the subunits was analyzed by polyacrylamide gel electrophoresis in sodium dodecyl sulfate followed by staining with Coomassie blue. Left panel. Aliquots (60 μl) of non-induced cultures (N) or of cultures after two hr of IPTG induction (I) were analyzed. Right panel. Bacteria were harvested two hr after IPTG induction, disrupted by sonification and centrifuged at 10,000 x g for 20 min at 4°C. The inclusion bodies (0.8 to 1.5 μl) (P) and the supernate fractions (10 to 12 μl) (S) were analyzed. Taken from (9).

Table 1. Recovery of Recombinant CKII from *Drosophila* and Human Expressed in *E. coli*.

Subunits of CKII	Total CKII in Bacterial Extracts[a]	CKII in Inclusion Bodies[b]
	%	%
Drosophila α	20	85
Drosophila β	20	100
Human α	30	60
Human β	30	75

[a] % of total protein in bacterial extracts
[b] % of total CKII in bacterial extracts

Denaturation and Renaturation of CKII

Complete denaturation of the α and β subunits in inclusion bodies is required for renaturation to occur. Denaturation is accomplished by incubation of inclusion body protein in 8 M urea and 0.05 to 0.3 M dithiothreitol for 2 hr. Renaturation is carried out in the presence of 0.4 M arginine, a fixed ratio of dithiothreitol to glutathione (oxidized) and 5 μg/ml casein. At room temperature, renaturation of the subunits and the holoenzyme occurs rapidly; 50% renaturation is observed at about 20 min, as measured by the assay for protein kinase activity. Maximal activity is reached within 1 hr and can be maintained at room temperature for at least 24 hr. The native holoenzyme, purified from rabbit reticulocytes and diluted with renaturation buffer, is also stable for at least 24 hr, with little or no loss of activity. Denatured/renatured holoenzyme exhibits about 75% of the activity of the native holoenzyme (9).

To examine the critical conditions required for the renaturation of CKII, each component in the renaturation buffer was analyzed with regard to recovery of activity (Table 2). Omission of the protein substrate for CKII decreases the renaturation efficiency by 60%. The chaotrophic agent, arginine, is optimal at 0.4 M and reduction or removal of arginine reduces or eliminates the activity. Thus, substrate and arginine are crucial for the optimal renaturation of CKII. The ratio of dithiothreitol and oxidized glutathione is important for maintaining the redox state and of major importance in renaturation of the recombinant holoenzyme from *Drosophila*, as compared to that from human (18).

The renatured α subunit binds to the Mono S column tightly and elutes around 450 mM NaCl, a position identical to that of the soluble α subunit (5). The β subunit binds to Mono Q and elutes at < 500 mM NaCl. The minor contaminants in inclusion bodies are removed by these chromatography steps (9).

Table 2. Conditions for Renaturation of Casein Kinase II

Additions	Activity	
	units/ml	%
GSSG/DTT, 0.4 M arginine, casein	418,461	100
GSSG/DTT, 0.4 M arginine	172,158	41
GSSG/DTT, 0.2 M arginine	50,603	12
GSSG/DTT	23,611	6
None	11,713	3

Casein kinase II from rabbit (4 μl, 2248 units) was denatured in 8 M urea and 0.1 M dithiothreitol (DTT) for 2 hr at room temperature, and renatured in Buffer A containing the indicated additions. Oxidized glutathione (GSSG) was 4 mM, DTT was 3 mM and casein was 5 μg/ml. Taken from Lin and Traugh (9).

PURIFICATION OF RECOMBINANT CKII

The recombinant holoenzyme is reconstituted from inclusion bodies containing the α and β subunits in a 1:1 molar ratio by denaturation and renaturation. A quick initial assay for formation of the holoenzyme is the enhanced catalytic activity over that of the α subunit alone. This assay is used as an indicator of proper renaturation of the holoenzyme and thus also of the α and β subunits. The protein kinase activity is stimulated 4- to 6-fold compared to that of the renatured α subunit alone (Table 3). The β subunit has no phosphotransferase activity.

The reconstituted holoenzyme is purified to apparent homogeneity by one-step chromatography on a heparin-TSK column. This step also concentrates the enzyme. Any excess β subunit does not adhere to the heparin column, while excess α subunit adheres to the column and may copurify with the holoenzyme. Any excess α subunit can be removed by an additional FPLC step on Mono Q. Under these conditions, the α subunit does not adhere to the resin, while the reconstituted holoenzyme and β subunit bind.

Table 3. Reconstitution of CKII from Recombinant Subunits Expressed in *E. coli*

Renatured Subunits	^{32}P Incorporated	Activity
	cpm	-fold
CKIIα	22,771	1.0
CKIIβ	0	0
CKIIα + CKIIβ	91,724	4.0

Inclusion bodies (11 μg), containing either the α or β subunits alone or in combination (total 22 μg), were denatured, renatured and assayed as described in Protocols. Taken from Lin and Traugh (9).

The *Drosophila* enzyme has been purified 3.5-fold from inclusion bodies on the heparin column, with a 75% recovery in activity (9). The specific activity of the recombinant holoenzyme is 849,000 pmol/min/mg, which is 5- to 10-fold higher than that of the α subunit alone. The activity is about 60% of that reported for the holoenzyme from *Drosophila* assayed under similar conditions (19).

One potential problem is the presence of one or more major breakdown products of the subunits in the inclusion bodies. For instance, a fragment of the human α subunit (~35 kD) is present which has similar chromatographic properties on heparin-TSK gel to those of the α subunit and holoenzyme. To obtain subunit or holoenzyme free of breakdown product, the denatured/renatured inclusion bodies were chromatographed through a small phosphocellulose column to concentrate the recombinant holoenzyme, purified by FPLC on Mono Q, then chromatographed on heparin-TSK gel. This provided pure holoenzyme, free of any breakdown product with a recovery of greater than 50% of the total activity (Table 4).

Table 4. Purification of Human CKII Renatured and Reconstituted from Inclusion Bodies

	Protein	Activity	Specific activity	Yield
	mg	nmol/min	nmol/min/mg	%
Renatured protein	25.0	1500	60	100
Phosphocellulose	10.0	1122	112	75
Mono Q	4.0	811	201	54
Heparin-TSK gel	3.3	793	238	53

Inclusion bodies containing equimolar subunits of α and β subunits of human CKII were denatured, renatured and the holoenzyme was concentrated by chromatography on phosphocellulose and purified by FPLC on Mono Q and Heparin-TSK.

CHARACTERIZATION OF RECONSTITUTED CKII

Recombinant CKII can be examined with regard to a number of parameters: molecular weight and subunit composition, sensitivity to inhibition by acidic compounds, stimulation by polyamines, and optimal salt concentration for activity. The values obtained for the recombinant holoenzyme, as compared to those for the α subunit alone, enable determination of whether the correct confirmation is obtained. The subunit composition of the *Drosophila* and human enzyme can be determined by gel filtration at a high salt concentration (600 mM NaCl). The high salt is utilized to prevent aggregation of enzyme which occurs at lower salt concentrations (19,20). The correctly-formed tetramer migrates at 160,000 as determined for the native holoenzyme, while any nonreconstituted subunits chromatograph at a lower molecular weight (Figure 2). Analysis of the subunit ratio in the holoenzyme following polyacrylamide gel electrophoresis in sodium dodecyl sulfate shows the α and β subunits are present at approximately equal concentrations, indicating formation of the $\alpha_2\beta_2$ tetramer. Stimulation of the catalytic activity by reconstitution with the β subunit is 4- to 6-fold for *Drosophila* and human. Heparin inhibits both the renatured recombinant α subunit and the reconstituted holoenzyme to a similar extent ($\geq$ 90% at 1 μg/ml). The polyamine spermine stimulates the catalytic activity of the native holoenzyme to the same extent as that observed with native CKII. The extent of stimulation is dependent on the Mg^{2+} concentration; at lower concentrations, stimulation by spermine is greater (21). The isolated α subunit is not stimulated, as would be expected since polyamines act through the β subunit (5,9,10). The optimal salt concentration for the recombinant holoenzyme is 150 to 200 mM KCl, while that for the α subunit is 50 mM KCl.

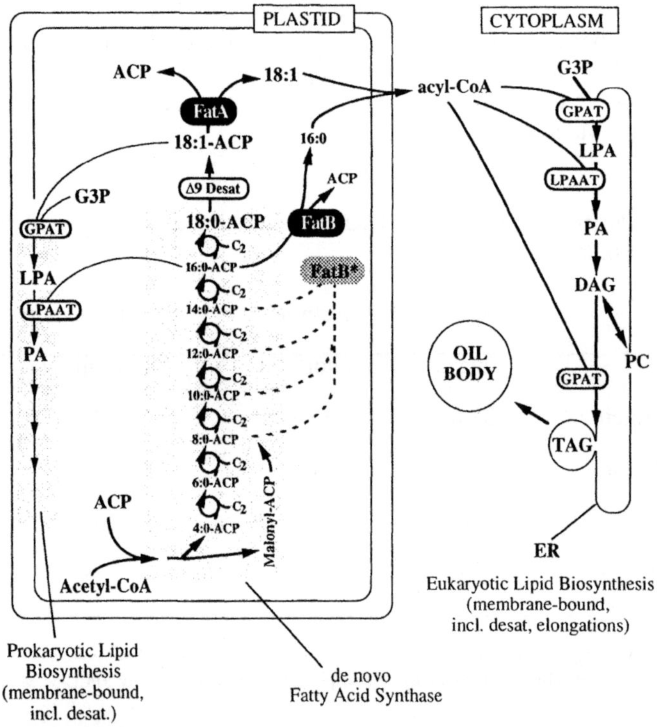

Figure 2. Analysis of reconstituted CKII by gel filtration. Inclusion bodies containing *Drosophila* α and β subunits in a 1:1 ratio were renatured and purified by FPLC on heparin-TSK gel. CKII was analyzed following gel filtration on a Superose 12 column. The positions of the heterotetramer, α subunit and the molecular weight markers are identified. Taken from (9).

The properties described for the recombinant holoenzyme are identical to those observed for the native holoenzyme. Thus, the characteristics of the recombinant holoenzyme mirror those of native CKII. Gel filtration identifies the correct molecular weight, and a comparison of the activity of α alone and α reconstituted with β shows significant stimulation. Activity ratios at 50 mM and 200 mM NaCl show differences between α and reconstituted $\alpha_2\beta_2$ activity, as does stimulation by polyamines.

CONCLUDING REMARKS

Recombinant subunits of wild-type and mutant CKII can be reconstituted to form active holoenzyme. Soluble subunits of CKII expressed in *E. coli* can be purified by various chromatography procedures and usually require at least three chromatography steps (5,10). However, at high levels of expression, the majority of the expressed protein is present primarily in inclusion bodies. The recovery of active recombinant CKII from inclusion bodies of *E. coli* by the denaturation/renaturation procedure described above, allows purification and analysis of insoluble recombinant proteins. Since the recombinant subunit of *Drosophila* CKII is insoluble, it is possible to study the role of the individual subunits in the function and regulation of CKII, as well as the function of CKII in metabolism. The described procedure permits answering the question of whether there is sufficient homology to produce cross-species recombinant holoenzymes. In this regard we have been able to show that CKII from human and *Drosophila* form highly active holoenzyme, even though the β subunit of *Drosophila* requires a different redox ratio for appropriate renaturation (18). The denaturation/renaturation procedure also allows for utilization of mutations present only as insoluble protein to be utilized in analysis of the molecular structure and functional properties of the enzyme. It has been used to characterize the nucleotide binding site in the α subunit of CKII and to analyze the function of the autophosphorylation site on the β subunit (15,19,22). Finally, the procedure can be adapted to a variety of different proteins, including other protein kinases.

Since the critical components required to achieve the native structure *in vivo* are lacking *in vitro*, correct folding can be achieved only by adjusting the solvent conditions. Disulfide bond formation is enhanced by modulating the ratio of dithiothreitol/oxidized glutathione which provides a disulfide reshuffling system. The chaotropic agent arginine acts by destabilizing the nonspecific interactions between partially-folded intermediates, thus enhancing correct folding (11,23). Arginine is shown to be critical in obtaining active heterotetramers of CKII. The substrate casein is also important in renaturation by binding to the active site and thus stabilizing the correct folding intermediates. Renatured and reconstituted under these conditions, CKII has identical characteristics to native holoenzymes and to holoenzyme formed from soluble recombinant subunits. With other proteins, similar levels of arginine and oxidized glutathione/dithiothreitol can be utilized, although a substrate specific for the enzyme to be renatured should be utilized rather than casein.

ACKNOWLEDGMENTS

This work was supported by a grant from the U.S. Public Health Service, GM26738.

REFERENCES

1 Tuazon, P.T. and Traugh, J.A. (1991) in Advances in Second Messenger and Phosphoprotein Research (Greengard, P. and Robison, G.A., eds.), Vol. 23, pp. 123-164, Raven Press, New York, NY.

2 Pinna, L.A. (1990) Biochim. Biophys. Acta 1054, 267-284.
3 Issinger, O.-G. (1993) Pharmac. Ther. 59, 1-30.
4 Hanks, S.K., Quinn, A.M. and Hunter, T. (1988) Science 241, 42-52.
5 Lin, W.-J., Tuazon, P.T. and Traugh, J.A. (1991) J. Biol. Chem. 266, 5664-5669.
6 Meggio, F., Donella-Deana, A., Brunati, A.M. and Pinna, L.A. (1982) FEBS Lett. 141, 257-262.
7 Hu, E. and Rubin, C.S. (1990) J. Biol. Chem. 265, 20509-20615.
8 Grankowski, N., Boldyreff, B. and Issinger, O-G. (1991) Eur. J. Biochem. 198, 25-30.
9 Lin, W-J. and Traugh, J.A. (1993) Protein Express. Purific. 4, 256-264.
10 Jakobi, R. and Traugh, J.A. (1992) J. Biol. Chem. 267, 23894-23902.
11 Buchner, J. and Rudolph, R. (1991) Bio/Technology 9, 157-162.
12 Hathaway, G.M., Lundak, T.S., Tahara, S.M. and Traugh, J.A. (1979) Methods Enzymol. 60, 495-511.
13 Palen, E. and Traugh, J.A. (1991) Biochemistry, 30, 5586-5590.
14 Jakobi, R. and Pyerin, W. (1989) Eur. J. Biochem. 183, 227-233.
15 Lin, W-J., Sheu, G-T. and Traugh, J.A. (1994) Biochemistry 33, 6998-7004.
16 Studier, F.W., Rosenberg, A.H., Dunn, J.J. and Dubendorff, J.W. (1990) Methods Enzymol. 185, 60-89.
17 Tabor, S. and Brent, R. (1990) in Current Protocols in Molecular Biology (Ausubel, F.A., Brent, R., Kingston, R.E., Moore, D.D., Seidman, J.G., Smith, J.A. and Struhl, K., eds) pp. 16.2.1-16.2.11, Greene Publishing and Wiley-Interscience, New York, NY.
18 Lin, W-J., Jakobi, R. and Traugh, J.A. (1994) J. Prot. Chem. 13, 217-225.
19 Glover, C.V.C., Shelton, E.R. and Brutlag, D.L. (1983) J. Biol. Chem. 258, 3258-3265.
20 Hathaway, G.M. and Traugh, J.A. (1979) J. Biol. Chem. 254, 762-768.
21 Hathaway, G.M. and Traugh, J.A. (1984) Arch. Biochem. Biophys. 233, 133-138.
22 Jakobi, R. and Traugh, J.A. (1995) Eur. J. Biochem. 230, 111-117.
23 Rudolph, R. (1990) in Modern Methods in Protein and Nucleic Acid Analysis. (Tschesche, H., ed.), pp. 149-171, Walter de Gruyter, Berlin and New York, NY.

PLANT ACYL-ACP THIOESTERASES: CHAIN-LENGTH DETERMINING ENZYMES IN PLANT FATTY ACID BIOSYNTHESIS

Toni Voelker

Calgene, Inc.
1922 Fifth Street
Davis, CA 95616

INTRODUCTION

De novo plant fatty acid biosynthesis takes place in an organelle, the plastid. Acyl chains are elongated by 6 to 7 dissociable enzymes, together constituting a so-called type II fatty acid synthase. During the extension cycles the acyl chains are covalently bound, *via* a thioester, to the prosthetic group of a small, soluble, acyl carrier protein (ACP; 78 amino acid residues) (1,2). Taken together, these features make the plant fatty acid synthase distinct from the cytoplasmic, multi-subunit fatty acid synthase of animals and fungi (type I), but quite similar to the one common in prokaryotes (3). This functional similarity is one of the many vestiges of the prokaryotic ancestry of the plastids, which descended from cyanobacterial endosymbionts (4).

Acyl-chain elongation in plastids can be terminated by transferring the newly-formed acyl chain directly from ACP to glycerol-3-phosphate, thereby retaining the acyl groups in the plastidial, "prokaryotic", membrane glycerolipid synthesis pathway (Figure 1). Alternatively, specialized acyl-ACP thioesterases can intercept the prokaryotic pathway of plastids by hydrolyzing the newly-formed acyl-ACP into free fatty acids and ACP. The free fatty acids exit the plastids by an unknown mechanism and become esterified to CoA. The cytoplasmic CoA-esters supply the "eukaryotic" lipid biosynthesis pathway (5). Phospholipids, triglycerides and other neutral lipids of this pathway are assembled by enzymes located in endoplasmatic reticulum membranes (Figure 1). Acyl-ACP thioesterases, therefore, by catalyzing the first committed step in the eukaryotic lipid biosynthesis pathway in all plant cells, play a crucial role in the distribution of *de novo* synthesized acyl-groups between the two pathways (6), reviewed in (1, 2). The acyl-ACP thioesterase is considered to be an essential enzyme of all plant cells. Since this is the first review dedicated solely to the subject of plant acyl-ACP thioesterases, the author wishes to begin with a brief historic outline.

Acyl-ACP thioesterase activity in plants was first detected by Shine et al. (7) in avocado mesocarp extracts and subsequently partially purified and characterized by Ohlrogge et al. (8). With the use of different acyl-ACP substrates *in vitro*, acyl-ACP

Genetic Engineering, Vol. 18, Edited by J.K. Setlow
Plenum Press, New York, 1996

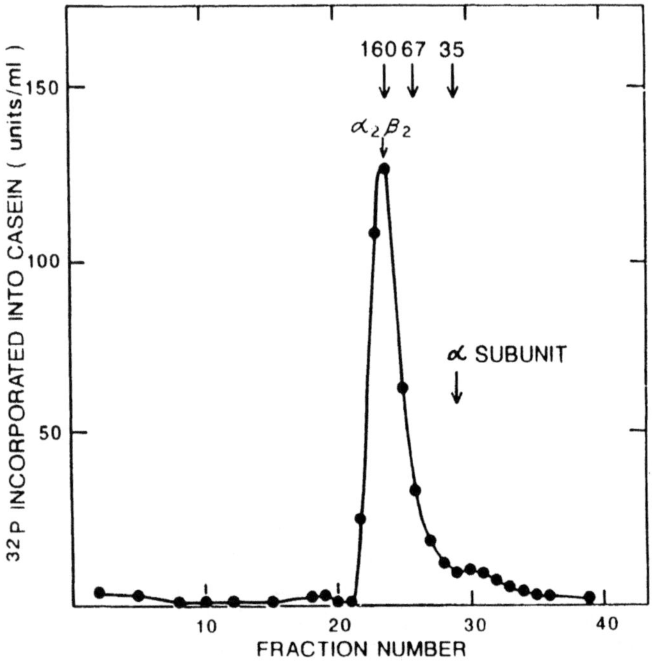

Figure 1. Plant Lipid Biosynthesis. Emphasis is given to the "two pathway" aspect of lipid biosynthesis; only selected enzymes (boxed) and metabolites are shown. Each 4-step chain elongation cycle of the *de novo* fatty acid synthesis is symbolized by a circle. Oil bodies are the storage organelles for TAGs, in principle oil droplets surrounded by a specialized, half-unit membrane. Membrane-bound components and substrates of the lipid synthesis pathway are shown "embedded" in the respective membranes. **ACP**, acyl carrier protein; **DAG**, 1,2- diacylglycerol; **DAGAT**, DAG acyltransferase; **Δ9**, stearoyl-ACP Δ9 desaturase; **ER**, endoplasmic reticulum; **FatA/B**, the mnemonic of the two forms of acyl-ACP thioesterases; **FatB***, derived, medium-chain FatBs of specialized tissues, pathways dotted; **G3P**, glycerol-3-phosphate; **GPAT**, G3P acyltransferase; **LPAAT**, lysophosphatidic acid acyltransferase; **PA**, phosphatidic acid; **PC**, phosphatidyl choline; **TAG**, triacylglycerol.

thioesterase-enriched fractions from three plant tissues (avocado mesocarp, coconut endosperm and jojoba seeds) were found to be most active on 18:1-ACP, much less so on 18:0-ACP and 16:0-ACP, and only minimally on 12:0-ACP and 14:0-ACP. This discovery supplied one essential "mosaic stone" for the formulation of the "two-compartment concept" of plant lipid biosynthesis (9), as outlined in the previous paragraphs. Since oleate (18:1 (Δ9 *cis*)), together with its derivatives, usually represent the bulk of the *de novo* fatty acyl carbon flux through the eukaryotic pathway (2), and 16:0 and its derivatives make up the remainder [between 5 and 20% (10)], the preferences of the acyl-ACP substrates were in good accordance for normal plant tissues. However, the activity profile obtained with the coconut endosperm was puzzling, since this specialized tissue produces almost exclusively so-called medium-chain (C8-C14) fatty acids for deposition in triglycerides for storage (coconut oil) (10).

During the decade following their discovery, only a few thioesterases were purified to near-homogeneity. The first example was the oleoyl thioesterase from maturing safflower seeds, which was reported to be a dimer of M_r 41 kD subunits and to have an activity profile (11) as described for the partially purified enzymes (8). The mechanism for the production of medium-chain fatty acids resisted biochemical investigation (12), and remained an enigma for many years. In 1988, Harwood listed various possible mechanisms, but no direct evidence was provided for any one (13). Only more recently have several thioesterases been purified sufficiently to allow protein sequencing. This information facilitated the cloning of the respective cDNAs, which in turn, after appropriate expression in heterologous hosts, allowed the functional test of the encoded polypeptides. In a next stage these few cDNAs were used for the isolation of more than 30 cDNAs and genes from more than 10 different species. In summary, the role of acyl-ACP thioesterases in the determination of chain length is now well described. It also became clear that plants harbor acyl-ACP thioesterase genes of two distinct, but related classes, now labeled *FatA* and *FatB*, from **F**atty **A**cyl-ACP **T**hioesterase (14). The *FatA* genes encode oleoyl-ACP specific enzymes (FatA), while *FatB* genes encode thioesterases with preferences for saturated substrates (FatB) of various chain lengths (Figure 1). In order to describe the species from which the respective enzyme or gene is derived, the first letters of the genus and species, respectively, are combined and placed in front of the "Fat" mnemonic (14) (see Table 1 for examples).

PROTEIN SEQUENCE-BASED THIOESTERASE CLONING

Since acyl-ACP thioesterases were perceived to be tools useful for the genetic engineering of seed oil composition [proposed by (15,16)], commercial interest drove the cloning of most cDNAs and genes encoding these enzymes. This made it necessary to include the patent literature for this review. For example, it was speculated that a gene for a then still-to-be discovered 12:0-ACP thioesterase could be introduced into a temperate zone oil crop, thereby enabling this transformant to produce the detergent feed stock lauric acid instead of the usual C18 fatty acids (15). Medium-chain producing seeds of several species were investigated, but no evidence of medium-chain thioesterase activity was detected (8,12,17,18). In 1991, Pollard et al. (19) reported the first evidence of such an enzyme in maturing seeds of California bay trees (*Umbellularia californica*), which produce a laurate-rich oil. Extracts from medium-chain producing cotyledons yielded significant 12:0-ACP hydrolytic activity when compared to immature seeds of the same species (19). This activity was subsequently purified approximately 40,000-fold to near-homogeneity (20). After SDS-PAGE a polypeptide doublet was observed with an M_r of 34 kD. The native thioesterase migrated in gel filtration chromatography with a MW of 42,000, suggesting a monomer. Sequences from the N-terminus of these polypeptides, as well as internal peptides, were determined. Based on these limited amino acid sequences, a mixed-oligonucleotide PCR (21) strategy was used for the amplification of specific transcript fragments from first-strand cDNA of maturing bay seeds. After maturing bay seed cDNA libraries were screened with the PCR fragments, cDNAs from two distinct but related transcripts of about 1500 bases were obtained (14,22,23). The more prominent cDNA class, derived from what is now named *Uc FatB1* (see Table 1 for nomenclature) has an open reading frame encoding a 382 amino acid polypeptide matching most of the short peptide sequences obtained from the purified protein. The N-terminus of the purified protein was located at residue 85. A molecular mass of 34 kD could be calculated for the remainder of the polypeptide, in good accordance with that of the purified protein. Since any poly(A)RNA-derived protein has to be imported into the plastid

Table 1. Sources of Sequences, Specificities

Species	Mnemonic	Previous Designations	References
Allium porrum (leek)	*Ap FatA1*	A-1	(79)
	Ap FatB1	9-1	(39,79)
	Ap FatB2	8-2	(39,79)
	Ap FatB3	NL6-4	(79)
Arabidopsis thaliana	*At FatA1*	TE 1-7	(42)
	At FatB1	TE 3-2	(42)
Brassica napus (rapeseed)	*Bn FatA1*	NL2	(32)
	Bn FatA2	NL3	(32)
Brassica rapa	*Br FatA1*	pCGN3266	(14)
Carthamus tinctorius (safflower)	*Ct FatA1*	2-1	(29)
	Ct FatA2	5-2	(29)
Cinnamonum camphorum (camphor)	*Cc FatB1*	pCGN5219	(75)
Coriandrum sativum (coriander)	*Cs FatA1*		(34)
Cuphea hookeriana	*Ch FatA1*	CLT7	(40,87)
	Ch FatB1	CMT9	(14,87)
	Ch FatB2	CMT7	(39,40,52)
	Ch FatB3	CMT13	(39,40)
	Ch FatB4	CMT10	(39,40)
Cuphea lanceolata	*Cl FatB1*	*Cl*te13; *Cl*teg7	(43,46)
	Cl FatB2	*Cl*te5; *Cl*teg4	(43,46)
	Cl FatB3	*Cl*te2; *Cl*teg1; *Cl*TEg100	(43,46,53)
	Cl FatB4	*Cl*teg16; *Cl*TEg200	(43,46,53)
Cuphea palustris	*Cp FatB1*		(54)
	Cp FatB2		(54)
Cuphea wrightii	*Cw FatB1*		(82,88)
	Cw FatB2		(82,88)
Glycine max (soybean)	*Gm FatA1*	2	(27,28)
	Gm FatA2	4	(27,28)
Mangifera indica (mango)	*Mi FatA1*	MI-06	(39,89)
	Mi FatB1	MI-09	(39,89)
	Mi FatB2	M4-23	(39,89)
Ulmus americana (elm)	*Ua FatB1*		(14,39,87)
Umbellularia californica (California bay)	*Uc FatB1*	BTE	(23,39)
	Uc FatB2		(14)

in order to access the plant fatty acid synthase, a transit peptide was predicted (chloroplastic import reviewed by 24). Gene bank searches yielded no significant matches, indicating that the bay thioesterase belongs to a novel class of genes. When expressed in *E. coli* as a LacZ-fusion product (from residue 85), *Uc* FatB1 showed an extreme preference for 12:0-ACP (25) (Figure 2A). The full-length *Uc FatB1* reading frame was combined with a napin seed storage protein promoter (26), and used to transform *Arabidopsis thaliana*. Maturing seeds of transformants showed a dramatically elevated 12:0-ACP hydrolytic activity when compared to the control, and laurate represented up to 20% of the acyl groups in mature seeds (23). Thus, by partially redirecting a long-chain fatty acid synthase to laurate production *in vivo*, this work demonstrated for the first time that acyl-ACP thioesterases can determinate acyl chain length *in vivo*. It also confirmed a mechanism for medium-chain fatty acid synthesis in plants and showed the feasibility of creating an annual laurate oil crop.

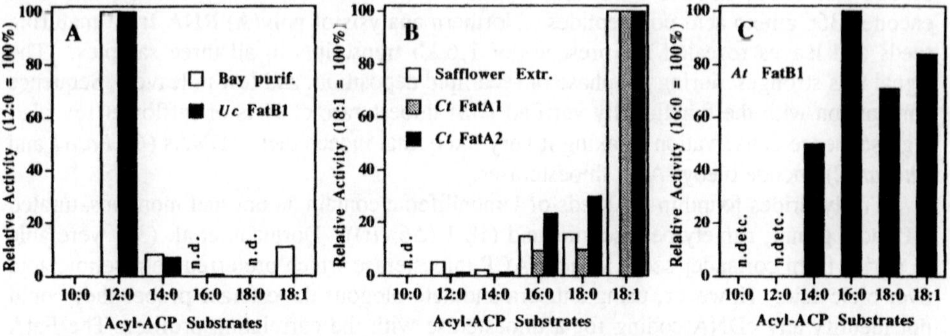

Figure 2. Substrate Specificities of Protein-Purified and cDNA-Expressed Fat Enzymes. Respective enzyme extracts were assayed with different acyl-ACP substrates. In order to allow comparisons of specificity, the respective highest activity for each preparation was set at 100%. Enzyme sources were either purified plant proteins (open bars) or *E. coli* extracts of expressed cDNAs (filled bars; mature portion, see text). **n.d.**, not determined; **n.detc.**, not significantly above bacterial background; data for Figure 1A are from (25), Figure 1B from (29), Figure 1C from (34).

Two groups succeeded in cloning oleoyl thioesterase cDNAs using limited peptide sequences and tested the expressed enzymes *in vitro* [soybean (27,28), safflower (29)]. These enzymes were purified from maturing seeds. The safflower thioesterase preparation used for protein sequencing consisted of 40 kD and 34 kD polypeptides which yielded identical N-termini and internal peptide sequences. Screening of 10^5 plaques of a maturing seed cDNA library yielded 5 clones of a type termed 2-1 (which matched all the peptide sequences, now *Ct FatA1*) and one clone, 5-2 (which matched no peptide sequences completely, now *Ct FatA2*). Full-length cDNAs were about 1500 bp and encoded 80% identical polypeptides of 385 and 389 amino acids, respectively. The N-terminus of the purified protein (40 and 34 kD polypeptides) was located at residue 61 of the derived *Ct* FatA1 sequence, most likely defining a 60 amino acid transit peptide and a 325 amino acid mature protein with a calculated MW of 37,254. After plasmid-based synthesis of the mature *Ct* FatA1 in *E. coli*, a 40 kD polypeptide was detected. This matched the apparent molecular weight of the partially purified safflower thioesterase (11). The 34 kD form was most likely a C-terminal-processed artifact of the purification procedure. The activity profiles of *E. coli*-expressed *Ct* FatB1 and *Ct* FatB2 were rather similar, both preferred 18:1-ACP (29) (Figure 2B). This defined the two safflower cDNAs as encoding oleoyl-ACP thioesterases, now termed FatA.

After 8500-fold purification, the soybean oleoyl-ACP thioesterase had a native M_r of 75 kD, and showed an M_r of 39 (minor) and 34 (major) kD in SDS-PAGE, suggesting a dimer *in vivo*. Peptide sequence-based cDNA cloning yielded two nearly identical (97%) sequences encoding 367 amino acid polypeptides (27). N-terminal sequencing of the M_r 34 kD form defined a 55 amino acid transit peptide. Production of LacZ fusions of the mature enzymes in *E. coli* allowed the determination of the activity profiles, which were very similar to the safflower cDNAs. Soybean, tobacco, and canola transformants harboring various promoter constructs showed that 18:1-ACP hydrolytic activities of tissue extracts could be increased and decreased. Significant changes in fatty acyl composition in the transformed tissues were reported later (28,30).

An oleoyl thioesterase was also purified from maturing seeds of *Brassica napus* (31). The enzyme consisted of immunologically related 38 and 33 kD polypeptides; however, only the 38 kD form was detected in an immunoblot from direct seed extracts. Based on internal peptide sequence information, Loader et al. (32) were able to isolate two classes of cDNAs. Both cDNAs were nearly identical (98% in the reading frame), and both encoded 366 amino acid polypeptides. Northern analysis of poly(A) RNA from maturing seeds and leaves revealed the presence of 1.6 kb transcripts in all three samples. The signal was strongest during the phase of seed lipid deposition, and low in leaves. Sequence comparison with the functionally verified *FatA* thioesterase cDNAs of safflower revealed high sequence conservation, making it very likely that indeed these cDNAs (*Bn FatA1* and *Bn FatA2*) encode oleoyl-ACP thioesterases.

Triglycerides found in the seeds of Umbelliferae contain an unusual monounsaturated C18 acyl group, namely petroselinic acid (18:1 (Δ6 *cis*)). Dörmann et al. (33) were able to purify from coriander seeds an acyl-ACP thioesterase which preferred petroselinic acid over oleic acid. However, using antisera and heterologous thioesterase probes they could not identify any cDNA coding for a thioesterase with the correlating profile. The FatA thioesterase from coriander (*Cs* FatA1) was found to prefer 18:1 (Δ9 *cis*) over other 18:1 substrates (34). After *E. coli* expression the *Cs* FatA1 mature protein (N-terminus inferred by homology with safflower FatA sequences) accumulated as a 40 kD polypeptide, which after exposure to coriander endosperm extracts was processed to a 34 kD form.

In summary, the protein purification of 18:1-ACP thioesterases and the subsequent cloning and expression of cDNAs of several dicot species revealed that most species investigated carry one or two related nuclear genes. Species-to-species conservation is about 75% for the mature enzyme portion and the respective coding area. Almost no sequence conservation is found in the transit peptide sequences, which is normal (24). The 360 to 380 amino acid pre-proteins are synthesized in the cytoplasm from relatively rare (~ 1 in 10,000) 1500 to 1600 nucleotide poly(A) mRNA. An ~ 60 amino acid N-terminal transit peptide is removed after import into the stroma of the plastids, where the mature enzyme accumulates as a dimer consisting of polypeptides of approximately 325 amino acid length, with an M_r of ~ 40 kD. A portion of the C-terminus is obviously proteolytically sensitive and is easily lost during purification, leading to the ~ 34 kD form. *FatA*-encoded oleoyl-ACP thioesterases are somewhat active on 16:0-ACP and 18:0-ACP (5 to 20% of 18:1), and therefore could explain satisfactorily the acyl-ACP hydrolytic activities observed in crude extracts (8) (Figure 2B), and could as well be responsible for the small amount of palmitate and stearate deposited in "eukaryotic" lipids of most tissues. As first reported (29), the FatA thioesterases are related to the bay 12:0-ACP thioesterase, but to a somewhat lesser degree than found between oleoyl-ACP thioesterases (50% sequence conservation). This could reflect the phylogenetic relationships of the respective species; Lauraceae (California bay) are only very distantly related to the other dicot species (35). However, the emergence of many more *Fat* cDNAs proved this interpretation to be incorrect, as discussed below. None of the plant acyl-ACP thioesterase sequences had significant matches with the yeast or animal thioesterases (36,37), nor do they contain the serine-active site consensus sequence. In contrast to the serine-active thioesterases of other kingdoms, the plant FatA and FatB thioesterases appear to utilize an active-cysteine mechanism (20,29).

DNA-SEQUENCE BASED THIOESTERASE cDNA CLONING

Fat sequences from few species facilitated the isolation of the homologous genes or cDNAs from many different angiosperm species, thereby shunting protein purification procedures. Hitz and Yadav (27) used the soybean cDNAs to detect RNA transcripts in tomato, *Brassica*, sunflower and *Arabidopsis thaliana*, and isolated genomic fragments of a *FatA* gene from *Brassica napus* by direct hybridization. In an alternative route, primers designed from randomly chosen short amino acid sequences of the soy thioesterases were used in PCR to amplify short *FatA* sequences from genomic DNA of *Cuphea lanceolata*, a medium-chain producing species. In an independent direct-hybridization library screening, Knutzon was able to obtain a *FatA* cDNA from *Brassica rapa* using safflower cDNA probes (*Br FatA1)*, reported in (14).

The major thrust of these activities was directed towards the cloning of sequences encoding thioesterases with unique specificities, especially for medium chains and 18:0, and several strategies proved to be successful. The most extensive cloning efforts were undertaken at Calgene, Inc. The primary approach was based on the rationale that some of the sequences completely conserved between the safflower oleoyl thioesterases and the bay medium-chain thioesterase (e.g., Figure 3) should be present in the thioesterases of other species. In a first approach, the maturing seeds of *C. hookeriana* [seed reserve triglycerides contain 60 mol% 8:0 and 25% 10:0; (38)], and maturing seeds of elm [seed reserve triglycerides with 80 mol% 10:0; (10)], were used as sources of template. Thioesterase sequences were amplified from primary cDNA using a mixed-oligonucleotide PCR strategy (21). Primers were designed with 5 to 7 residue amino acid sequences conserved between the safflower 18:1-ACP thioesterases *Ct* FatA1 and *Ct* FatA2 and the California bay 12:0-ACP thioesterase *Uc* FatB1. Two combinations were successful, and amplified at least two distinct transcripts in *C. hookeriana* and one transcript from elm, all similar to *Uc FatB1*. The amplified fragments facilitated the cloning of several *FatB* cDNAs (*Ch FatB1, Ch FatB2, Ch FatB3, Ch FatB4, Ua FatB1*) (14,39,40) (Table 1). After a comparison of these sequences with the oleoyl thioesterases of safflower, soybean and *Brassica* and the medium-chain thioesterase of bay, it was found that all elm and *Cuphea* sequences were much more related to the latter than to the oleoyl thioesterases. Phylogenetic relationships of the respective species could be ruled out as the cause for the similarity pattern, since Ulmaceae (elm) and Lythraceae (*C. hookeriana*) are much more closely related to the Brassicaceae, for example, then to the Lauraceae (California bay) (35). This allowed Jones et al. to propose that probably all higher plants carry genes of two thioesterase sequence classes, *FatA* and *FatB*, independently of the presence or absence of medium chains in seed storage oils (14). The first proof of this hypothesis emerged in the course of random sequencing of an *A. thaliana* developing-silique library, in which a partially sequenced cDNA appeared to code for a *Uc FatB1* homologue (41). With this fragment as a probe, a full-length *FatB* cDNA (*At FatB1*) was subsequently isolated and characterized (42).

Thioesterase cDNAs and genes were also cloned from maturing seeds of *Cuphea lanceolata* [seed triglycerides have 83% 10:0, (38)]. Using *Uc* FatB1 as a sole source of sequence information for PCR amplification, Töpfer et al. (43) selected short sequences

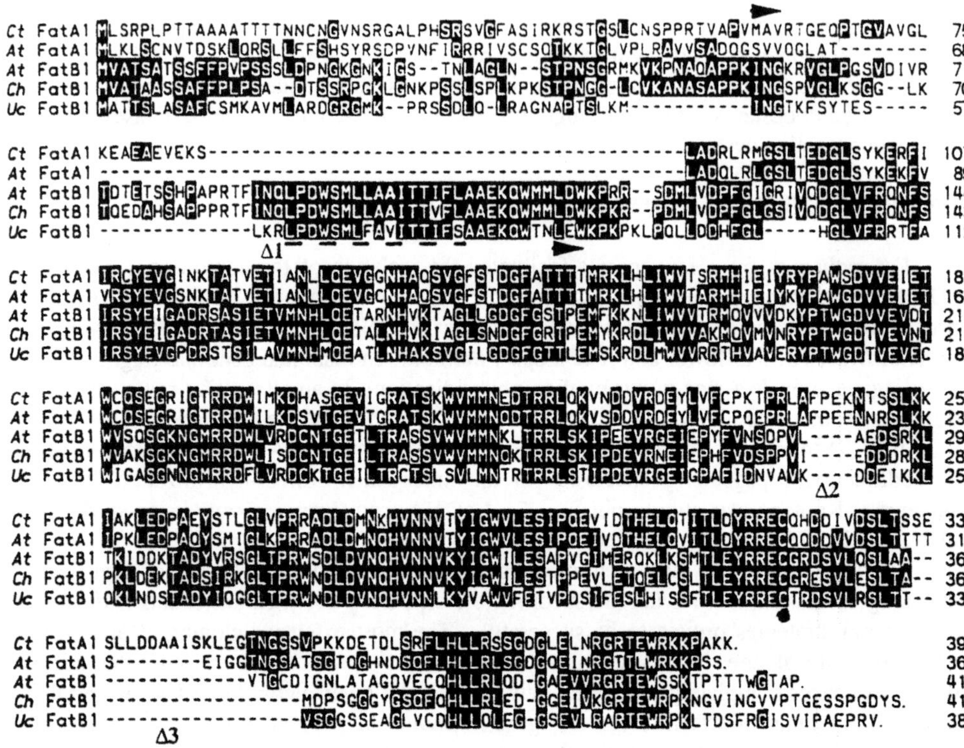

Figure 3. Alignment of Derived Polypeptide Sequences of FatA and FatB Thioesterases. Derived amino acid sequences of selected Fat representatives (see Table 1 for sources), were aligned and the consensus shaded. The arrows depict the respective N-termini of the purified *Ct* FatA1 and *Uc* FatB1 enzymes. Δ1, Δ2, Δ3, point to the three major sequence-length variations indicative for FatA *versus* FatB ancestry. The black dot is located at the only cysteine residue conserved in all the known Fat sequences.

(preferentially resulting in low-degeneration sequence after reverse translation) for the design of primers in a 3′ RACE protocol (44). One single 8 amino acid sequence from *Uc* FatB1 (residues 277 to 284) was reported to be successful. Subsequently, the amplified fragment was used as a probe for the isolation of four closely related *FatB* cDNAs (*Cl FatB1, Cl FatB2, Cl FatB3, Cl FatB4*) (Table 1). Töpfer et al. carried out the only exhaustive cloning effort for *FatB* genomic fragments reported to date. Altogether, 23 genomic clones of *C. lanceolata* were characterized. They fell into 4 classes corresponding to the 4 different cDNAs. The haploid genome of *C. lanceolata* consists of 6 chromosomes. Since 8 is considered the basal chromosome number (x) for the genus, it follows that each of the 4 *FatB* genes are members of a gene family, and not simply alleles of one gene in a polyploid genome, as could be argued for the *FatB* genes of *C. hookeriana* (x ~ 5), for example (45). All 4 genes appear to produce transcripts with 6 introns, one of which is in the 5′ untranslated region. The coding exons are spread over 2 to 2.3 kb of the genome. Not unexpectedly, the location and approximate size of the introns are conserved between the *FatB* genes of *Cuphea* and *Brassica* (27). This gives some guidance for estimating the *FatB* gene family size based on genomic Southern analysis. After sequence comparison with the *Fat* sequences available, it was concluded that plant thioesterases fall into two sequence-classes, which were termed not quite correctly "long chain" and "medium-chain" (46). More recently, *FatA* and *FatB* cDNAs from several more species were isolated, either by evolutionary PCR or by hybridization with heterologous probes (Table 1). To illustrate the main features distinguishing FatA versus FatB, Figure 3 shows the derived amino acid sequences of several representatives from different plant families.

SEQUENCE COMPARISONS, PHYLOGENY

Twelve *FatA* cDNA sequences, isolated from seven different plant families (Table 1), encode 363 to 390 residue pre-proteins. The length variation is almost exclusively restricted to the N-terminal region, which is normal, since these portions most likely represent stromal transit peptides (24). In contrast, the mature portions of the polypeptides are highly conserved in length and sequence; a region homologous to residues 89 to 327 in *Ct* FatA1 (Figure 3) is conserved at least 75% between all FatA representatives.

Twenty-two *FatB* cDNAs were isolated from six different plant families (Table 1). Pre-proteins range from 382 to 418 residues, and, excepting representatives isolated from a single species or genus, such as *U. californica* or *Cuphea*, the sequences of an approximately 60 to 90 amino acid N-terminal portion are conserved at a very low level, if at all. Significant similarity starts abruptly with a near-completely conserved region (*Uc* FatB1 residues 60 to 82) (Figure 3), and conservation averaging 70 to 80% continues nearly to the C-terminus. All FatB sequences share the Δ1, Δ2 and Δ3 length variations relative to FatA (Figure 3), which defines them as markers for class-specificity. Conserved deletion/insertion events have been considered to be reliable indicators of phylogenetic relationships; e.g., (47).

The advent of more FatB representatives provided an explanation for a previously puzzling feature of *Uc* FatB1. The mature N-terminus of purified *Uc* FatB1 was originally determined by amino acid sequencing as residue 85 (23), arrow in Figure 3. However, this would place the most N-terminal region conserved among all FatB representatives (residues 60 to 80 for *Uc* FatB1, Δ1 in Figure 3) into the transit peptide. This sequence would be unusual for stromal transit peptides, not only for its high degree of sequence conservation. It contains a hydrophobic region which contains all the properties of a eukaryotic transmembrane domain (Figure 4), and resembles a conserved feature of thylakoid transit peptides (24). This conserved region is now believed to belong to the N-terminal portion of the processed mature protein (14). Indeed, a 40 kD form of *At* FatB1 has been detected in *A. thaliana* tissues by immunoblotting. It co-migrated with an *E. coli*-expressed *At* FatB1 with an engineered N-terminus at residue 88, just upstream of the conserved hydrophobic domain [Figure 3; (42)].

Interestingly, FatA polypeptides do not have the hydrophobic domain (Δ1, Figures 3 and 4), making this feature a Fat-class indicator. Since it appears to be nonessential for enzyme activity *in vitro* and *in vivo* [Figures 2 and 5; (51)] its role remains unexplained. Perhaps FatB is loosely anchored to a membrane or complexed with the other enzymes of the fatty acid synthase *via* this domain. Only one cysteine (close to the C-terminus, indicated by a dot, Figure 3) is conserved between all Fat sequences to date, making it a likely candidate for the active-site residue, which supports the cysteine-inactivation studies (19,29).

For the generation of the phylogenetic tree of Figure 5A, we compared a total of 34 thioesterase polypeptide sequences from 15 species of 9 different angiosperm families (Table 1). In the analysis only the ~300 residue region of the "mature" portion of the derived polypeptides was used (residues 81 to 402 for *Ch* FatB1, Figure 3, for example). In this region, the sequences of the sample are practically colinear and scores of 40 to 98%

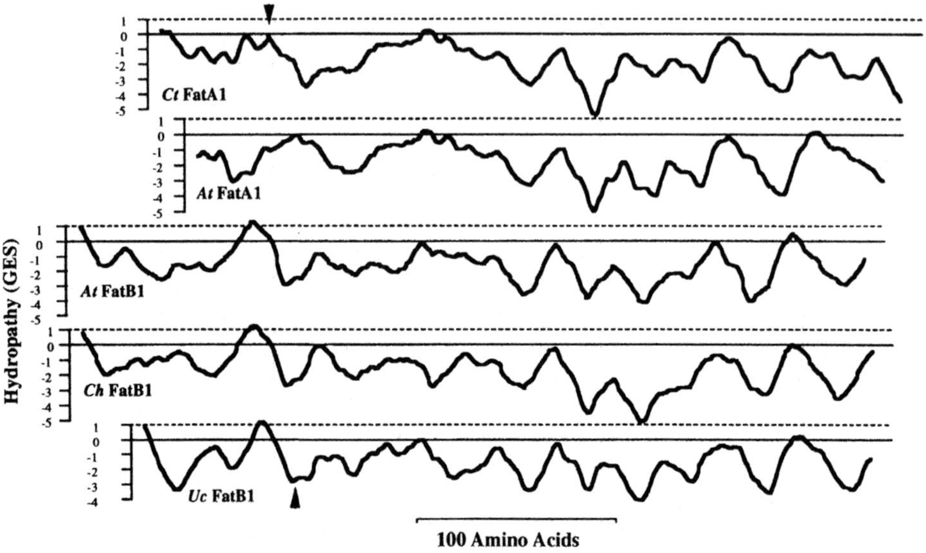

Figure 4. Hydropathy Plots of *Fat*-Derived Polypeptides. The hydropathy profiles of all sequences aligned in Figure 3 were determined by the default setting of TopPredII (90) with the use of the GES scale. The arrows depict the respective N-termini of the purified *Ct* FatA1 and *Uc* FatB1 enzymes. Hydrophobic regions extending beyond the dotted line are predicted to be membrane-spanning domains.

sequence identity were obtained after the introduction of a moderate number of gaps for better alignment (not shown). A parsimonious search produced the tree shown in Figure 5.

As previously reported (14,46) for much smaller samples, a deep branching at the origin of the tree separates FatA from FatB sequences (Figure 5). Evidently this represents the first evolutionary separation of the encoding genes. Note that representatives of both sequence classes were isolated from several species, namely from *A. thaliana* (*At*), *C. hookeriana* (*Ch*), leek (*Ap*) and mango (*Mi*). This represents the strongest support for the hypothesis that the basic division into *FatA* and *FatB* sequences reflects neither taxonomic relationships nor seed acyl compositions. Thus *FatA* and *FatB* genes are paralogous, i.e., are derived from an ancient gene duplication (48). The creation and subsequent conservation of two distinct sequence classes of thioesterase in all species investigated suggests that the encoded enzymes have different functions. Since the separation of *FatA* and *FatB* must pre-date the evolutionary separation of angiosperms into monocot and dicots, both *Fat* classes must have existed before the radiation of the angiosperms, which is dated 150 to 300 million years ago; reviewed in (49), see also (50). The sum of the sequence data clearly demonstrates that the *FatA/B* gene duplication is much older than the monocot-dicot divergence, which makes it likely that both thioesterase classes are present in the gymnosperms and possibly also in more distant taxa of plants (49). Furthermore, *FatA* and *FatB* genes should be present in all species of these taxa unless secondarily lost.

BIOCHEMICAL CHARACTERIZATION OF cDNA-EXPRESSED FAT ENZYMES

In many cases, the acyl-ACP hydrolytic profiles of *Fat* -encoded polypeptides were determined *in vitro* after expression in *E. coli*. For this purpose the reading frames coding for the "mature" portion of the respective pre-proteins were either fused to the N-terminal portion of LacZ (14,25,27), or supplied with an initiation codon (29,42). Both strategies

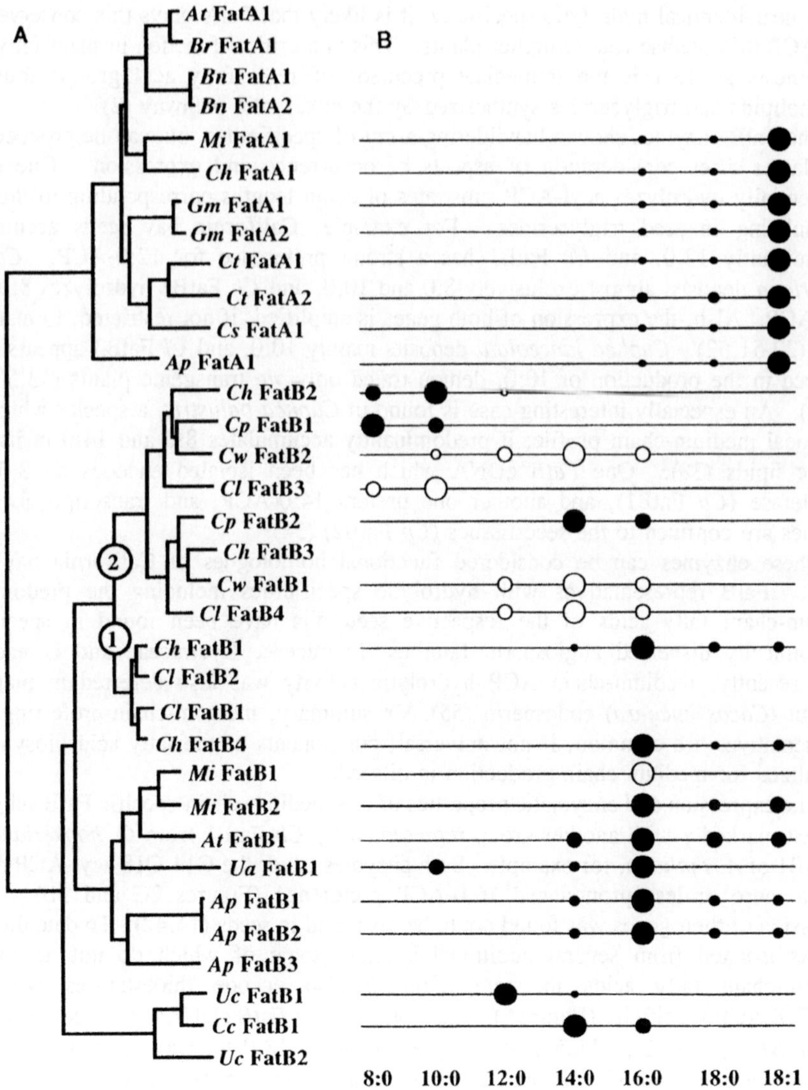

Figure 5. Phylogenetic Relationships and Substrate Specificities of Fat Thioesterases. **5A**, The derived amino acid sequences of all *Fat* genes and cDNAs available to the author (Table 1) were aligned as described in Figure 3. After selecting a region of true homology between all members of the sample (equivalent for residues 86 to 387 of *Ct* FatA1, see Figure 3), the alignment was processed by the heuristic version of PAUP (91), which calculates phylogenetic trees based on parsimony. The seven trees obtained showed variation in the connection of a few nodes with short branch lengths; the most representative tree is shown; see also (35). **5B**, Compilation of Fat specificities from the sources listed in Table 1. The dots on the horizontal line to the right of each Fat member symbolize the respective enzyme specificity in a semi-quantitative way by circles with 4 different diameters. Largest circle, highest preference; second largest circle, 50 to 85% of highest activity; second smallest circle, 10 to 45% of highest preference; smallest circles, detectable, but below 10% of highest activity. Filled circles: measured *in vitro* after expression in *E. coli*; empty circles, deduced from oil composition data of transformed plants. Acyl-ACP substrates are indicated at the bottom.

were successful. All FatA enzymes tested appeared to be 18:1-specific, with minor activities on 18:0 and 16:0 substrates (Figure 5B). This suggests that FatA is involved predominantly in the hydrolysis of 18:1-ACP. Since all isolates from several plant families gave a near-identical hydrolytic specificity, it is likely that FatA plays this conserved 18:1-ACP thioesterase role in higher plants. This is a critical function in plant fatty acid biosynthesis as 18:1 is the immediate precursor of most fatty acyl groups found in phospholipids and triglycerides synthesized by the eukaryotic pathway (2).

The FatB enzymes show a bewildering array of specificities, but can be grouped into two classes after consideration of aspects of occurrence and expression. One group preferentially hydrolyzes acyl-ACP substrates of chain lengths corresponding to the ones accumulating in seed triglycerides. For example, California bay seeds accumulate predominantly 12:0, and *Uc* FatB1 has a strong preference for 12:0-ACP. *Cuphea hookeriana* deposits almost exclusively 8:0 and 10:0, and *Ch* FatB2 hydrolyzes 8:0- and 10:0-ACP. Also, the expression of both genes is amplified, if not restricted, to maturing seeds (20,51,52). *Cuphea lanceolata* deposits mainly 10:0, and *Cl* FatB3 appears to be involved in the production of 10:0, demonstrated only *via* transgenic plants (43,53; see below). An especially interesting case is found in *Cuphea palustris*, a species which has a bimodal medium-chain profile; it predominantly accumulates 8:0 and 14:0 in its seed storage lipids (38). One *FatB* cDNA which has been isolated encodes an 8:0/10:0 thioesterase (*Cp* FatB1), and another one prefers 14:0-ACP, and transcripts for both enzymes are confined to the seed tissues (*Cp* FatB2) (54).

These enzymes can be considered functional homologues of California bay's *Uc* FatB1. FatB representatives with hydrolytic specificities including the predominant medium-chain fatty acids of the respective seed oils have been found in species of taxonomically dispersed angiosperm families (Lauraceae, Lythraceae and Ulmaceae). More recently, medium-chain ACP hydrolytic activity was also detected in maturing coconut (*Cocos nucifera*) endosperm (55). In summary, medium-chain-preferring FatB representatives are common, if not universal, components of the fatty acid biosynthesis specialized for medium-chain production in oilseeds.

The expression and enzymatic properties of this medium-chain-specific FatB subgroup contrast markedly with another group represented by *Ch* FatB1 from *C. hookeriana* and *At* FatB1 of *A. thaliana*, for example. Both enzymes cover the C14-C18 acyl-ACP range, with a more or less pronounced 16:0-ACP preference (Figures 2C and 5B). Also, expression of their genes was found not to be restricted to seeds (14,42). To date the *FatB* cDNAs isolated from several additional species (some of which do not accumulate medium-chain fatty acids in storage lipids) also encode thioesterases with this C14-C18-ACP specificity (Figure 5). Examples are the *FatB* cDNAs from leek epidermis (*Ap FatB1, Ap FatB2*), which produces predominantly 18:0 for wax formation. Similar examples are *Mi FatB1, Mi FatB2* from maturing mango seeds, which accumulate predominately 18:0 for deposit in storage triacylglycerols. The only *FatB* member isolated from maturing seeds of camphor [10:0 and 12:0 accumulating; Lauraceae (10)], with a preference shifted to 14:0 instead of 16:0, is slightly different. It could be an evolutionary intermediate between the typical, 16:0-ACP-preferring, FatB and a specialized, laurate-specific FatB like *Uc* FatB1. In an alternative scenario, it is conceivable that the typical FatB can undergo some "specificity drift" during evolution of the species in order to fulfill slightly different requirements. This might have happened during the evolution of camphor. The FatB representative from elm (*Ulmus americana*) seeds (80 mol% 10:0), *Ua* FatB1, is rather unusual, and shows features of both groups. This thioesterase clearly has activity on 10:0-ACP, correlating with the seed acyl composition but about equal or higher activity is seen with 16:0-ACP. This makes this enzyme a hybrid between both groups. It is conceivable that such bimodal enzymes (see also the broad specificities of *Cw* FatB1 and *Cw* FatB2) can support medium-chain production as well as function in the role of the typical FatB, depending on the status of the resident fatty acid synthase of the respective tissues.

The typical 16:0-ACP-preferring FatB is probably present in all major plant parts, if not cells, and is not restricted to medium-chain-producing species. Based on expression pattern, enzyme specificity, and presence in all angiosperm species investigated, this type of FatB was considered to be the evolutionary ancestor of the specialized medium-chain thioesterases, which obviously evolved several times during the radiation of the angiosperms, as the phylogenetic tree shows (Figure 5). Therefore, it was concluded that all higher plant tissues not only contain an oleoyl-ACP thioesterase, but also a somewhat less specific, 16:0-centered thioesterase which was labeled palmitoyl thioesterase (14). Prior to its discovery *via* the cloning of its cDNA, very little, and mostly indirect, biochemical evidence had been accumulated for the existence of a 16:0-ACP hydrolyzing activity that was separate from the oleoyl thioesterase (28,56). Also, extracts of palm mesocarp, a tissue naturally producing high-palmitate oil, showed a marked elevation of 16:0-ACP hydrolysis compared to other plant tissues (57). This could be caused by the upregulation of a *Ch* FatB1-like enzyme, as was demonstrated with transgenic *Brassica napus* and soybean (14,30).

ENGINEERING BACTERIAL FATTY ACID BIOSYNTHESIS WITH FATA/B

As discussed in the introduction, the plastid-localized *de novo* fatty acid synthase of plants is functionally very similar to the system found in *E. coli*. When mature versions of FatB were synthesized in *E. coli* (14,25,42,51), the bacterial fatty acid biosynthesis could be redirected from supplying acyl chains for glycerolipids to producing predominantly free fatty acids. The chain lengths of the accumulated fatty acids always correlated with the enzymes' specificities. The amounts of free fatty acids accumulated were especially dramatic in *E. coli* strains deficient in the utilization of fatty acids (58). The interception rate of the bacterial long-chain biosynthesis was found to be dependent on the culture age and the respective FatB enzyme. The extent of interception increased to near-hundred percent when the cultures were reaching stationary phase (25), thereby depleting the *in vivo* acyl-ACP pools >C10 (59). This demonstrated compatibility between elements of the fatty acid synthesis pathway of the two kingdoms, and allows the use of *E. coli* as a very useful heterologous system for the study of FatB action *in vivo*.

During the extension of 8:0-ACP to C10-ACP, the acyl flux through *E. coli*'s fatty acid synthase splits into saturated and unsaturated pathways (58). Therefore saturated and unsaturated C10-C16 acyl-ACPs are available for hydrolysis by the introduced FatB enzyme. Indeed, unsaturated fatty acids accumulated in cultures harboring different FatB enzymes (Figure 6). This is direct evidence that medium-chain FatB enzymes can hydrolyze unsaturated medium-chain acyl-ACP substrates *in vivo* (such substrates being currently unavailable for *in vitro* activity measurements). It is interesting to note that the chain-length preference for unsaturated substrates is always shifted to longer chains relative to the preference for saturated substrates, as if unsaturation makes FatB substrates appear "shorter" to the enzyme. It supplies circumstantial *in vivo* evidence that the evolutionary ancient Fat enzyme, which was proposed to have had a specificity comparable to *At* FatB1 (Figure 2C) was probably capable of facilitating plastidial export of saturated and unsaturated long chains (14).

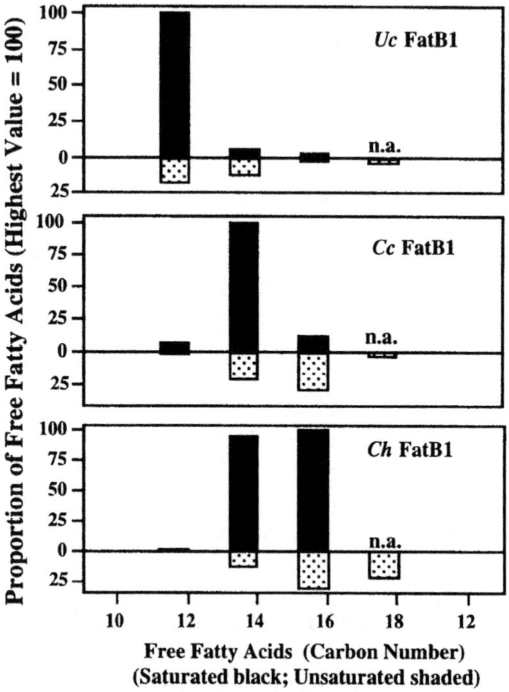

Figure 6. Engineering of *E. coli* Fatty Acid Biosynthesis with FatB. The portions of *FatB* cDNAs coding for the mature enzymes were expressed in an *E. coli* strain deficient in the degradation of free fatty acids. In this strain the impact of Fat on the bacterial fatty acid biosynthesis can be measured directly, since its acyl-products cannot be re-metabolized and accumulate in the medium (25). Since *E. coli* extends saturated and monounsaturated acyl-ACPs [from C10 on; reviewed in (58)], the hydrolytic potential of a given Fat enzyme for unsaturated medium chains can be evidenced indirectly. In order to allow more direct comparisons between experiments, the measurement of the respective fatty acid with the highest accumulation in the respective culture (nmol of free fatty acid/ml culture) was assigned the value 100. **n.a.**: the FatB potential on hydrolysis of 18:0-ACP cannot be assessed, because *E. coli* does not produce this substrate. Saturated fatty acids (black bars) above the x-axes, unsaturated fatty acids (shaded bars) are below the x-axes. Original data for *Uc* FatB1 from (25); *Cc* FatB1, Voelker, unpublished data; *Ch* FatB1 (14).

ENGINEERING PLANT FATTY ACID BIOSYNTHESIS WITH *FATA/B*

As first demonstrated by Voelker et al. in 1992 (23), Fat enzymes can redirect plant fatty acid biosynthesis to medium chains. Usually, cDNA coding areas for the respective pre-proteins were placed behind seed storage protein promoters and the chimeric genes were introduced into the genome of the target species. There are also two reported cases in which *FatB* genes of *C. lanceolata* (4 kb genomic fragments) were used for transformation, and the regenerated plants displayed modified seed oil compositions, although to a moderate extent (43,53). To date, at least half a dozen different *Fat* engineering projects have been successful, i.e., the introduced genes induced significant changes in the acyl compositions of seed oils. The engineered long-chain producers were transformed to produce medium chains ranging from C8 to C14. Palmitate (16:0) levels could be increased drastically by expressing FatB enzymes (14). Suppression of FatB in seeds of transgenic soybean *via* genetic engineering with a soybean *FatB* cDNA reduced the palmitate proportion from 16 mol% to less than 5% (30). This represents the only reported study in which the role of a FatB thioesterase was directly probed in its native species, and the results support the theory that FatB activity is responsible for at least a fraction of the palmitate production in seeds of higher plants. Complementing these results, the overexpression of soybean *FatB* with seed storage gene promoters in transgenic soybeans doubled the 16:0 portion, while the analogous overexpression of soybean *FatA* elevated the 18:1 level by 50% and 16:0 portions were somewhat reduced (30). In general, the chain-length profile of Fat-induced changes in acyl-group compositions reflected the enzyme profile determined *in vitro*. In addition, these results demonstrated that single-enzyme engineering for altered vegetable oils is feasible [reviewed most recently by (53)].

Did such chain-length engineering lead to oils with compositions comparable to the ones found in the species from which the thioesterases originated, i.e., can the engineering with medium-chain FatB enzymes replicate the production of oils found in natural medium-chain producers? Figure 7 shows that this is only partially the case, since the engineered oils differ somewhat in quantitative and qualitative aspects. Medium chains usually make up 80 to 90% or more of the respective seed oil composition (10), while engineering seed oils to more than ~ 60 mol% medium chains has not been achieved. The triglyceride positional composition of engineered high-laurate oil has been reported, and even in oils in which positions 1 and 3 of its triglycerides were almost completely occupied with medium chains, the position 2 contained almost exclusively unsaturated C18 chains (60). It has been previously shown *in vitro* that the triacylglycerol biosynthesis machinery of conventional oilseeds selects against saturated acyl groups for this position, e.g., (61), with lysophosphatidic acid acyl transferase (LPAAT, Figure 1) probably being the responsible agent (62). This *in vivo* and *in vitro* evidence demonstrates that the stringencies of triglyceride biosynthesis of conventional oil seeds could be possible factors contributing to the observed plateau at ~ 66 mol% that saturates after the introduction of a FatB enzyme. In addition, it is possible that the final saturated proportion of the oil is determined by more than just the mixture of fatty acids exported from the plastids. Eccleston et al. (63) have shown that expression of *Uc FatB1* with a 35 S promoter in transgenic plants can lead to moderate laurate accumulation in seeds (7 mol%), but no laurate was detected in leaves. These results did not reflect gene expression, since much higher 12:0-ACP hydrolytic activities were detected in leaf hydrolysates than in maturing seeds, and it was demonstrated that a large fraction of the *de novo* fatty acid biosynthesis in leaf chloroplasts was diverted to laurate production. This raises the question of whether a recycling of laurate may be also possible in the maturing seed, effectively compensating for selective acyl-CoA utilization by the acyl transferases of the Kennedy pathway. This possibility is currently under investigation (64).

FatB enzymes did produce medium chains in transgenic plants, but their action resulted in an altered chain-length profile when compared with that found in species in which they
naturally reside (Figure 7). The engineered profiles were found to be shifted to longer or shorter chain lengths. For example, no 8:0 accumulates in seeds of *C. lanceolata*, but the 8:0 portion of medium chains in oil from *Cl FatB3*-engineered plants is high. Inversely, California bay seeds accumulate 10:0, but none is produced in *Uc FatB1*-engineered plants.

One could argue that this simply reflects the cloning and expression of the "wrong" *FatB* genes, but Ohlrogge (64) believes that this is not always the case. It is possible that some of the discrepancies between natural and engineered oils reflect the respective FatB's competition with condensing enzymes of the different resident fatty acid synthases for acyl-ACP substrates (Figure 1). During plant fatty acid biosynthesis, acyl-ACPs are extended by ß-ketoacyl-ACP synthases, which in "normal" tissues are very active on substrates shorter than C16 or C18, respectively [reviewed in (65)]. This might be the reason for the observed similar and low levels of medium-chain ACP pools (65), potentially requiring high concentrations of thioesterases in order to interfere with extensions. The medium-chain-length distribution in engineered oils (Figure 7) usually reflects the respective enzyme's specificities as measured *in vitro* (compare Figures 2, 5B and 7), which suggests there are comparable pool sizes for the C8-14 acyl ACPs in conventional fatty acid synthases *in vivo*. This accords with the limited data resulting from direct acyl-ACP measurements in maturing oilseeds (65). In natural medium-chain producers the fatty acid synthase might be modified in respect of its chain extension preferences. Applying this hypothesis, the *C. hookeriana* fatty acid synthase would have

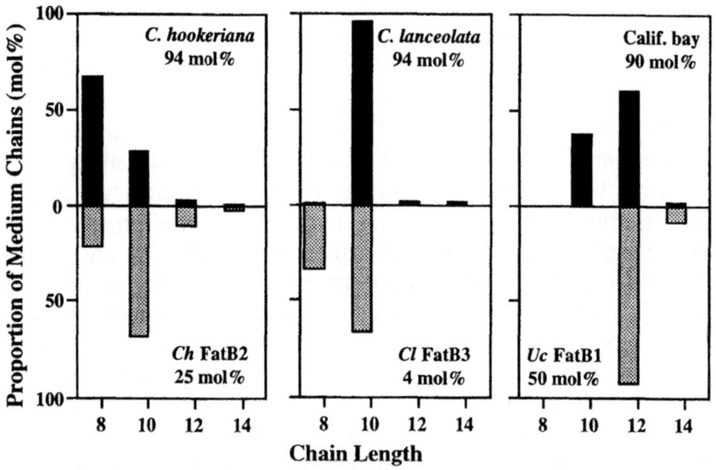

Figure 7. Comparison of Seed Acyl Compositions of Natural and Engineered Medium-Chain Producing Plants. The medium-chain fatty acyl composition from seeds of three naturally medium-chain producing plant species are shown, compared with the proportion of the induced fatty acyl groups after expression of *FatB* cDNAs from these species in maturing seeds of canola. Natural compositions: California bay (19); *Cuphea* species (38); engineered canola: California bay (60), *C. hookeriana* (52); *C. lanceolata* (43,53). Mol% numbers in each plot indicate the proportion of medium-chains of the total acyl composition. Untransformed canola, a long-chain producer, which does not accumulate any significant amounts of medium chains, is not shown.

a low elongation capacity for acyl-ACP ≥8:0, and an increased pool of 8:0-ACP would be available to *Ch* FatB2 for hydrolysis. Most of the substrate flux would be intercepted, leaving a very diminished pool for *Ch* FatB2's action on 10:0-ACP. Inversely, if the *C. lanceolata* ß-ketoacyl-ACP synthases extend efficiently up to 10:0-ACP, but elongate this substrate only poorly, *Cl* FatB3 would not compete well at the C8 stage, but would easily hydrolyze the large 10:0-ACP pool. This would result in an oil rich in 10:0. There is indirect evidence for specialized fatty acid synthases in maturing seeds of medium-chain producers (66,67). Furthermore, there is one reported claim that partially purified condensing enzymes of *C. lanceolata* seed extracts hardly elongate 10:0-ACP (68), but to date enzymes with such specificities have eluded investigators. In maturing seeds of *C. lanceolata* pools of acyl-ACPs ≥ C12 are very low, which demonstrates a high rate of interception (69). Employing computer simulations, Davies (55) developed a pathway kinetic model in which a thioesterase competed with condensation reactions during two successive extension cycles. The model demonstrates that, depending on the competing enzymes' kinetic parameters, even very large differences are possible between thioesterase specificities (as measured under equal substrate concentrations *in vitro*) and fatty acids produced *in vivo*. Strong evidence for the validity of this model for type I fatty acid synthases was found during *FatB*-expression studies in *E. coli* (25,51).

SUMMARY, PERSPECTIVES FOR FUTURE RESEARCH

The recent dramatic advances in the field, made possible by the application of an integrated approach fusing biochemistry and genetic engineering, are uncovering the multi-faceted biology of plant acyl-ACP thioesterases. *De novo* fatty acid biosynthesis of probably all higher plants employs two related but functionally different acyl-ACP thioesterases, FatA and FatB, encoded by nuclear genes. FatA representatives are usually synthesized from 1 to 2 very related transcripts as pre-proteins, which have to be imported into the stroma of plastids in order to function. They accumulate as soluble 40 kD polypeptides in the stroma and are responsible for shuttling predominantly 18:1 to the eukaryotic pathway. FatB usually is encoded by 1 to 4 genes, and the FatB thioesterases of tissues harboring conventional long-chain acyl biosynthesis have a relatively broad specificity, centered around 16:0-ACP, and there is evidence that FatB is normally involved in the production of 16:0, though additional functions cannot be excluded. *At* FatB1 displays what can probably be regarded as the "typical" profile (Figures 2C and 5) for FatB. Several lines of evidence, based on detection of transcripts and immunoblots, indicate that in plant tissues FatB gene products are generally 5 to 10 times more frequent than FatA (34,70). This is in accordance with the emergence in the *A. thaliana* random sequencing project of expressed sequence tags (71) coding for Fat (Genbank to date: 10 ESTs are *At FatB1*, 2 are *At FatA1*). To date, the rice EST project (72) caught only 3 Fat cDNAs, and all are FatB. It is unresolved as to why FatB activities usually escape detection in crude tissue extracts, even though it was shown that kinetic parameters of FatA and FatB enzymes are comparable (34).

There is compelling evidence that Fat hydrolyzes thioesters by an active-cysteine mechanism (20,29), contrasting with the serine-active thioesterases of other phyla (73). To date Fat cannot be integrated into a larger enzyme family, since recent database searches uncovered no significant matches with non-Fat entries (51). In seeds specialized in the production of saturated fatty acids in the C8-C14 range, FatB enzymes usually are present with hydrolytic profiles corresponding to seed acyl composition. Such specialized FatB enzymes evolved independently several times during the radiation of the angiosperms [discussed in (14) in more detail; see also Figure 5]. Expression of *FatB* cDNAs or genes in seeds of transgenic plants normally induces changes in the acyl group composition towards more saturates. This represents functional proof that such enzymes are important in the determination of chain length and overall levels of saturates, but the transgenic data point to at least two other key players: ß-ketoacyl-ACP synthases as suppliers and competitors for substrates, thereby influencing chain length, and specialized lysophosphatidic acid acyltransferases facilitating the access of saturated fatty acids to position 2 of the triglycerides (60,74).

What aspects of plant acyl-ACP thioesterase biology warrant further study? Several topics of interest are listed.

Structure Function. Obviously, Fat enzymes can discriminate between substrates with respect to chain length, unsaturation, and location of the double bond. This implies that these enzymes must have a quite precise "ruler" for measuring ACP-bound acyl chains, but the ruler's specificities can be vastly modified during evolution (Figure 5B). The first reported steps towards understanding Fat structure-function relationships were facilitated by the very closely-related Lauraceae representatives *Uc* FatB1 and *Cc* FatB1, which prefer 12:0-ACP and 14:0-ACP respectively (Figure 5). Swapping segments of coding areas between both cDNAs showed that the length-determination capacity resides in the C-terminal half of the mature enzymes (75). Most dramatically, replacing only three selected amino acid residues of *Uc* FatB1 with the homologous ones from *Cc* FatB1 transformed *Uc* FatB1 into a 14:0-ACP-preferring thioesterase (75). While such studies will allow the identification of important residues, for a complete description of the active site the tertiary structures of several Fat enzymes with different chain-length preferences need to be solved. As demonstrated by Yuan et al. (75), changes in primary structure which modify chain-length preferences can be rather subtle; also many different solutions might be possible [e.g., shown for lipoxygenases; (76)]. Generally, such changes will hardly be detectable in the evolutionary "noise" accumulated during the separate history of the genes.

Missing Specificities. There are several plant species which synthesize in specialized tissues substantial amounts of stearate, either for carbon storage in triglycerides, or wax production. For example 18:0 represents 35% of seed acyl groups of the cacao plant, but no biochemical evidence for a 18:0-ACP thioesterase was found (77). An 18:0-ACP-preferring extract from leek epidermis appeared to include a tentatively assigned FatB thioesterase [by immunological cross reactivity (78)]. To date only *FatA* and *FatB* cDNAs encoding enzymes with typical specificities were reported from 18:0-producing tissues of mango and leek (79), leaving the nature of the preferential 18:0-ACP hydrolysis unidentified. It is conceivable that in such tissues the thioesterases are not specialized, but the Δ9 desaturase (Figure 1) is suppressed. Indeed, cocoa butter-like oils could be engineered by antisense repression of canola's Δ9 desaturase (80). Most likely the minor 18:0-ACP activities of conventional FatA and FatB enzymes (Figure 5) were able to hydrolyze the drastically increased quantities of this intermediate. Biochemical evidence exists for a petroselinic-ACP thioesterase, but exhaustive cloning and expression efforts based on *Fat*-specific probes and antisera could not identify a matching Fat representative (33,34,64). Broad-range (C8-C18) acyl-ACP thioesterases were reported from *C. lanceolata* and *C. wrightii* (81), but no cDNA encoding such an enzyme has been found in these species (43,53,82). Maybe the enzyme preparations contained a mixture of several *Cl* FatB members, which could probably cover this range (see Figure 5). It is possible that such enzyme profiles are produced by still unknown Fat enzymes, but the existence of evolutionarily unrelated enzymes with convergent, similar functions cannot be excluded.

Enigmatic FatB N-Terminal Domain. All reports on FatA show that this enzyme is synthesized as a pre-protein, and a conventional stromal transit signal is removed after import into the plastids. The situation with FatB is much less clear. Unfortunately, the N-terminus of only one purified FatB was determined, and later it was found to be a proteolytic artifact. Using sequence conservation between FatB representatives from different families, Jones et al. (14) considered the putative mature N-terminus to be located just upstream of the first conserved region (Figure 3), and assigned the predicted membrane-spanning domain (Figure 4) to the mature portion. It is therefore conceivable that the native FatB resides in or at a membrane, which would partition it from the soluble FatA (Figure 4). Loose association of plant fatty acid synthase with thylakoids has been observed, e.g., (83). More recently, the designation of the FatB mature N-terminus according to (14) was challenged by *in vivo* data resulting from *Uc FatB1* expression in transgenic canola. In maturing seeds of these plants the prominent form of *Uc* FatB1 accumulates with an M_r predicted for the pre-protein (60), opening the possibility that FatB does not have a conventional stromal transit peptide. There is precedent for an imported stromal protein lacking a cleaveable transit peptide (84).

Evolution of Medium-Chain FatB Enzymes. As discussed earlier, medium-chain-specific enzymes evolved several times from the typical FatB. It would be important to trace whether for example the development of 10:0-ACP thioesterases involved several intermediates, or whether such enzymes are direct descendants from the typical version. A prime subject for such studies is represented by the genus *Cuphea*, since it harbors species with a wide diversity of chain lengths (38). As Figure 5 shows, 4 clades of *Cuphea* FatBs have already diverged before speciation, e.g., all 4 different FatB members of *C. hookeriana* have orthologous (48) members in *C. lanceolata* (and possibly also in the other two species included in the tree, but not all members may yet be cloned). Most orthologous enzymes appear to have very similar specificity (Figures 5A and 5B, not all specificities known), which makes it very likely that the development of the respective specificities preceded speciation. It is interesting to note that both clades containing enzymes with derived specificities (i.e., C8/10 and C14, Figure 5) diverged *via* a single branch (labelled "2") from the clade containing the typical FatB enzymes *Ch* FatB1 and *Ch* FatB4 (branch labelled "1"). This indicates that the two clades with derived enzymes are descendants of a single duplicated gene copy, which subsequently was duplicated again, allowing further diversity (by sequence and function). This in turn raises the question whether orthologs of several or all of the 4 clades exist also in *Cuphea* species which do not accumulate medium chains (38,45). If they exist, do they display the respective specialization, but do not lead to an oils phenotype because of low expression and/or a non-specialized fatty acid synthase? Unfortunately, all current FatB sequences come from rather closely-related species of the large genus (~ 260 species) which is divided into more than 10 quite diverged sections (45). Therefore, the current supply of *Cuphea* FatB sequences gives only few clues about events early in the explosive Fat-evolution of this genus. It is interesting to note in this respect that there is one report of 10:0 accumulation of a Lythraceaen species outside of *Cuphea* (45), arguing for the development of 10:0-ACP preferring FatB to have preceded the radiation of *Cuphea*. Sequences and specificities of FatB representatives from many more *Cuphea* or Lythraceaen species have to be described before a more detailed phylogenetic reconstruction can be attempted.

Origins of Fat and the FatA/B Split. As discussed extensively (14), the fact that the prokaryotic fatty acid biosynthesis of cyanobacteria (the ancestors of plastids) does not employ acyl-ACP thioesterases or a Δ9 stearoyl-ACP desaturase [Figure 1; (85,86)] suggests that both enzymes probably have evolved from unknown ancestors following the establishment of the prokaryotic endosymbiont in the eubacterial cell [endosymbiontic evidences reviewed in (4)]. It was further concluded (14) that the ancient plant acyl-ACP thioesterase, Fat, probably had a broad specificity which remains somewhat conserved in the unspecialized FatB enzymes such as *At* FatB1. Much later, Fat split into two lineages, FatA and FatB. FatB preserved the ancient function, and FatA developed into the oleoyl-ACP thioesterase, making it the more recently-emerged form. It would be a challenging and exciting project to investigate the Fat status of extant species from algae to higher plants. This would probably allow the reconstruction of the phylogenetic events crucial for the development of a relatively recent metabolic pathway unique to plants: the plastidial fatty acid export system which bridges prokaryotic and eukaryotic lipid biosynthesis.

ACKNOWLEDGMENTS

This review is dedicated to H. Maelor Davies, who, with skill, patience and an excellent team forced the bay thioesterase through many columns, and over a period of two years tamed the wild bronco from the canyons of California. Once broken in, we took its *Fat* and "just followed the yellow Brassica road", in order to make the world's first genetically engineered commercial oil crop. The fallout from this amazing workhorse continues to fertilize plant biotechnology and the science of plant fatty acid metabolism.

I would like to thank all colleagues who contributed unpublished information: K. Dehesh, J. Fillatti, D.J. Hawkins, D.S. Knutzon, J.C. Kridl, L. Yuan, all Calgene, Inc.; J.B. Ohlrogge, Michigan State University; M.B. Slabaugh, J.M. Leonard and S.J. Knapp, Oregon State University, and T. Kinney, duPont de Nemours & Co. Several colleagues at Calgene, Inc. read the manuscript critically, DeeDee Walker and Jennifer Yeager edited the manuscript.

REFERENCES

1 Ohlrogge, J. and Browse, J. (1995) Plant Cell 7, 957-970.
2 Gibson, S., Falcone, D.L., Browse, J. and Somerville, C. (1994) Plant, Cell and Environment 17, 627-637.
3 Ohlrogge, J.B. (1982) TIBS 7, 386-387.
4 Gray, M.W. (1989) Trends Genet. 5, 294-295.
5 Roughan, P.G. and Slack, C.R. (1982) TIBS 9, 383-386.
6 Löhden, I. and Frentzen, M. (1988) Planta 176, 506-512.
7 Shine, W.E., Mancha, M. and Stumpf, P.K. (1976) Arch. Biochem. Biophys. 172, 110-116.
8 Ohlrogge, J.B., Shine, W.E. and Stumpf, P.K. (1978) Arch. Biochem. Biophys. 189, 382-391.
9 Stumpf, P.K. (1981) TIBS 6, 173-176.
10 Hilditch, T.P. and Williams, P.N. (1964) The Chemical Constitution of Natural Fats. Chapman and Hall, London.
11 McKeon, T.A. and Stumpf, P.K. (1982) J. Biol. Chem. 257, 12141-12147.
12 Oo, K.-C. and Stumpf, P.K. (1979) Lipids 14, 132-143.
13 Harwood, J.L. (1988) Annu. Rev. Plant Physiol. Plant Mol. Biol. 39, 101-138.
14 Jones, A., Davies, H.M. and Voelker, T.A. (1995) Plant Cell 7, 359-371.
15 Knauf, V.C. (1987) Trends Biotechnol. 5, 40-46.
16 Battey, J.F., Schmid, K.M. and Ohlrogge, J.B. (1989) Trends Biotechnol. 7, 122-126.
17 Slabas, A., Roberts, P. and Osmesher, J. (1982) in Biochemistry and Metabolism of Plant Lipids (Wintermans, J.F.G.M. and Kuiper, P.J.C., eds.), pp. 251-256, Elsevier, Amsterdam.
18 Pollard, M.R. and Singh, S.S. (1987) in The Metabolism, Structure and Function of Plant Lipids (Stumpf, P.K., Mudd, J.B. and Ness, W.D., eds.), pp. 455-463, Plenum Press, New York, NY.
19 Pollard, M.R., Anderson, L., Fan, C., Hawkins, D.J. and Davies, H.M. (1991) Arch. Biochem. Biophys. 284, 306-312.
20 Davies, H.M., Anderson, L., Fan, C. and Hawkins, D.J. (1991) Arch. Biochem. Biophys. 290, 37-45.
21 Gould, S.J., Subramani, S. and Scheffler, I.E. (1989) Proc. Nat. Acad. Sci. U.S.A. 86, 1934-1938.
22 Davies, H.M., Pollard, M., Voelker, T.A. and Thompson, G.A. (1991) International Patent Application No. WO-91-16421. Published October 31, 1991.
23 Voelker, T.A., Worrell, A.C., Anderson, L., Bleibaum, J., Fan, C., Hawkins, D.,

Radke, S. and Davies, H.M. (1992) Science 257, 72-74.
24 deBoer, A.D. and Weisbeek, P.J. (1991) Biochim. Biophys. Acta 1071, 221-253.
25 Voelker, T.A. and Davies, H.M. (1994) J. Bacteriol. 176, 7320-7327.
26 Kridl, J.C., McCarter, D.W., Rose, R.E., Scherer, D.E., Knutzon, D.S., Radke, S.E. and Knauf, V.C. (1991) Seed Sci. Res. 1, 209-219.
27 Hitz, W.D. and Yadav, N.S. (1992) International Patent Application No. WO 92/11373. Published July 9, 1992.
28 Yadav, N., Wierzbicky, A., Knowlton, S., Pierce, J., Ripp, K., Hitz, W., Aegerter, M. and Browse, J. (1993) in Biochemistry and Molecular Biology of Membrane and Storage Lipids of Plants (Murata, N. and Somerville, C.R., eds.), Vol. 9, pp. 60-66, The American Society of Plant Physiologists, Rockville, MD.
29 Knutzon, D.S., Bleibaum, J.L., Nelsen, J., Kridl, J.C. and Thompson, G.A. (1992) Plant Physiol. 100, 1751-1758.
30 Kinney, T. (personal communication).
31 Helleyer, A., Leadlay, P.F. and Slabas, A.R. (1992) Plant Mol. Biol. 20, 763-780.
32 Loader, N.M., Woolncr, E.M., Hellyer, A., Slabas, A.R. and Safford, R. (1993) Plant Mol. Biol. 23, 769-778.
33 Dörmann, P., Frentzen, M. and Ohlrogge, J.B. (1994) Plant Physiol. 104, 839-844.
34 Dörmann, P., Kridl, J.C. and Ohlrogge, J.B. (1994) Biochim. Biophys. Acta 1212, 134-136.
35 Chase, M.W., Soltis, D.E., Olmstead, R.G., Morgan, D., Les, D.H., Mishler, B.D., Duvall, M.R., Price, R.A., Hills, H.G., Qiu, Y.-L., Kron, K.A., Rettig, J.H., Conti, E., Palmer, J.D., Manhart, J.R., Sytsma, K.J., Michaels, H.J., Dress, W.J., Karol, K.G., Clark, W.D., Hedrén, M., Gaut, B.S., Jansen, R.K., Kim, K.-J., Wimpee, C.F., Smith, J.F., Furnier, G.R., Strauss, S.H., Xiang, Q.-Y., Plunkett, G.M., Soltis, P.S., Swensen, S.M., Williams, S.E., Gadek, P.A., Quin, C.J., Eguiarte, L.E., Golenberg, E., Jr, G.H.L., Graham, S.W., Barrett, S.C.H., Dayanandan, S. and Albert, V.A. (1993) Annu. Missouri Bot. Gard. 80, 528-580.
36 Amy, C.M., Witkowski, A., Naggert, J., Williams, B., Randhawa, Z. and Smith, S. (1989) Proc. Nat. Acad. Sci. U.S.A. 86, 3114-3118.
37 Chirala, S.S., Kuziora, M.A., Spector, D.M. and Wakil, S.J. (1987) J. Biol. Chem. 262, 4231-4240.
38 Graham, S.A., Hirsinger, F. and Röbbelen, G. (1981) Amer. J. Bot. 68, 908-917.
39 Voelker, T.A., Yuan, L., Kridl, J. and Hawkins, D. (1995) International Patent Application No. WO-95-13390. Published May 18, 1995.
40 Fillatti, J.A., Knutzon, D. and Dehesh, K. (1995) Poster at the NPLC workshop "Biochemistry and molecular biology of plant fatty acids and glycerolipids", June 1-4, 1995, South Lake Tahoe, CA.
41 Grellet, F., Cooke, R., Raynal, M., Laudié, M. and Delseney, M. (1993) Plant Physiol. Biochem. 31, 599-602.
42 Dörmann, P., Voelker, T.A. and Ohlrogge, J.B. (1995) Arch. Biochem. Biophys. 316, 612-618.
43 Töpfer, R., Martini, N. and Schell, J. (1995) International Patent Application No. WO 95/06740. Published March 9, 1995.
44 Frohman, M.A., Dush, M.K. and Martin, G.R. (1988) Proc. Nat. Acad. Sci. U.S.A. 85, 8998-9002.
45 Graham, S.A. (1989) Crit. Rev. Food Sci. Nutr. 28, 139-173.
46 Martini, N., Schell, J. and Töpfer, R. (1995) in Plant Lipid Metabolism (Kader, J.-C. and Mazliak, P., eds.), pp. 495-498, Kluwer Academic Publishers, Netherlands.

47 Morden, C.W. and Golden, S.S. (1989) Nature 337, 382-384.
48 Fitch, W.M. (1970) System. Zool. 19, 99-113.
49 Sanderson, M.J. and Donoghue, M.J. (1994) Science 264, 1590-1593.
50 Martin, W., Gierl, A. and Saedler, H. (1989) Nature 339, 46-48.
51 Voelker, T.A. (unpublished data).
52 Dehesh, K., Jones, A., Knutzon, D.S. and Voelker, T.A. (unpublished data).
53 Töpfer, R., Martini, N. and Schell, J. (1995) Science 268, 681-686.
54 Dehesh, K., Edwards, P., Hayes, T., Cranmer, A.M. and Fillatti, J. (1995) Plant Physiol. 109 (in press).
55 Davies, H.M. (1993) Phytochemistry 33, 1353-1356.
56 James, D.W.J. and Dooner, H.K. (1990) Theoret. Appl. Genet. 80, 241-245.
57 Sambanthamurti, R. and Oo, K.-C. (1990) in Seed Storage Compounds (Quinn, P.J. and Harwood, J.L., eds.), pp. 166-168, Portland Press, London.
58 Cronan, J.E. and Rock, C.O. (1987) in *Escherichia coli* and *Salmonella typhimurium*: Cellular and Molecular Biology (Neidhardt, F.C., Ingraham, J.L., Low, K.B., Magasanik, B., Schaechter, M. and Umbarger, H.E., eds.), Vol. 1, pp. 474-497, Amer. Soc. Mol. Biol., Washington, D.C.
59 Ohlrogge, J., Savage, L., Jaworski, J., Voelker, T. and Post-Beittenmiller, D. (1995) Arch. Biochem. Biophys. 317, 185-190.
60 Voelker, T.A., Hayes, T.R., Cranmer, A.C., Turner, J.C. and Davies, H.M. Plant J. (in press).
61 Bafor, M., Stobart, A.K. and Stymne, S. (1990) J. Amer. Oil Chem. Soc. 67, 217-225.
62 Oo, K.-C. and Huang, A.H.C. (1989) Plant Physiol. 91, 1288-1295.
63 Eccleston, V., Voelker, T.A. and Ohlrogge, J. (1995) Planta 197 (in press).
64 Ohlrogge, J.B. (personal communication).
65 Ohlrogge, J.B., Jaworski, J.G. and Post-Beittenmiller, D. (1993) in Lipid Metabolism in Plants (T.S. Moore, Jr., ed.), pp. 3-32, CRC Press, Baton Rouge, LA.
66 Deerberg, S., Twickel, J.v., Förster, H.-H., Cole, T., Fuhrmann, J. and Heise, K.-P. (1990) Planta 180, 440-444.
67 Heise, K.-P. and Fuhrmann, J. (1994) Prog. Lipid Res. 33, 87-95.
68 Schuch, R., Brummel, M. and Spener, F. (1993) International workshop "Biotechnological Aspects of Plant Lipids", May 26-29, 1993, Carmona, Spain, p. 31.
69 Singh, S.S., Nee, T.Y. and Pollard, M.R. (1986) Lipids 21, 143-149.
70 Dehesh, K., Hawkins, D.J., Knutzon, D.S., Kridl, J.S. and Yuan, L. (personal communication).
71 Newman, T., Bruijn, F.J.d., Green, P., Kende, H., Mcintosh, L., Ohlrogge, J., Raikhel, N., Somerville, S., Thomashow, M., Retzel, E. and Somerville, C. (1994) Plant Physiol. 106, 1241-1255.
72 Sasaki, T., Song, J.Y., Kogaban, Y., Matsui, E., Fang, F., Higo, H., Nagasaki, H., Hori, M., Miya, M., Murayamakayano, E., Takiguchi, T., Takasuga, A., Niki, T., Ishimaru, K., Ikeda, H., Yamamoto, Y., Mukai, Y., Ohta, I., Miyadera, N., Havukkala, I. and Minobe, Y. (1994) Plant J. 6, 615-624.
73 Lawson, D.M., Derewenda, U., Serre, L., Ferri, S., Szittner, R., Wei, Y., Meighen, E.A. and Derewenda, Z.S. (1994) Biochemistry 33, 9382-9388.
74 Davies, H.M., Hawkins, D.J. and Nelsen, J.S. (1995) Phytochemistry 39, 989-996.
75 Yuan, L., Voelker, T.A. and Hawkins, D.J. (1995) Proc. Nat. Acad. Sci. U.S.A. 92, 10639-10643.

76 Sloane, D.L., Leung, R., Craik, C.S. and Sigal, E. (1991) Nature 354, 149-152.
77 Griffiths, G., Walsh, M.C. and Harwood, J.L. (1993) Phytochemistry 32, 1403-1405.
78 Liu, D. and Post-Beittenmiller, D. (1995) J. Biol. Chem. 270, 16962-16969.
79 Yuan, L., Hawkins, D. and Kridl, J. (personal communication).
80 Knutzon, D.S., Thompson, G.A., Radke, S.E., Johnson, W.B., Knauf, V.C. and Kridl, J.C. (1992) Proc. Nat. Acad. Sci. U.S.A. 89, 2624-2628.
81 Dörmann, P., Spener, F. and Ohlrogge, J.B. (1993) Planta 189, 425-432.
82 Leonard, J.M., Slabaugh, M.B. and Knapp, S.J. (unpublished data).
83 Walker, K.A. and Harwood, J.L. (1985) Biochem. J. 226, 551-556.
84 Rathinasabapathi, B., McCue, K.F., Gage, D.A. and Hanson, A.D. (1994) Planta 193, 155-162.
85 Nishida, I., Imai, H., Ishizaki-Nishizawa, O., Tasaka, Y., Shirashi, H., Higashi, S., Hayashi, H., Beppu, T., Matsuo, T. and Murata, N. (1993) in Biochemistry and Molecular Biology of Membrane and Storage Lipids of Plants (Murata, N. and Somerville, C., eds.), pp. 79-88, The Society of American Plant Physiologists, Rockville, MD.
86 Wada, H., Gombos, Z., Sakamoto, T., Higashi, S., Los, D.A., Heinz, E., Schmidt, H., Nishida, I. and Murata, N. (1993) in Biochemistry and Molecular Biology of Membrane and Storage Lipids in Plants (Murata, N. and Somerville, C., eds.), pp. 67-78, The American Society of Plant Physiologists, Rockville, MD.
87 Voelker, T.A., Davies, M.H. and Knutzon, D.S. (1994) International Patent Application No. WO-94-10288. Published May 11, 1994.
88 Slabaugh, M., Leonard, J., Huestis, G., Crane, J. and Knapp, S. (1995) in Plant Lipid Metabolism (Kader, J.-C. and Mazliak, P., eds.), Kluwer Academic Publishers, Netherlands.
89 Hawkins, D. and Kridl, J. (personal communication).
90 Claros, M.G. and Heijne, G.v. (1994) Comput. Appl. Biosci. 10, 685-686.
91 Swofford, D.L. (1991) Version 3.0, Distributed by the Illinois Natural History Survey, Champaign, IL.

GENETIC ENGINEERING OF AN INSECT PARASITE

Randy Gaugler and Sarwar Hashmi

Department of Entomology
Rutgers University
New Brunswick, NJ 08903-0231

INTRODUCTION

Growing public concern over chemical insecticides, particularly groundwater contamination, residues on food and turfgrass, resistance development, and wildlife kills, have fueled an intense search for safer alternatives. The use of natural enemies, particularly parasitoids and insect pathogens, as biological insecticides have been considered by many to be the most viable pest management alternative. Nevertheless, biological insecticides are presently contributing little to replacing chemicals, making up less than 1.5% of all insecticides used in the U.S. (B. Carlton, personal communication). However, at present, greater than 90% of all biologicals used are some derivative of *Bacillus thuringiensis*. If endotoxins produced by *B. thuringiensis* were regarded as chemical insecticides, and there are persuasive arguments to this effect (1), then the most important true biologicals are insect-parasitic or entomopathogenic nematodes.

Entomopathogenic nematodes are lethal obligatory parasites of insects. They are ubiquitously distributed and comprise two families, the Steinernematidae and Heterorhabditidae. The families are not closely related phyogenetically but share similar life histories through convergent evolution (2). The only free-living stage is the infective juvenile which occurs in the soil, seeks out hosts, enters through natural body openings or thin cuticle, and penetrates into the host hemocoel. The nematodes carry symbiotic bacteria (*Xenorhabdus* or *Photorhabdus* spp.) which are released into the host, usually causing host death in 24 to 48 hr. The nematodes feed upon the bacteria and liquefying host contents, mature, mate, and produce one or more generations (Figure 1) which emerge from the cadaver as infective juveniles in search of new hosts (3).

The potential of nematodes for insect pest suppression has been recognized for more than 60 years. Attempts during the 1930s to exploit *Steinernema glaseri* against the Japanese beetle, *Popillia japonica*, failed because early researchers were unaware of the nematode's bacterial symbiote (4). Subsequent rearing on artificial media containing antimicrobial agents to repress contaminants eliminated the symbiote, precluding field success.

There are presently nearly 60 laboratories in 38 countries working on entomopathogenic nematodes, and the number is growing. This explosion of interest since

Genetic Engineering, Vol. 18, Edited by J.K. Setlow
Plenum Press, New York, 1996

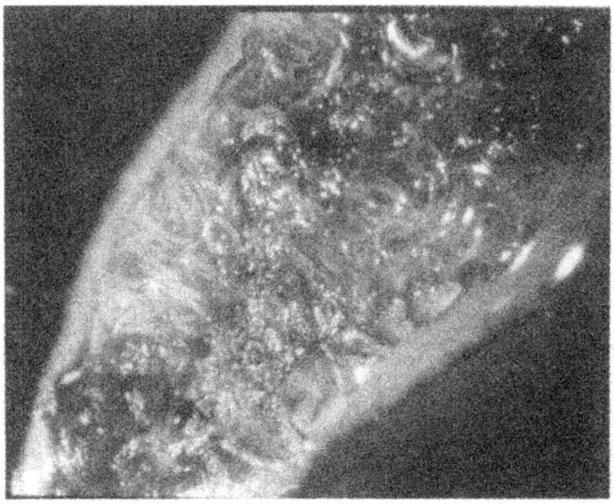

Figure 1. *Galleria mellonella* larvae infected with *Heterorhabditis bacteriophora* HP88.

the mid-1980s is a function of their impressive attributes as biological insecticides (3). Whereas virtually all biological control agents are constrained by narrow host specificity, the experimental host range of these unique parasites includes hundreds of insect species. This is due to the associated bacteria, which kill hosts so quickly that parasitic adaptation to the host life cycle is not required. Despite their lethality to insects, lack of pathogenicity to mammals (5) led the U. S. Environmental Protection Agency to exempt all species and strains belonging to *Steinernema* and *Heterorhabditis* and their associated bacteria from registration requirements under the Federal Insecticide Fungicide Rodenticide Act (FIFRA). Mass production is easily accomplished for some species with standard liquid fermentation methods in tanks up to 150,000 liters (6). Continuous fermentation improvements offer promise that insecticidal nematodes will be able to compete economically with chemicals against many insect pests. Nematodes are commercially produced in the U.S., Europe and China, and are being successfully applied against pests of turfgrass (mole cricket, fleas), artichokes (artichoke plume moth), mushrooms (sciariid flies) and ornamentals (black vine weevil). Despite their biocontrol attributes and commercial successes, worldwide retail sales for 1995 are estimated at no more than $15 million (P. Grewal, personal communication).

Perhaps the most significant reason why insecticidal nematodes are not having an impact beyond niche markets is their lack of tolerance to environmental stress, particularly temperature extremes (7), solar radiation (8), and desiccation (9). Poor stability results in inadequate shelf-life and field persistence. Of the six nematode species that have been commercialized, only *S. carpocapsae* can be stored for more than a few weeks at room temperature, with the use of a novel granule formulation that induces nematode cryptobiosis (10). This formulation breakthrough provides a maximum shelf-life for *S. carpocapsae* of six months, far short of the two-year industry standard for conventional insecticides. Extreme temperatures in the warehouse or during transport have had disastrous results, causing complete nematode mortality before the nematodes even reach the user. Curran (11) showed that as much as 90% of infective juveniles applied to the soil disappear within a week. Such limited field persistence is certainly responsible in part for the "efficacy gap" between chemical and nematode-based insecticides (12).

There are over twenty species of *Heterorhabditis* and *Steinernema* (13). Hundreds of isolates have been collected from every inhabited continent (14) and many of these nematodes have been tested for their insecticidal activity. Despite the high degree of natural genetic variability among these many species and strains, realization of the full potential of nematodes for biological control will require improvements in their virulence, host range and particularly their environmental tolerance. Genetic improvement has been advocated for overcoming these limitations (15). Entomopathogenic nematodes are exceptionally well suited for genetic studies due to their simplicity, transparency, ease of laboratory culture, short generation time and small genome size. Furthermore, *Heterorhabditis* is a self-fertilizing hermaphrodite, thus mutations become homozygous automatically, pure lines can be maintained and single gene mutants can be isolated. Depending on culture conditions, the development of males can be induced, allowing the transfer of gene markers for cross-breeding (16). These are all attributes that the genetically little known entomopathogenic nematodes share with another nematode, the free living *Caenorhabditis elegans*, the most completely understood metazoan.

APPROACHES TO GENETIC IMPROVEMENT

Genetic manipulation is a well-established approach for improving crops and livestock. Classical selective breeding has been applied to improve insect predators and parasitoids, most notably by Hoy (17,18) in increasing the pesticide resistance of phytoseiid mites. Similar efforts have been made to genetically improve entomopathogenic nematodes as bioinsecticides. Selection of *S. carpocapsae* for increased host-finding of scarab larvae achieved a several-fold improvement over the wild type (19-21). A key element of this selection scheme was provision of adequate genetic diversity through strain hybridization. Relaxation of selection pressure resulted in a rapid decline in response to host cues. The enhancement was found to result from increased chemosensitivity to host-released carbon dioxide (22). Nevertheless, selection was unable to overcome the extremely poor adaptation of *S. carpocapsae* to scarabs, and the selected S20 strain did not provide enhanced field efficacy (23). Subsequent attempts to improve nematode field efficacy *via* classical selection have been disappointing or marginally successful (24,25).

Gaugler et al. (23,26) suggest that selection is a valid approach to strain improvement, but caution that the approach is limited because it is most likely to succeed when the selected nematode is already known to attack the target insect. Additionally, laboratory selection can result in unpredictable genetic changes. Some fitness parameters such as dispersal and overwintering ability may be inconsequential for nematodes that are primarily intended for use as bioinsecticides; others are directly (e.g., pathogenicity) or indirectly (e.g., reproductive potential) relevant, and their impairment would reduce effective use. An evaluation of fitness of the selected S20 strain by Gaugler et al. (27) revealed that nematode pathogenicity, sex ratio and vigor remained unchanged, while reproductive capacity increased and storage stability declined. Finally, selective breeding is effective only if the desired alleles are present in the population subjected to selection. Insufficient genetic variation in resistance to ultraviolet (UV) radiation led Gaugler et al. (20) to reject selective breeding as an option to improve UV tolerance.

An alternative approach to selective breeding for introducing genetic changes is through induction of mutations. Unlike selective breeding, a genetically homozygous population of mutants can be established from a single individual of *Heterorhabditis*. Moreover, single-gene mutations can be induced, eliminating the cumulative polygenic changes associated with selection. With a collection of single-gene mutations all affecting the same trait, one can dissect the trait into its specific molecular components, map the corresponding genes, assess their relation to the desired phenotype, and, by controlled crosses, construct novel strains with specific combinations of desired traits. Adapting the mutagenesis protocols described by Brenner (28), the initial goal in our laboratory and others has been the isolation and characterization of morphological mutants of *H. bacteriophora* (16,29,30,39). These initial successes encouraged insect nematode laboratories across Europe to propose a cooperative effort to develop *Heterorhabditis* linkage maps (I. Glazer, personal communication). In addition to morphological mutants, we have recently isolated a behavioral mutant. Of special significance has been the reported isolation of a putative desiccation-tolerant mutant of *Heterorhabditis* (A. Burnell, personal communication). Nevertheless, the current enthusiasm for conventional mutagenesis is more likely to contribute critically-needed genetic analyses than to produce improved strains.

Genetic engineering or recombinant DNA technology is the most powerful and revolutionary tool for genetic improvement. This technique offers significant advantages over selective breeding or mutagenesis, making it possible to induce small defined changes into established genotypes without destabilizing effects. Ordinarily, the use of recombinant DNA methods to manipulate and improve organisms as poorly understood at both the genetic and molecular levels as steinernematid and heterorhabaditid nematodes could not be seriously contemplated (31). However, the close relationship of these nematodes with *C. elegans* (same superfamily), the model organism in molecular genetics for nearly three decades, provides an extraordinary early opportunity to exploit the new technology. This chapter discusses how advances made with *C. elegans* and other organisms can be applied or adapted to improve the biological control potential of insect nematodes by genetic engineering.

IDENTIFYING USEFUL GENES

The first step in genetic engineering is identifying genes that encode for useful traits, e.g., genes that regulate processes such as nematode development, cold acclimatization and response to stress in these nematodes. Most genes currently available for insertion into entomopathogenic nematodes originate from *D. melanogaster* or *C. elegans* (Table 1). However, changes in one or more attributes in nematodes may not result in improved field performance (23). Furthermore, desirable traits such as anhydrobiosis and chemoreception may be encoded by multiple genes and thus be more difficult to manipulate by genetic transformation than traits such as pesticide resistance which may be regulated by single genes (32). For instance, metallothionein genes (*mtn*), which appear to function in homeostasis of copper and cadmium and in their detoxification, have been cloned from *Drosophila* and other organisms (33). Although genes from *Drosophila* have yet to be transferred into entomopathogenic nematodes, it might be useful to introduce the *mtn* gene to promote resistance to copper-containing fungicides. The *opd* gene, which encodes resistance for organophosphate-based insecticides, may also add potentially valuable traits to entomopathogenic nematodes. This gene was isolated from the bacterium, *Pseudomonas diminuta*. Phillips et al. (34) put *opd* under the control of a heat shock (*hsp70*) promoter from *Drosophila*, injected the plasmid and expressed the *opd* gene in *D. melanogaster*. A transgenic nematode with the *opd* gene may be resistant to organophosphate-based insecticides. Two heat-shock protein genes from *C. elegans* have been transferred to entomopathogenic nematodes (35,36). Further progress in the areas of molecular biology and biochemistry will lead us to identify genes for traits such as shortening developmental time, enhancing progeny production, extending temperature tolerances and altering host or habitat preferences. For example, antifreeze protein genes cloned from the wolffish

Anarhichas lupus, have been expressed in transgenic *Drosophila* (37,38) and have potential for nematodes.

Table 1. Examples of Cloned Genes for Useful Traits for Genetic Manipulation of Entomopathogenic Nematodes

Genes	Sources	References
Roller (phenotypic marker *rol6* [*su1006*])	*C. elegans*	(74)
Twitcher (phenotypic marker) *unc22*	*C. elegans*	(75)
Small heat-shock protein gene *hsp16*	*C. elegans*	(103,104)
Heat-shock protein gene *hsp70*	*C. elegans*	(121)
Acetylcholinesterase (*Ace*) (pesticide resistant)	*D. melanogaster*	(126)
mdr49 & *mdr65*	*D. melanogaster*	(127)
metallothionein (*mtn*) (copper resistant)	*D. melanogaster*	(33)

The isolation and characterization of genes encoding for important traits such as cold tolerance and desiccation resistance from entomopathogenic nematodes may increase the frequency of transformation because these genes might pass more efficiently into other entomopathogenic nematodes. Also, cloning genes from one species of entomopathogenic nematode and inserting them into the same or another species of these nematodes will likely lessen regulatory restrictions. So far, no genes for specific traits are available from entomopathogenic nematodes. However, one possibility centers on the ability of infective juvenile *S. glaseri* to produce an anti-immune protein(s) which suppresses the host immune system, thus allowing the nematode to avoid encapsulation and ensure bacterial release and establishment (Wang and Gaugler, unpublished data). Analysis of such anti-immune proteins may lead to the identification of one or more genes that could be incorporated into these nematodes for the overexpression of this trait. The nematode *S. glaseri* produces an enzyme which enables penetration into the host hemocoel (39). A study of this enzyme might lead us to understand the molecular basis of penetration in these nematodes and, thereby, the identification of genes. Thus, two critical areas for future research will be: (i) the identification, isolation, and cloning of useful genes (traits) and ii) the regulation of foreign gene expression in entomopathogenic nematodes.

METHODS OF GENETIC TRANSFORMATION

Rapid advances have been made in the development of diverse techniques for transforming widely different organisms. Foreign DNA is introduced into cells by disruption of the cell surface. Successful methods depend upon measures that limit the severity of disruptions in the cell surface, embryo, or whole organism.

Several gene delivery methods have been described that use the natural gene transfer abilities of bacterial (40,41) and viral pathogens (42) to mediate foreign DNA transfer, but these are generally suitable only for plant systems. A notable exception has been the use of baculoviruses (which are highly pathogenic to insects and promising biological insecticides in their own right) as vectors to carry foreign DNA into insect cell lines (43-

45). Baculovirus-based expression systems have been employed for the commercial production of proteins of biomedical and agricultural importance. Many of these pathogens are known to infect nematodes (46), including insect nematodes, and it is intriguing to consider their potential use in gene transfer. However, we will restrict the present discussion to physical methods of gene transfer that possess more immediate potential for genetic modification of entomopathogenic nematodes.

Electroporation

Electroporation involves the use of electrical impulses of high field strength to permeabilize biomembranes reversibly, thereby opening the cell wall for the introduction of genetic material into cells (47). Conditions required to transfer DNA into mouse cells electrically have been reported, and a simple apparatus was used to transfer DNA into mouse and human cells by electroporation (47). Bellofato and Cross (48) were able to show a transient expression of the bacterial *chloramphenicol acetyl transferase* (CAT) gene after electroporation into *Leptomonas*. Laban and Wirth (49) obtained similar results with the CAT gene after electroporaton into *Leishmania*. Successful transient and stable transformation of *Trypanosoma brucei* were obtained by electroporation (50). The optimization of the gene delivery system to plant tissue with electroporation has mainly involved eliminating explant released nucleases, prolonging the DNA/explant incubation time, expanding the pulse time, wounding the explant either enzymatically or mechanically or adjusting the explant in a specific orientation in the electroporation chamber. Electroporation has not been demonstrated for nematodes, presumably because of their large size ($>$ 1 mm) and thick cuticle. Conceivably, these limitations might be overcome by dissolving the cuticle of gravid hermaphrodites with alkaline hypochlorite to release the eggs (51), and exposing the embryos to electroporation.

Microprojectile Acceleration

Acceleration of DNA-coated microprojectiles was developed for gene delivery into various plants (52-55). Different approaches have been used to accelerate these particles including explosion from a 0.22 caliber cartridge (hence the derivation of the "gene gun") (52), electric arch discharge (54) and controlled gas pressure (56,57). The microprojectiles have sufficient momentum to penetrate the plant cell wall and membranes, and can carry DNA or other substances into the interior of the bombarded cells. Baldarelli and Lengyel (58) obtained transient expression of DNA in *Drosophila* embryos after ballistic introduction with DNA carrying tungsten particles and suggested that a modified version of this method could yield stable germ-line transformation in *Drosophila*. Zelenin et al. (59) introduced and expressed foreign DNA into the developing embryos of loach (*Misgurnus fossilis*), rainbow trout (*Salmo gairdneri*) and zebra fish (*Brachydanio rerio*) with tungsten microprojectiles coated with galactosidase (pRSV-β-gal) and aminoglycoside 3′-phosphotransferase II (pSV3-neo) genes. About 70% of the eggs survived the bombardment and showed the expression of the injected sequence (59).

The optimal target of the particles to ensure the production of transgenic nematodes is unclear. Because the particles are only slightly smaller than the oocyte nucleus, it would appear that a direct hit on the nucleus would cause irreparable damage. Therefore, the introduction of DNA into the adjacent cytoplasm or into the eggs of gravid females or hermaphrodites may be the preferred target for successful transformation.

Efforts to transform *H. bacteriophora* in our laboratory employed a BioRad Biolistic Transformation System® that uses helium gas pressure. Tungsten and gold microparticles with sizes ranging from 0.5 to 1.5 micrometers were coated with plasmid vectors (pRF4 and pPCZ1) carrying *C. elegans* genes coding for Rol 6 and the 16 kD heat-shock protein, respectively. Lipid agar plates were overlaid with 30,000 nematode embryos, bombarded at a helium pressure of 1100 PSI and a target distance of 6 cm, and then incubated. Less than 10% of the embryos survived the treatment and no transformants were identified. Due to the laborious nature of screening, however, we are likely to have missed transgenic nematodes unless efficiency were at least 0.5%. We predict that optimizing particle delivery, particularly for pressure, distance and DNA concentration, would ultimately be successful. Nevertheless, microparticle acceleration is best suited for plants where transformation is easily detected. Simpler and far less expensive methods are available for nematode transformation.

Fiber-Mediated DNA Delivery

Agitation of yeast cells (*Saccharomyces cervisiae*) with glass beads and plasmid DNA with the use of a vortex mixture resulted in successful genetic transformation (60). Kaeppler et al. (61) subsequently reported silicon-carbide fiber-mediated DNA delivery into plant cells. In their protocol, a suspension of tobacco and maize cells was vortexed in the presence of silicon-carbide fibers and plasmid DNA encoding glucuronidase (GUS). Transient expression was recorded. This simple procedure relies on the DNA-coated fibers to be carried into the cells when collisions occur. We are exploring the use of fiber-mediated DNA delivery using both nematode embryos and gravid hermaphrodites of *H. bacteriophora*. Although this approach is in its early stages, the results so far are not promising.

Microinjection

Microinjection has been the most common method of gene transfer into oocytes (62,63), eggs (64,65) and embryos (66). The feasibility of injecting RNA was first demonstrated in *C. elegans* when a nonsense suppresser tRNA was injected into a strain carrying an amber mutation (67). Total tRNA was isolated from the *sup7* strain that suppresses amber mutations (68,69).

Advances in genetic transformation of *C. elegans* have demonstrated that integrative transformation can be achieved reproducibly after DNA is microinjected directly into maturing oocyte nuclei (70). In the transformed animals that have so far been analyzed, one to ten copies of the injected DNA are found integrated at a particular chromosomal site (70). When populations derived from injected animals are screened by Southern hybridization (71), they contain hundreds of copies of the transforming DNA in a long, extrachromosomal tandem array. Heritable extrachromosomal DNA transformation in *C. elegans* was first described by Stinchcomb et al. (72) after microinjecting DNA into the gonad cytoplasm. DNA molecules injected into the central core of cytoplasm in each arm of the hermaphrodite gonad of *C. elegans* underwent a transient period of reactivity. This resulted in the formation of large heritable extrachromosomal structures that experienced very little further rearrangement (73). Germ cell nuclei in *C. elegans* develop initially in a syncytium and, when cell membranes later envelop them, the exogenously added DNA is packaged into the oocyte. Transforming DNA is generally not integrated into the chromosomes but rather is maintained as a concatamer of introduced sequences. The stability of this extrachromosomal array is variable, depending somewhat upon the size of the concatamer, with larger arrays maintained better than smaller ones. The copy number of introduced sequences can be on the order of a few hundred. It is not clear which experimental factors determine whether integrants or extrachromosomal tandem arrays are obtained, although selection scheme and point of injection seem to be important.

Microinjection of *C. elegans* typically involves DNA co-injected with a plasmid that encodes a marker gene conferring a recognizable phenotype. Two commonly used markers are *rol6* (*su1006*), a dominant allele that causes transformed strains to roll [this is encoded in plasmid pRF4 (74)] and a construct that synthesizes an antisense complement to a portion of the *unc22* gene, thereby causing animals to twitch [this marker is encoded on plasmid pPD10.41 (75)]. We used microinjection to generate *H. bacteriophora* lines carrying *hsp16-lacZ* fusion as a marker to identify transformants (35). DNA for microinjection is prepared by mixing 50 μg/ml of marker DNA and 50 μg/ml of DNA to be assayed. The concentration of DNA introduced can vary between 50 and 200 μg/ml. Only 10 μl total volume is required as needles are filled with very small amounts of DNA. In the microinjection protocol, the needles are filled with DNA and loaded into a micromanipulator and the injection needle is attached to a pressure source (nitrogen gas). Adult hermaphrodite nematodes containing a few eggs are injected. The needle is aligned next to the nematode gonad, pressed firmly against the cuticle and inserted into the gonad. After injection, a small drop of M9 buffer (28) is placed onto the nematode, and the nematode is transferred to a fresh agar plate seeded with bacteria.

Microinjection methods for transferring foreign DNA into *C. elegans* have helped advance basic studies of gene expression and have permitted the introduction of genes of interest for genetic studies (70,72,73). Using a microinjection method, Fire (70) injected a cloned *C. elegans* amber suppressor tRNA gene, *sup7*, into *C. elegans* with an amber termination mutation in gene *tra3* whose function is required for fertility. Transient expression of *sup7* was demonstrated by the presence of fertile nematodes in the generation after injection. In a few cases, the fertile nematodes gave rise to stable suppressor lines carrying injected DNA sequences. The suppressor activities were mapped to chromosomal loci, indicating that the exogenous DNA had integrated into the genome (70).

Mello et al. (73) used a cloned mutant collagen gene, *rol6* (*su1006*) (74), as a dominant genetic marker for DNA transformation and showed that large extrachromosomal arrays assembled directly from the injected molecules, and that homologous recombination drives array assembly. These workers suggested that the size of the assembled transgenic structures determines whether or not they will be maintained extrachromosomally or lost. Low copy number extrachromosomal transformation can be achieved by adjusting the relative concentration of DNA molecules in the injection mixture.

Microprobes

Microinjection has been used for genetic transformation of plants, animals, and microorganisms, and remains the exclusively used method for *C. elegans* transformation. Nevertheless, this technique is laborious, requires expensive specialized equipment, and considerable expertise. As an alternative to microinjection for introducing foreign DNA into entomopathogenic nematodes, we have developed a remarkably simple, inexpensive and highly efficient transformation system which makes use of arrays of microprobes fabricated by silicone micromachining technology. The microprobes are fabricated on a silicon wafer overlaid with a large number of pyramidal projections (Figures 2A, 2B). Each microprobe is extraordinarily sharp and will pierce cells and transmit DNA and the depth of penetration can be controlled with great precision.

Because of our previous success with the *hsp16* gene, we wanted to determine the efficiency of the microprobe array using the same plasmids (36). In our protocol, adult hermaphrodite nematodes were collected at the 4 to 6 egg stage from *in vitro* culture and

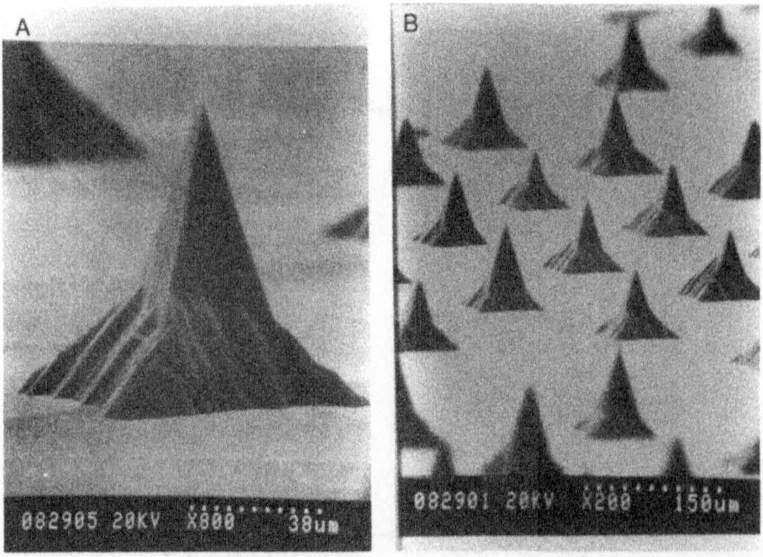

Figure 2. Scanning electron micrograph of a microprobe array. A) Each individual microprobe is approximately 80 μm high; B) The tips are in square array separated by 150 μm, resulting in an areal density of about 4400 microprobes per cm^2.

washed with distilled sterile water. The array was placed on the surface of a lipid agar plate preseeded with *H. bacteriophora*' s symbiotic bacteria, *P. luminescens*. A DNA mixture was made by 5 μg DNA (plasmid vector pPCZ1 containing *C. elegans* 16 kD heat shock promoter fused to *E. coli* β-galactosidase, 12.5 μl of 2.5 M $CaCl_2$ and 5 μl of 0.1 M spermidine-free base. Ten μl of the DNA mixture was pipetted onto the array to coat the tips of the probe with DNA. After five minutes, 10 μl of concentrated nematode suspension (approximately 200 to 250 nematodes) were pipetted onto the top of the array and left at room temperature for 5 to 6 minutes (Figure 3). The array was then removed from the agar plate and the plate with nematodes was incubated at 25° C until the nematodes produced progeny. Eight percent of the progeny tested showed expression of *hsp16-lacZ* fusion in the F1 generation. Moreover, expression was apparent in the next three generations, demonstrating that transformation was heritable.

The ovary of the adult heterorhabditid hermaphrodite consists of a core of cytoplasm surrounded by a sheath of germ nuclei. It is likely that DNA coated on the tips of the microprobes is carried through the cuticle into the gonad cytoplasm as the nematodes crawl on the microprobes. Expression of β-galactosidase in the progeny implies that some probe tips passed through the nematode cuticle to allow DNA delivery into the gonad cytoplasm. Because of the presence of many microprobes in a small area, multiple injections are possible. There was little evidence of damaged nematodes, except in the few nematodes where the tip of the probe was observed to have penetrated deeply into the body. This means of DNA transfer appears ideally suitable for nematodes, especially as further refinements change the geometry of the microprobes to increase the number of site-directed DNA insertions.

The microprobe's simplicity of use and high degree of efficiency make it an extremely useful tool that may speed up studies involving transgenic plants or animals. Our success in transforming *H. bacteriophora* is spurring efforts to use microprobe-mediated DNA

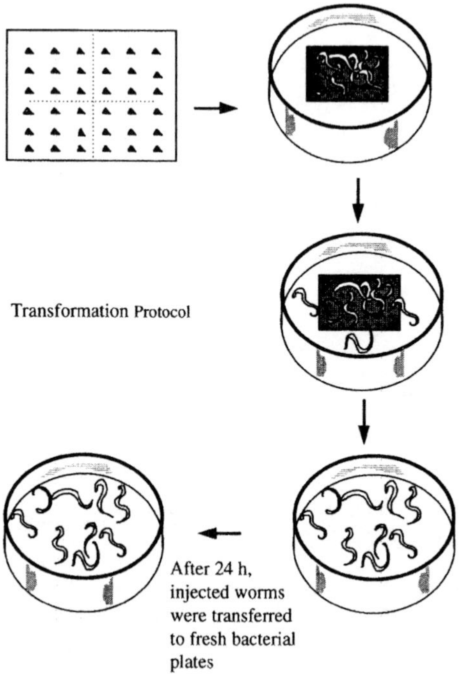

Figure 3. A schematic drawing of microprobe-mediated gene transfer method. The microprobes are integral with the support that are prepared by etching a silicon wafer. There are 200 to 250 microprobes on a 5 x 8 mm substrate. The nematodes were placed on a DNA-coated microprobe array.

delivery systems for plants, and we believe microprobes can be adapted for use with popular model systems such as *C. elegans* and *Drosophila*. One key to future success will be increasing the frequency of foreign DNA integration by exploring processes that favor chromosomal integration of the vectors.

TRANSPOSONS

Transposons are DNA sequences able to move from one chromosomal locus to another. Early investigations of these "jumping genes" demonstrated their ability to affect the regulation of gene activity, and to cause mutations and chromosomal rearrangements in corn (76). Because of this ability to cause insertion, deletions and other DNA rearrangements, transposons have been identified as a major source of spontaneous mutations in many organisms including *Drosophila* (77,78) and *C. elegans* (79,80).

The genome of *C. elegans* contains multiple conserved copies of the transposable element Tc1. The 1.6 kb DNA sequence of Tc1 is present in 30 to 500 copies per haploid genome, depending on the nematode strain (79,81). Structure and sequence analyses have shown that Tc1 elements have 54 bp perfect inverted terminal repeats and an internal open reading frame of 819 nucleotides (82,83). The structure of Tc1 resembles that of P elements of *Drosophila* and Ac/Ds elements of corn in having short, terminal inverted repeats (81). The detection of an element related to *C.elegans* Tc1 in bacteria, invertebrates and vertebrates indicates that this family of transposable elements exists in a larger variety of hosts than previously expected (84). This makes Tc1, a well-characterized transposon in *Caenorhabditis* spp., a potential tool for similar studies to be developed for other organisms, including entomopathogenic nematodes. Six transposable element families have been described in *C. elegans*: Tc1 through Tc6 (83,85).

The use of Tc1 allows the cloning of genes defined solely by genetic and phenotypic criteria (79). For instance, a method for inducing mutations with the Tc1 transposable element has facilitated the cloning of genes identified by mutations (86). Many *C. elegans* genes have been cloned by "transposon tagging" following the isolation of the Tc1 allele at a particular locus (87). The Tc1 transposon induces frequent deletion of the flanking DNA, apparently resulting from Tc1 excision followed by imprecise DNA repair. Zwaal et al. (88) used target-selected gene inactivation after transposon insertion. In their protocol, they established a frozen library of 5000 *C. elegans* lines mutagenized by Tc1 insertion, from which insertion mutants of genes of interest were recovered. Nematodes were then screened by polymerase chain reaction (PCR) to detect derivatives in which Tc1 and 1000 to 2000 bp of flanking DNA were deleted. Zwaal et al. (88) were able to isolate Tc1 insertions in 16 different genes and obtained deletion derivatives for six.

Most research on germ line transformation methods with transposons as vectors has been accomplished with *D. melanogaster*. Spradling and Rubin (89) and Rubin and Spradling (90) discovered that a transposable element, the P element, could be genetically manipulated to serve as a vector to carry foreign DNA into the chromosomes of germ line cells of *Drosophila*. An intact P element consists of 2907 bp of DNA with a coding sequence that produces a single polypeptide with transposase activity. Intact P elements have 31 bp inverted repeats flanking the coding region and these inverted repeats are required if the P element is to transpose. Efficient techniques for genetic transformation and the availability of a well-characterized genome of *C. elegans* offer promise in adapting these tools for entomopathogenic nematodes.

Abad et al. (91) examined the distribution of Tc elements among 19 nematode species of the class Secernentea including the entomopathogenic nematode *S. carpocapsae*. These researchers found a wide distribution of Tc3 elements among nematode species. We have initiated a search for transposable elements in entomopathogenic nematodes using Tc series specific probes from *C. elegans* (92). We are screening our library (> 150 strains) of cryopreserved *Steinernema* and *Heterorhabditis* species and strains with the Tc-specific probes. The DNA of these nematode species or strains is digested with different restriction enzymes and screened by Southern blot hybridization with Tc probes.

The isolation and characterization of active transposons from entomopathogenic nematodes will open new doors to strain improvement. The use of transposons as insertion vectors into genomic DNA will increase the efficiency, and of particular interest, the stability, of transformation. Moreover, interruption of genes by transposon insertion (site-directed mutagenesis) can provide new information on gene function. In short, transposon studies will form the starting point for exploration of the genome.

MARKERS TO MONITOR GENE EXPRESSION

Several markers are available for monitoring gene expression and protein localization within cells. The *E. coli uid* A gene encoding the enzyme β-glucuronidase (GUS) is a widely used scorable marker gene in plant transformation experiments (93). The *E. coli lacZ*-encoded enzyme, β-galactosidease (94), has been used as a reporter molecule for studies of gene expression in a variety of prokaryotic and eukaryotic systems (95), including entomopathogenic nematodes (35,36). The *E. coli* β-galactosidase gene is used as a marker enzyme for studies of gene expression because both the enzyme and the gene have well characterized properties (96). Rapid and precise histochemical staining procedures have made β-galactosidase a particularly good reporter gene for assays of tissue-specific expression in transgenic animals (97). However, these markers and others such as firefly luciferase (*luc*) (98) and bacterial luciferase (*lux*) (99), require exogenously added substrates and cofactors which kill the cells or organism (95). To select a living organism with a particular expressed gene, it is essential to have a marker to monitor gene expression and protein localization in living organisms. Chalfie et al. (100) recently used *Aequoria victoria* green fluorescent protein (*gfp*) as a marker for gene expression in *C. elegans*. Green fluorescent protein, expressed in prokaryotes and eukaryotic cells, is capable of producing a strong green fluorescence when excited by blue light (450 to 490 nm) with a long-pass emission filter (100). Because exogenous substrates are not required for the fluorescence, *gfp* can be used to monitor gene expression in living organisms.

Screening to identify transformed individuals is simple when a marker such as *gfp* is available. To test the utility of *gfp* as a marker in entomopathogenic nematodes, we injected plasmid DNA (pGF/mec4) with a *C. elegans mec4* promoter fused to *A. victoria gfp*. We observed the expression of *mec4/gfp* in the touch receptor cells in the tail region of *H. bacteriophora* (101). The expression of *gfp* in entomopathogenic nematodes constitutes a powerful tool that can be used to address a variety of interesting molecular and even ecological questions. This will be a particularly valuable tool in risk assessment studies on the environmental fate and horizontal and vertical transmission that must be conducted before field release of genetically engineered nematodes is contemplated.

Behavioral markers have frequently been used to identify living transformed *C. elegans*. The twitcher marker is a construct that synthesizes an antisense complement to a portion of the *C. elegans unc22* gene, causing nematodes to twitch. Roller, *rol6* (*su1006*), a dominant allele, causes transformed nematodes to roll about their longitudinal axis, resulting in circular movement. We attempted to use the roller allele as a marker in initial transformation experiments with *H. bacteriophora*, but we did not observe the roller behavior. Possibly *rol6* collagen does not interact appropriately with the basement membrane, to cause the same changes in *H. bacteriophora* as it does in *C. elegans*.

In the absence of a selectable marker, transformation can be confirmed through Southern blot hybridization. The identical analysis can be done by PCR (102) with specific primers designed for the injected sequence.

HEAT-SHOCK GENES AND THEIR EXPRESSION IN *H. bacteriophora*

The goal for genetic engineering of entomopathogenic nematodes is to improve their biological control potential through incorporation of genes encoding useful traits. The desired traits should be encoded by as few genes as possible, ideally one. Fortunately, some traits that could enhance tolerance to environmental stress and are conferred by single genes have been isolated and cloned from *C. elegans* and *Drosophila*. Among the suitable genes available from *C. elegans* are a family of genes encoding 16 kD heat-shock protein (*hsp16*s) (103,104). Our first goal was to study *hsp16* expression in our model organism: the HP88 strain of *H. bacteriophora*. We selected *H. bacteriophora* for our studies, because in addition to its many biocontrol attributes, its developmental biology is similar to *C. elegans* in having self-fertilizing hermaphrodites (16,105). Thus, progeny can be easily screened for the presence of transgenes.

Small Heat-Shock (*hsp16*) Genes

All organisms appear to express a set of proteins in response to heat-induced stress that enable them to acquire transient resistance to otherwise lethal temperatures (107,108). The heat-shock response (109,110) was originally discovered in *Drosophila* (111), and is characterized by the induction of a unique set of polypeptides: the heat-shock proteins (Hsp). Generally, Hsp can be subdivided into three families: 20 to 30, 70 and 80 to 90 kD proteins. Small heat-shock proteins exist as high molecular weight complexes in vertebrates, *Drosophila*, yeast and plants. The small heat-shock proteins in *C. elegans* consist of proteins ranging from 16 to 25 kD (112,113). Two cDNA clones encoding 16 kD heat-shock proteins have been isolated by screening a cDNA library constructed from RNA of heat-shocked nematodes with labeled and size-fractionated probe $(A)^+$ heat-shock RNA (112). Using these cDNA clones as probes, Russnak and Candido (103) and Jones et al. (104) defined the genomic organization of 16-kD heat-shock genes. The four major *C. elegans hsp16* genes reside at two loci designated *hsp16A* and *hsp16B*, and are arranged as two divergently transcribed pairs (103,104).

We used microinjection to generate *H. bacteriophora* lines carrying *hsp16-lacZ* fusion as a marker to identify transformants (35). Plasmid vector pPCZ1 containing a *C. elegans* 16 kD heat-shock promoter was fused to *E. coli* β-galactosidase (106). DNA was then injected into the central core of cytoplasm in each arm of the hermaphrodite gonad at a DNA concentration of 100 μg/ml. The worms were then permeabilized by lyophilization and acetone treatment, stained with 3%X-galactosidase, and incubated at 25°C (114). Nematodes were screened for the presence of transgenes by heat shocking for two hours at 33°C (106). This effort yielded heritable transformation in 6 to 7% of the progeny (Table 2). Approximately 13 to 22 transformed progeny were obtained from each injected hermaphrodite. Staining was not observed either in uninjected control worms or in the

Table 2. Frequency of Transformation in *Heterorhabditis bacteriophora* HP88

pRF4 + pPCZ1 (1:3)	Number[a] injected	Total F1 assayed	Total	Per injected animal[b]	% expression of total F1
100 μg/ml	18	1290	91	13	7
	21	1435	70	22	6
	14	1225	69	18	6

[a] Results of three experiments with different numbers of nematodes injected. Each injected nematode was grown separately on a single bacterial plate; progeny obtained from a plate were stained separately for locating β-galactosidase.
[b] Number of F1 expression is the average of an experiment.

progeny of the injected worms not subjected to heat-shock treatment before the staining procedure. In the latter case, this indicated that the expression of *hsp16* transgenes was heat-shock dependent. The reproduction of injected animals was followed for several generations. Although transformants were retained, their frequency decreased after the third and fourth generations. We presume that, as in *C. elegans*, introduced DNA is not integrated into the genome and thus tends to be diluted in the population over time.

The frequency of foreign DNA integration can be increased by increasing the DNA concentration or by exploiting processes that favor genomic integration. We investigated this in *H. bacteriophora* by injecting DNA at concentrations ranging from 100 to 300 μg/ml and screening for β-galactosidase expression in the progeny. At concentrations of 100 μg/ml or above, the frequency of transformation remained unchanged. Additional factors may also be adjusted to optimize efficiency. For example, we suspect that genes originating from entomopathogenic nematodes rather than *C. elegans* might be passed more efficiently to progeny with increased levels of expression.

We find a high degree of homology between the *hsp70* genes of *C. elegans* and *H. bacteriophora* (115). Thus, a *C. elegans* gene may provide easy access in *H. bacteriophora*. The only essential requirement will be the ability to identify and clone the genes of interest in *H. bacteriophora* or other entomopathogenic nematode species. Such studies are in progress. Our success in transformation of *Heterorhabditis* opens new avenues to develop strains of insecticidal nematodes with enhanced traits for biological control. Transformation of *Heterorhabditis* with the use of a promoter from *C. elegans* should be useful to others who wish to determine whether their favorite *C. elegans* genes work in entomopathogenic nematodes.

Thermotolerance of Transgenic *H. bacteriophora*

In many organisms a temperature increase above the normal physiological temperature causes a heat-shock response which consists of a profound alteration of gene expression. The best correlation of this type is observed between the major 70 kD heat-shock proteins (Hsp70) and increased thermotolerance (116,117). The heat-shock proteins have been described in many classes of organisms, and the most closely studied and highly conserved of these is Hsp70.

DNA sequence analysis of heat-shock genes isolated from several eukaryotic species, including *C. elegans*, has revealed the presence in promoter regions of consensus sequences closely related to that of *Drosophila* (118). The various *hsp70* related genes are organized into an *hsp70* multigene family and have been identified in many organisms including *Drosophila* (110), human (119), *Saccharomyces cervisiae* (120) and *C. elegans* (121,122). The genes have been characterized in *C. elegans* and several have been isolated (121). It is believed that as a class, the Hsp70 family of proteins catalyze the unfolding of entire native or denatured proteins. In the latter capacity, they may enable the cell to eliminate or renature proteins damaged by high temperature.

Because heat-shock proteins are responsible for thermotolerance, we tested whether overexpression of these genes would trigger heat-shock induction in *H. bacteriophora* sufficient to enable the infective juvenile to survive otherwise lethal temperatures. We used microinjection to introduce a plasmid vector, pUC18, containing a 6.6 kb *Eco*R1 fragment of the *hsp70*A gene from *C. elegans* (received from D.L. Baillie). Clonal pools of animals from the injected nematode's generation were assayed for the presence of the introduced sequence of the *hsp70* gene with the use of PCR and by Southern blot hybridization. We found that progeny of the injected nematodes had taken up the injected DNA (data not shown). A heat-exposure protocol was designed to determine the acquisition of thermotolerance in the progeny derived from the injected worms. The infective juveniles produced from injected worms were subjected to two hours of exposure to 35°C, acclimatized for one hour at 25°C, and then incubated for one hour at 40°C followed by 25°C overnight. The nematodes surviving 40°C were transferred individually to a lipid agar plate previously seeded with *Photorhabdus luminescens* bacteria, and incubated at 25°C. The resulting F2 infective juveniles were again subjected to the heat-exposure assay. Survival rate was more than 90%, whereas only 4 to 5% of progeny collected from the uninjected controls were capable of exposure to 40°C (Figure 4). Nematodes surviving 40°C were used to infect an insect host, *Galleria mellonella*, and further heat exposures were conducted on infective juveniles emerging from the host cadaver two to three generations later.

Induction of Hsp70 protein is believed to enable cells or organisms to recover from previous thermal stress, providing a degree of transient protection from subsequent heat challenge, and might have a fitness cost if expressed at normal temperatures (123). However, such a fitness cost has not been reported by the induction of Hsp70 above the normal physiological temperature (124). By genetic transformation, we have obtained a stable line of transgenic *H. bacteriophora* expressing the *hsp70* gene above normal growth temperature and conferring tolerance to heat stress. We have used the eight-cycle heating protocol described above to examine the frequency of selection for nematodes that express the *hsp70* gene. We emphasize that the heat resistance demonstrated in transgenic *H. bacteriophora* is not a consequence of repeated heat treatment. Transgenic *H. bacteriophora* show several copy numbers of the introduced *hsp70* genes compared with a single copy in wild-type populations in Southern blot analysis (unpublished data). Moreover, the survival of our transgenic nematodes in each heating cycle was 15- to 18-fold greater than in wild types. By incorporating the *hsp70* gene into *H. bacteriophora*, we have produced a transgenic nematode with enhanced high-temperature tolerance. More importantly, we have opened the door to the introduction of other genes of interest.

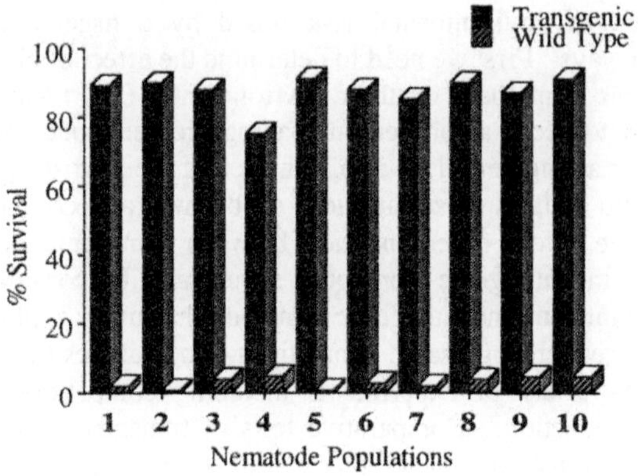

Figure 4. Expression of *C. elegans hsp70* gene confers thermal resistance to *H. bacteriophora* HP88. F1 generation of both wild-type and transformed infective juveniles of the nematode were exposed to 35°C for two hr, acclimated for 1 hr at 25°C and then exposed to 40°C for one hr, and survivals were determined after 24 hr.

TRANSGENIC NEMATODES AND RISK

We have developed three lines of transgenic *H. bacteriophora*: RRGEx1[gen-1(*hsp16*)], RRGEx2[gen-2(*hsp70*)] and RRGEx3[gen-3(*mec4/gfp*)]. These are the first transgenic insect parasites. The heat-shock protein gene *hsp70* of *C. elegans* was introduced to increase the heat tolerance of these nematodes. Laboratory studies to compare the heat-shock response of transformed and wild-type nematodes indicated that transformation has rendered these nematodes eighteen times more tolerant of high temperatures (40°C) than the untransformed wild-type strain (HP88). Because *hsp* genes may also impart cold-hardiness (125), overexpression of *hsp70* in our strains may provide dual temperature tolerance. We anticipate that these transgenic nematodes will show increased field persistence and efficacy in an increased range of environments. We are presently conducting efficacy experiments in a P3 containment greenhouse to test this hypothesis, but true validation will require field releases. In short, despite the exemption from government registration these parasites enjoy in most countries, an assessment of risk is required before any release of transgenic nematodes can be seriously considered.

Characterization will be the first stage in assessing transgenic strains. That is, how do the transgenic nematodes differ from the parental HP88 strain? Does the genetic modification affect nematode growth, survival and host range? Special emphasis must be placed on determining whether infectivity and host range have been altered. It is also important to understand precisely the relationship between the introduced gene and the resulting effect on the transformed nematode in terms of fitness cost. For example, heat-shock protein genes (*hsp70*) express at temperatures above the normal physiological temperature in many organisms to provide them protection from heat. This protein might have a fitness cost if expressed at normal temperatures. Feder et al. (123) reported that overexpression of Hsp70 at normal temperature conferred a deleterious effect to *Drosophila* cell growth in tissue culture. The expression of Hsp70 at normal growth temperatures reduces development but the overexpression at elevated temperature might be useful for certain organisms. Understanding the nature of the cost of a particular gene product at the molecular, biochemical and physiological level is clearly essential to the production of an improved nematode.

The assessment of environmental risk posed by transgenic nematodes should be approached in two ways. First we need to determine the effect of alterations to the biology of entomopathogenic nematodes on their environmental fate. Although horizontal gene transfer is unlikely to occur among entomopathogenic nematode species, this could be a rare possibility over a long period of time, considering their broad spectrum of insecticidal activity. We should address these questions on behavioral, ecological and physiological bases. Second, the effect of competition between transgenic nematodes and natural populations of entomopathogenic nematodes should also be considered. The successful introduction of *gfp* into entomopathogenic nematodes has many applications. For instance, it will allow us to create recognizable populations of nematodes (nematode tagging) which can be used in a variety of experiments including competition assays for survival, infectivity and reproduction. Comparative tests of transgenic and wild-type nematodes would be the primary role of *gfp* for this work. However, it will also enable classical mark/recapture type experiments which were impossible without such a marker.

Concern exists about the possible dangers from spread of transgenes to other species in the environment. However, the excellent safety record of entomopathogenic nematodes and considerable interest in reducing use of chemical pesticides has increased acceptance of new approaches.

CONCLUSIONS

Recombinant DNA technology shows early promise for the rational improvement of entomopathogenic nematodes. Few technical obstacles exist in this field of research where efficient new transformation methods, effective visible markers, transposons and transgenic strains with enhanced stress tolerance have already been generated. The availability of *C. elegans* as a genetic and molecular model obviously provided the foundation for these advances. Progress will accelerate as additional laboratories envision the prospect of "tailoring" insect nematodes specific for a particular pest and environment. Because the full potential of this approach is dependent on the availability of cloned genes of interest, heavy initial emphasis is likely to be placed on isolating new genes, particularly those of entomopathogenic nematode origin, encoding for important traits (e.g., anti-immune proteins). The ability to insert transposons into particular sites along the chromosomes is certain to be developed, making site-directed mutagenesis feasible, as well as stable genomic transformation. All this, in turn, will inevitably lead to the need for more information on classical and molecular genetics. Moreover, the development of transgenic strains will provide instruments for the study of insect nematode biochemistry, physiology, and even ecology.

Despite our enthusiasm for the powerful new molecular tools now available for the genetic improvement of entomopathogenic nematodes, we recognize that recombinant DNA technology offers no panacea. One has to look no further than *B. thuringiensis*, where after more than a decade of intense research effort in university and industry laboratories, a wild-type strain, *B. thuringiensis* var. *kurstaki*, remains the commercial leader. We are also acutely aware that not all of the limitations faced by entomopathogenic nematodes are resolvable by genetic engineering. The effective field use of these fascinating parasites will continue to rely heavily on behavioral and ecological research that identifies optimal temporal and spatial opportunities for application (9). Grower attitudes concerning biologicals are not amenable to engineering. These challenges notwithstanding, initial success in the genetic engineering of entomopathogenic nematodes is a significant milestone in the sixty-year effort to develop insecticidal nematodes, and an extremely promising development for the future.

ACKNOWLEDGMENTS

We thank Drs. Ghazala Hashmi, Moeen AbuHatab, Ed Lewis, and Robin Stuart for their helpful suggestions.

REFERENCES

1 Martin, P.A.W. (1994) Amer. Ent. 40, 85-90.
2 Poinar, G.O., Jr. (1993) Fund. Appl. Nematol. 16, 33-338.
3 Gaugler, R. and Kaya, H.K. (eds.) (1990) Entomopathogenic Nematodes in Biological Control. CRC Press, Boca Raton, FL.
4 Gaugler, R., Campbell, J., Selvan, M. and Lewis, E. (1992) Biol. Contr. 2, 181-187.
5 Gaugler, R. and Boush, G.M. (1979) Environ. Entomol. 8, 658-660.
6 Grewal, P. and Georgis, R. (1994) pp. 126-130, Proc. IVth Intern. Colloq. Invertebr. Pathol., 28 August-2 September, Montpellier.
7 Molyneux, A.S. (1985) Revue de Nematol. 8, 165-170.
8 Gaugler, R. and Boush, G.M. (1979) J. Invertebr. Pathol. 33, 121-123.
9 Georgis, R. and Gaugler, R. (1991) J. Econ. Entomol. 84, 713-720.
10 Silver, S.C., Dunlap, B. and Grove, D.I. (1995) Intern. Patent WO 95/05077.
11 Curran, J. (1993) in Nematodes and the Biological Control of Insect Pests (Bedding, R., Akhurst, R. and Kaya, H., eds.) pp. 67-77, CSIRO, East Melbourne.
12 Gaugler, R. and Georgis, R. (1991) Bio. Contr. 1, 269-274.
13 Kaya, H.K., Bedding, R.A. and Akhurst, R.J. (1993) in Nematodes and the

Biological Control of Insect Pests (Bedding, R., Akhurst, R. and Kaya, H., eds.) pp. 1-10, CSIRO, East Melbourne.
14 Hominick, W.M., Reid, A.P., Bohan, D.A. and Briscoe, B.R. (1995) Biocontr. Sci. Technol. (in press).
15 Gaugler, R. (1988) Agric. Ecos. Environ. 24, 351-360.
16 Zioni (Cohen-Nissan), S., Glazer, I. and Segal, D. (1992) J. Nematol. 24, 359-364.
17 Hoy, M.A. (1990) in New Directions in Biological Control (Baker, R. and Dunn, P., eds.) pp. 405-417, UCLA Symp. Mol. Cell. Biol., New Series, Alan R. Liss Press, New York, NY.
18 Hoy, M.A. (1992) Exp. Appl. Acarol. 14, 393-416.
19 Gaugler, R., Campbell, J.F. and McGuire, T.R. (1989) J. Invertebr. Pathol. 54, 363-372.
20 Gaugler, R., McGuire, T.R. and Campbell, J.F. (1989) J. Nematol. 21, 247-253.
21 Gaugler, R. and Campbell, J.F. (1991) Environ. Entomol. 20, 700-706.
22 Gaugler, R., Campbell, J.F. and Gupta, P. (1991) J. Invertebr. Pathol. 57, 234-241.
23 Gaugler, R., Glazer, I., Campbell, J.F. and Liran, N. (1994) J. Invertebr. Pathol. 63, 68-73.
24 Grewal, P., Keil, C., Tomalak, M. and Gaugler, R. (1993) Ann. Appl. Biol. 123, 695-702.
25 Selvan, S., Gaugler, R. and Campbell, J. (1993) J. Econ. Entomol. 86, 353-359.
26 Gaugler, R. (1993) in Nematodes and the Biological Control of Insect Pests (Bedding, R., Akhurst, R. and Kaya, H., eds.) pp. 89-95, CSIRO, East Melbourne.
27 Gaugler, R., Campbell, J.F. and McGuire, T.R. (1990) J. Invertebr. Pathol. 56, 106-116.
28 Brenner, S. (1974) Genetics 77, 71-94.
29 Rahimi, F.R., McGuire, T.R. and Gaugler, R. (1993) Heredity 84, 475-478.
30 Koltai, H., Glazer, I. and Segal, D. (1994) J. Nematol. 26, 32-35.
31 Gaugler, R. (1987) in Biotechnological Advances in Invertebrate Pathology (Maramorosch, K., ed.) pp. 457-484, Academic Press, New York, NY.
32 Hoy, M.A. (1976) Environ. Entomol. 5, 833-839.
33 Theodore, L., Ito, A. and Maroni, G. (1991) Genet. Res. 58, 203-210.
34 Phillips, J.P., Xin, J.H., Kirby, K., Milne, C.P., Krell, P. and Wild, J.R. (1990) Proc. Nat. Acad. Sci. U.S.A. 87, 8155-8159.
35 Hashmi, S., Hashmi, G. and Gaugler, R. (1995) J. Invertebr. Pathol. 66, 293-296.
36 Hashmi, S., Ling, P., Hashmi, G., Reed, M., Gaugler, R. and Trimmer, W. (1995) BioTechniques 19, 766-770.
37 Rancourt, D.E., Peters, I.D., Walker, V.K. and Davies, P.L. (1990) Bio/Technology 8, 453-457.
38 Rancourt, D.E., Davies, P.L. and Walker, V.K. (1992) Biochim. Biophys. Acta 1129, 188-194.
39 AbuHatab, M., Selvan, S. and Gaugler, R. (1995) J. Invertebr. Pathol. 66, 125-130.
40 DeBlock, M., Herrera-Estrella, L., Van Montagu, M., Schell, J. and Zambryski, P. (1984) EMBO J. 3, 1681-1689.
41 Powell-Abel, P.A., Nelson, R.S., De, B., Hoffman, N., Rogers, S.G., Fraley, R.T. and Beachy, R.N. (1986) Science 232, 738-743.
42 Gasser, C.S. and Fraley, R.T. (1989) Science 244, 1293-1299.

43 Iatrou, K. and Meidinger, R.G. (1990) Proc. Nat. Acad. Sci. U.S.A. 87, 3650-3654.
44 Luckow, V.E. (1991) in Recombinant DNA Technology and Applications (Prokop, A., Bajpai, R.K. and Ho. C, eds.) pp. 97-152, McGraw Hill, New York, NY.
45 Yu, Z., Podgwaite, J.D. and Wood, H.A. (1992) J. Gen. Virol. 73, 1509-1514.
46 Poinar, G.O., Jr. and Jansson, H.-B. (eds.) (1988) Diseases of Nematodes. CRC Press, Boca Raton, FL.
47 Potter, H., Weir, L. and Leder, P. (1984) Proc. Nat. Acad. Sci. U.S.A. 81, 7161-7165.
48 Bellofatto, V. and Cross, G.A.M. (1989) Science 244, 1167-1169.
49 Laban, A. and Wirth, D.F. (1989) Proc. Nat. Acad. Sci. U.S.A. 86, 9119-9123.
50 Clayton, C.E., Fueri, J.P., Itzhaki, J.E., Bellofatto, V., Sherman, D.R., Wisdom, G.S., Vijayasarathy, S. and Mowatt, M.R. (1990) Mol. Cell. Biol. 10, 3036-3047.
51 Sulston, J. and Hodgkin, J. (1988) in The Nematode *Caenorhabditis elegans* (Wood, W.B., ed.) pp. 587-606, Cold Spring Harbor Laboratory, Cold Spring Harbor, NY.
52 Klein, T.M., Wolf, E.D., Wu, R. and Sanford, J.C. (1987) Nature 327, 70-73.
53 Sanford, J.C. (1988) Trends Biotechnol. 6, 229-302.
54 McCabe, D.E., Swain, W.F., Marinell, B.J. and Christou, P. (1988) Bio/Technology 6, 923-926.
55 Finer, J.J. and McMullen, M.D. (1990) Plant Cell Rep. 8, 586-589.
56 Morikawa, H., Lida, A. and Yamada, Y. (1989) Appl. Microbiol. Biotechnol. 31, 320-322.
57 Oard, J.H., Paige, D.F., Simmonds, J.A. and Gradziel, T.M. (1990) Plant Physiol. 92, 334-339.
58 Baldarelli, R.M. and Lengyel, J.A. (1990) Nucl. Acids. Res. 18, 5903-5904.
59 Zelenin, A.V., Alimov, A.A., Barmintzev, V.A., Beniumov, A.O., Zelenina, I.A., Krasnov, A.M. and Kolesnikov, V.A. (1991) FEBS Lett. 287, 118-120.
60 Costanzo, M.C. and Fox, T.D. (1988) Genetics 120, 667-670.
61 Kaeppler, H.F., Gu, W., Somers, D.A., Rines, H.W. and Cockburn, A.F. (1990) Plant Cell Rep. 9, 415-418.
62 Ozato, K., Inoue, K. and Wakamatsu, Y. (1989) Zool. Sci. 6, 445-457.
63 Tamiya, E., Sugiyama, T., Masaki, K., Hirose, A., Okoshi, T. and Karube, I. (1990) Nucl. Acids Res. 18, 1072.
64 McGrane, V., Carlson, J.O., Miller, B.R. and Beaty, B.J. (1988) Amer. J. Trop. Med. Hyg. 39, 502-510.
65 Morris, A.C., Eggleston, P. and Crampton, J.M. (1989) Med. Vet. Entomol. 3, 1-7.66.
66 Milne, C.P., Jr., Phillips, J.P. and Krell, P.J. (1988) J. Apicult. Res. 27, 84-89.
67 Kimble, J., Hodgkin, J., Smith, T. and Smith, J. (1982) Nature 299, 456-458.
68 Waterston, R.H. (1981) Genetics 97, 307-325.
69 Wills, N., Gesteland, R.F., Karan, J., Barnett, L., Botten, S. and Waterston, R.H. (1983) Cell 33, 575-583.
70 Fire, A. (1986) EMBO J. 5, 2673-2680.
71 Southern, E.M. (1975) J. Mol. Biol. 98, 503-517.
72 Stinchcomb, D.T., Shaw, J.E., Carr, S.H. and Hirsh, D. (1985) Mol. Cell. Biol. 5, 3483-3496.
73 Mello, C.C., Kramer, J.M., Stinchcomb, D. and Ambros, V. (1991) EMBO J. 10, 3959-3970.
74 Kramer, J.M., French, R.P., Park, E.C. and Johnson, J.J. (1990) Mol. Cell. Biol. 10, 2081-2089.

75 Fire, A., White-Harrison, S. and Dixon, D. (1990) Gene 93, 189-198.
76 McClintock, B. (1951) Cold Spring Harbor Symp. Quant. Biol. 16, 13-47.
77 Finnegan, D.J. and Fawcett, D.H. (1986) Oxford Surv. Eukaryotic Genes 3, 1-62.
78 Spradling, A.C. and Rubin, G.M. (1981) Annu. Rev. Genet. 15, 219-264.
79 Emmons, S.W., Yesner, L., Ruan, K.S. and Katzenberg, D. (1983) Cell 32, 55-65.
80 Liao, L., Rosenzweig, B. and Hirsh, D. (1983) Proc. Nat. Acad. Sci. U.S.A. 80, 3585-3589.
81 Emmons, S.W. (1988) in The Nematode *Caenorhabditis elegans* (Wood, W.B., ed.) pp. 47-79. Cold Spring Harbor Laboratory, Cold Spring Harbor, NY.
82 Moerman, D.G. and Waterston, R.H. (1989) in Mobile DNA (Berg, D.E. and Howe, M.M., eds.) p. 537. Amer. Soc. Microbiol., Washington, DC.
83 Rosenzweig, B., Liao, L. and Hirsh, D. (1983) Nucl. Acids Res. 11, 4201-4209.
84 Henikoff, S. (1992) New Biol. 4, 382-388.
85 Collins, J., Forbes, E. and Anderson, P. (1989) Genetics 121, 47-55.
86 Herman, R.K. (1988) in The Nematode *Caenorhabditis elegans* (Wood, W.B., ed.) pp. 17-45. Cold Spring Harbor Laboratory, Cold Spring Harbor, NY.
87 Moerman, D.G., Benian, G. and Waterston, R. (1986) Proc. Nat. Acad. Sci. U.S.A. 86, 2579-2583.
88 Zwaal, R.R., Broeks, A., Van-Meurs, J., Groenen, J.T. and Plasterk, R.H. (1993) Proc. Nat. Acad. Sci. U.S.A. 90, 7431-7435.
89 Spradling, A.C. and Rubin, G.M. (1982) Science 218, 341-347.
90 Rubin, G.M. and Spradling, A.C. (1982) Science 218, 348-353.
91 Abad, P., Quiles, C., Tares, S., Piotte, C., Castagnone-Sereno, P., Abadon, M. and Dalmasso, A. (1991) J. Mol. Evol. 33, 251-258.
92 Anderson, P., Emmons, S.W. and Moerman, D.G. (1992) in The Dynamic Genome (Federoff, N. and Botstein, D., eds.) pp. 319-333, Cold Spring Harbor Laboratory, Cold Spring Harbor, NY.
93 Jefferson, R.A., Kavanagh, T.A. and Bevan, M.W. (1987) EMBO J. 6, 3901-3907.
94 Helmer, G., Casadaban, M., Bevan, M., Kayes, L. and Chilton, M.D. (1984) Bio/Technology 2, 520-527.
95 Sihavy, T.J. and Beckwith, J.R. (1985) Microbiol. Rev. 49, 398-418.
96 Beckwith, J.R. and Zipser, D. (eds.) (1970) The Lactose Operon. Cold Spring Harbor Laboratory, Cold Spring Harbor, NY.
97 Goring, D.R., Rossant, J., Clapoff, S., Breitman, M.L. and Tsui, L.C. (1987) Science 235, 456-458.
98 Ow, D., Wood, K.V., Deluca, M., DeWet, J.R., Helinski, D.R. and Howell, S.H. (1986) Science 234, 856-859.
99 Koncz, C., Olsson, O., Langridge, W.H.R., Schell, J. and Szalay, A.A. (1987) Proc. Nat. Acad. Sci. U.S.A. 84, 131-135.
100 Chalfie, M., Tu, Y., Euskirchen, G., Ward, W.W. and Prasher, D.C. (1994) Science 263, 802-805.
101 Hashmi, S., AbuHatab, M. and Gaugler, R. (1995) p. 27. SIP 28 Annu. Meet. 16 July-21 July, Ithaca, NY.
102 Saiki, R.K. (1990) in PCR Protocols (Innis, M.A., Gelfand, D.H., Sninsky, J.J. and White, T.J., eds.) pp. 13-20. Academic Press, San Diego, CA.
103 Russnak, R.H. and Candido, E.P.M. (1985) Mol. Cell. Biol. 5, 1268-1278.
104 Jones, D., Russnak, R.H., Kay, R.J. and Candido, E.P.M. (1986) J. Biol. Chem. 261, 12006-12015.

105 Poinar, G.O., Jr. (1990) in Entomopathogenic Nematodes in Biological Control (Gaugler, R. and Kaya, H.K., eds.) pp. 23-61. CRC Press, Boca Raton, FL.
106 Stringham, E.G., Dixon, D.K., Jones, D. and Candido, E.P.M. (1992) Mol. Biol. Cell 3, 221-233.
107 Ashburner, M. and Bonner, J.J. (1979) Cell 17, 241-254.
108 Schlesinger, M.J., Ashburner, M. and Tissieres, A. (eds.) (1982) Heat Shock: From Bacteria to Man. Cold Spring Harbor Laboratory, Cold Spring Harbor, NY.
109 Lindquist, S. (1986) Annu. Rev. Biochem. 55, 1151-1191.
110 Lindquist, S. and Craig, E.A. (1988) Annu. Rev. Genet. 22, 631-677.
111 Ritossa, F. (1962) Experientia 18, 511-573.
112 Russnak, R.H., Jones, D. and Candido, E.P.M. (1983) Nucl. Acids Res. 11, 3187-3205.
113 Snutch, T.P. and Baillie, D.L. (1983) Can. J. Biochem. Cell Biol. 61, 480-487.
114 Fire, A. (1992) Gata 9, 151-158.
115 Hashmi, G., Hashmi, S., Selvan, S., Grewal, P. and Gaugler, R. (1995) pp. 26-27, SIP 28 Annu. Meet. 16 July-21 July, Ithaca, NY.
116 Lindquist, S. (1980) Develop. Biol. 77, 463-479.
117 Lindquist, S. (1980) J. Mol. Biol. 137, 151-158.
118 Pelham, H.R.B. (1985) Trends Genet. 1, 31-35.
119 Mues, G.I., Munn, T.Z. and Raese, J.D. (1986) J. Biol. Chem. 261, 874-877.
120 Craig, E.A. (1989) BioEssays 11, 48-52.
121 Snutch, T.P., Heschl, M.F.P. and Baillie, D.L. (1988) Gene 64, 241-255.
122 Heschl, M.F.P. and Baillie, D.L. (1989) DNA 8, 233-243.
123 Feder, J.H., Rossi, J.M., Solomon, J., Solomon, N. and Lindquist, S. (1992) Genes Dev. 6, 1402-1413.
124 Coleman, J.S., Heckathorn, S.A. and Hallberg, R.L. (1995) TREE. 10, 305-306.
125 Petersen, N.S., Young, P. and Burton, V. (1990) Insect Biochem. 20, 679-684.
126 Hall, L.M.C. and Spierer, P. (1986) EMBO J. 5, 2949-2954.
127 Wu, C.T., Budding, M., Griffin, M.S. and Croop, J.M. (1991) Mol. Cell. Biol. 11, 3940-3948.

THE STOP SIGNAL CONTROLS THE EFFICIENCY OF RELEASE FACTOR-MEDIATED TRANSLATIONAL TERMINATION

Warren P. Tate, Mark E. Dalphin, Herman J. Pel and Sally A. Mannering

Department of Biochemistry and the Centre for Gene Research
University of Otago
P.O. Box 56
Dunedin, New Zealand

INTRODUCTION

There are three important steps in protein synthesis where signals in the mRNA are critical for a successful outcome, namely the production of a functional protein. First the information in the nucleic acid which is to be translated into an amino acid sequence is signalled by successive triplet sense codons, second the frame is set by one sense codon, the initiation codon, which acts as the start of translation of the encoded information, and third the end of the information frame also has to be marked by a specific signal. The use of a range of different signals to mark each of these steps allows for differences in the efficiency with which different proteins are produced. In this review the focus is on the signal that marks the end of the frame, the translational termination signal. For a long time it was thought that termination would be the least interesting phase of protein synthesis but it has subsequently been found to have unexpected dimensions, providing a substratum of cellular regulation. The translational stop signal should now be thought of as a full stop in the large majority of cases, but as a pause in a fundamentally important minority of cases where alternative genetic events can occur.

CODON UTILIZATION

Initiation Signals

For the large majority of proteins the frame of translation is set by the one codon, AUG, and therefore other factors determine the efficiency of initiation. In prokaryotes the initiation codon selection by the ribosome is influenced by a polypurine domain (1) at an average of 7 nucleotides upstream from the codon (2). The spacing and the exact sequence vary significantly among mRNAs. Recently an optimal spacing between a reference point

---

Abbreviations: RF: Release Factor, eRF: Eukaryotic Release Factor, mRF: Mitochondrial Release Factor, 5′-DI: type I iodothyronine 5′-deiodinase, rRNA: Ribosomal RNA, mRNA: Messenger RNA, tRNA: Transfer RNA.

in the Shine-Dalgarno sequence and the initiation codon was determined. This optimal spacing probably reflects the preferred spacing between the anti-Shine-Dalgarno sequence in the rRNA and the tRNA in the P site (3). The critical point is that each mRNA has its unique brand of Shine-Dalgarno sequence and spacing and so initiation rate constants for individual mRNAs have the potential to vary significantly. Other *cis* and *trans* elements can influence initiation (4). Additionally, codons other than AUG may be used in rare cases to signal initiation. For example, AUU is found at the beginning of the coding information of the *infC* gene encoding initiation factor IF-3 and this unusual codon limits the expression of the protein (5).

In eukaryotes, the first AUG in the mRNA is generally the start codon and initiation rate constants are largely determined by events before the small ribosomal subunit sees this codon, for example the attachment of the small ribosomal subunit to the 5′ end of the mRNA and the scanning of the 5′ untranslated region. Nevertheless, there is an optimal context for the AUG to be recognized as the start of the coding information, and in suboptimal contexts (where the -3 nucleotide is not a purine and the base following the codon is not a guanine) there may be leaky scanning and the next AUG may be used (6).

Sense Codon Usage

Most amino acids have more than one codon specifying them and codon usage is neither uniform nor random. Organisms use synonymous codons (different codons representing the same amino acid) in particular ways. For example, in *Escherichia coli* highly-expressed genes have a marked bias in their sense codons, with those represented being the signals recognized by the most abundant tRNAs (7). Kurland believes that codon preference in this organism reflects a strategy to maximize the efficiency of translation at the fastest growth rates. The concentration of tRNAs that translate major codons increases while that of the others decreases at higher growth rates, so that a shift in the codon bias of the mRNAs is matched by isoacceptor tRNA bias. Indeed it may be the tRNA population, and therefore the ratio of unfavorable to favorable codons, that influences translation rates (8). While optimal codons are translated faster than others it does not necessarily reflect the overall rate of production of the particular protein. Clearly the rate of initiation of translation is a major determinant as to whether a particular mRNA is able to sequester a ribosome in a highly competitive manner, and this is irrespective of how many optimum sense codons it possesses. Collins et al. (9) have found that rare codons tend to be clustered in the first 10 codons, independent of the length of the protein and whether it is destined for export or is cytoplasmic, as though it is important generally to have a relatively slow initial rate of elongation (9). There may be more specialized use of rare codons to regulate against the inappropriate expression of genes; in these cases the upregulation of a decoding tRNA may activate expression at the appropriate time (10). Certainly clusters of the rare AGG arginine codon, particularly towards the beginning of the coding region, can significantly reduce the level of expression of a protein in *E. coli* (11).

Mutational biases reflecting the overall G+C content of the genome are also a major influence on the codon usage in an organism, particularly where this content is well away from ~50%. For example, in *Mycoplasma capricolum* almost all of the codons end in A or U and the G+C content of the genome is only 25%; conversely, *Micrococcus luteus* has a genomic G+C content of 75% and most of its codons end in G or C (7). In eukaryotes the codon usage patterns vary depending on the type of organism. For organisms with large effective population sizes, such as yeast and *Drosophila*, codon usage seems to be determined by translational selection rather like *E. coli,* whereas in humans codon usage reflects the base composition of the region of the chromosome in which the gene is found. This has been explained by the isochore theory which suggested that the human genome is a mosaic of regions of differing G+C content (12).

Hence, codon usage for the sense codons reflects both mutational biases among species and within a genome of a particular species, and translational selection. Mutational bias may be the dominant factor where the effective population size is modest.

The Stop Signal

The focus of our work has been on the translational stop signal. Just as several sense codons can specify the same amino acid, translational termination of protein synthesis is specified by one of three codons, UAA, UGA and UAG. An obvious question was whether there was any bias in how these three codons are used. This could be relatively easily answered, as discussed below, once the databases began to expand rapidly with sequence information from genes isolated from a number of different organisms. A more difficult question was whether the stop codons as displayed in the genetic code table represented the complete signal for translational stop. The triplet code was completed when the three codons that were unassigned to amino acids were shown to be signals for translational termination (13,14). Although the code table represents a compelling display of this genetic information in that it is accessible and easy to interpret, this format for displaying the stop signals precluded the possibility that they might be larger than the three bases specified. It soon became clear that stop codons were unlikely to be decoded by the simple triplet:triplet decoding scheme of the sense codons (15-17). The involvement of protein decoding molecules rather than tRNAs meant that the termination signal could be larger than a triplet and the wide variation in the efficiency by which stop codons were suppressed when introduced into different contexts hinted that the actual stop signal was not precisely defined by the triplet alone.

While the evidence became indisputable that the efficiency of translational stop codons could be influenced *in cis* by the surrounding sequence of the mRNA, at least outside of their natural contexts, it was not clear whether this was physiologically significant. Is this phenomenon simply indicating a variation in efficiency which does not interfere with normal stop codon function at natural sites? The suppression of a stop codon involves two competing events, aminoacyl-tRNA selection and peptidyl-tRNA hydrolysis, and it is hard to separate the effect of context on the tRNA-mediated event from the effect on the termination event. For example, the importance of the nucleotide immediately downstream of the codon was suggested from studies of the suppression of UGA (18) and of *supE,* a suppressor of UAG codons, while a further study of UAG and UGA suppression suggested the two bases downstream were important, and yet another study identified elements outside of the downstream nucleotides that were important (reviewed in 19).

Buckingham has carried out a systematic study of context changes introduced into a particular site containing a stop codon. The conclusions were that the nucleotide upstream of the codon, the nucleotide downstream of the codon, and the wobble nucleotide of the first downstream codon were all important (20). A critical question from these studies was whether the effects of sequence elements common to all sites were mediated through termination, whereas the effects of sequence elements specific for a particular site were mediated through the activity of the suppressor-tRNA (21). Pertinent to this question, Engelberg-Kulka and colleagues have shown that the efficiency of suppression of UGA stop codons is influenced by the base following, in the order A>G>C>U (18).

TRANSTERM DATABASE

We have approached the task of defining the stop signal by first establishing a database of the regions surrounding the stop codon in genes for which sequence information is available. The Translational Termination database (TransTerm), first established in 1993, initially consisted of the immediate context sequences around natural termination codons of 12,315 sequences from 45 organisms, together with the summary tables (22). Termination codon contexts were extracted with the use of the feature tables of Genbank entries for organisms which had over 40 valid sequences available. Frequencies of each stop codon from the different organisms were displayed, and since the data revealed an extraordinary bias in the base following the codon the frequencies of the tetranucleotide sequences based on a four-base termination signal model were included. Sequences of ten bases before (-10) and seven after (+10) the stop codon were also displayed. Sense codon bias of the coding region in each gene was estimated and listed according to the Codon Adaptation Index (CAI, a measure of sense codon bias and an indication of the level of expression) (23). Table 1 was constructed from data in TransTerm 1994. The 30 *E. coli* genes with the highest CAI values are listed along with their contexts at the stop codon. Displaying information in this way has enabled biases in stop codon and context to be obvious sometimes even at a cursory analysis. For example, in Table 1 it can be seen that TAA is the dominant stop codon used for the highest expression group and T at ~50% is the dominant +4 nucleotide (nucleotide immediately following the stop codon).

The TransTerm database was updated in 1994 to include initiation contexts and sense codon usage so that relationships could be investigated among the three types of signals. The 1994 database was constructed from 23,721 coding sequences from 93 organisms, and contained the sequence around the termination codon (-10 to +10), the sequence around the initiation codon (-20 to +10), the length, the G+C% of the third position of all codons (GC3), the CAI, and "the effective number of codons" statistic (Nc). Summary tables were included for each organism, which detail codon usage, stop codon and tetranucleotide stop signal usage, and matrices tallying base frequencies at each position around the initiation and termination codons (24). The TransTerm database is currently being updated again. Figure 1 illustrates the nucleotide biases around the initiation signals and termination signals for the selected set of *E. coli* genes with a CAI of >0.6 (representing ~100 genes or the top 5% according to expression levels). The nonrandomness, or χ^2 analysis, of the frequency of nucleotides in each position is shown; Figure 1A shows nucleotides -31 to +30 around the initiation signal, and Figure 1B, nucleotides -31 to +30 around the termination signal. The nonrandom distribution of nucleotides in the coding region is evident to the right of the start codon and to the left of the stop codon. The Shine-Dalgarno signal upstream from the initiation codon is clearly revealed by this analysis around the -10 position, as well as the bias in the nucleotide following the stop codon (the +4 position).

This database has been incredibly valuable to frame proposals about the translational stop signal that can then be tested experimentally. The first question to be asked was whether a particular stop codon was used more frequently and whether this varied from one organism to another. The analyses revealed that there was a bias in stop codon usage and that this bias varied from organism to organism. For example, in *E. coli* UAA is by far the predominant stop codon used (~60%), UGA is next (~30%) and UAG is underutilized (<10%) (25). The underutilization of UAG is accounted for in two ways, firstly TAG is scarce in the entire *E. coli* genome, and second there may be an evolutionary bias against its use as a termination signal. TAGN stop signals might be mistakenly altered to a sense codon by the mechanism of *vsr*-initiated DNA mismatched

Table 1 Selected *E. coli* data from TransTerm 1994[a]

STOP			LOCUS	CDS	CAI
ATACCGCAAG	TAA	TAGTACC	ECLIPOA	1	0.8488
TGAAGTTAAA	TAA	GCCAACC	ECOUW89	37	0.8410
CATCTCCAAA	TAA	GTTGAGA	ECGAP	1	0.8344
TTACCAGTTC	TAA	TCTCGAT	ECOOMPC	2	0.8195
AGTTCTGGGC	TAA	TTGCACG	ECOSTR3	2	0.8170
AACCAGCCTG	TAA	TCTCTGC	ECOCSPAA	1	0.8007
TGGTAAAATC	TAA	ATTTCCT	ECOD13187	1	0.7969
AGTTCTGAGC	TAA	TTGCCGA	ECOTGTUFB	1	0.7935
CGGCATGATG	TAA	TTGCCCT	ECGROESL	2	0.7863
TAAAGGCGAG	TAA	TTCTCTG	ECRSPA	2	0.7788
TCAATCTATG	TAA	TTAGATT	ECPFL	1	0.7734
AGAAGCTGAG	TAA	TAAGGCT	ECRPSB	1	0.7717
CGACGTTCTG	TAA	GATATTC	ECFDAPGK	6	0.7715
CAAGCAGTCT	TAA	TTATCAA	ECRPSB	2	0.7687
TTCTGTAAAC	TAA	TGCCTTT	ECOUW89	35	0.7675
GCCGCAGGCT	TAA	GTTCTCG	ECOMPA	2	0.7649
CGAAGCTGAA	TAA	CTTTTTC	ECORPLRPM	2	0.7421
TAAACAGGCT	TAA	GTCTGAC	ECOUW87	61	0.7419
GGGCAGCAAA	TAA	GACCGGA	ECOUW87	78	0.7385
CGTTCGCGCT	TAA	TTGCTGA	ECORPLY	1	0.7310
AGGCCGTAAA	TAA	GGAGCGT	ECRPOS10	8	0.7268
GAAAATCAAA	TAA	TTCTCGC	ECORPMBG	2	0.7265
CGCTAAGAAG	TAA	AAAATCA	ECFDAPGK	5	0.7265
AGTAGCTGAA	TAA	TTCGTTA	ECRPSFRI	4	0.7261
TCGTCTGTAC	TAA	TTCCCCG	ECORPSRPO	1	0.7260
CGTTTCTAAG	TAA	TAAAGCT	ECODNAAOP	1	0.7260
AGACAAAAAA	TAA	TCGCCCT	ECO110K	6	0.7171
TGTGAAGGCG	TAA	GGAGATA	ECRPOS10	3	0.7108
GGCGAAAAAA	TAA	TCATTTG	ECOUW87	51	0.7105
CCAGCAGGCG	TAA	TTTACGC	ECOTIG	1	0.7075

[a] The sequence data for the aligned *E. coli* termination regions were sorted on the CAI value for the whole coding region. The 30 sequences with the highest CAI values were selected and listed. The sequences were reduced to -10 to +10 nucleotides around the stop codon and spaces were added to the computer output to aid legibility. "STOP" indicates the stop codon. "LOCUS" is the GenBank Locus entry. "CDS" denotes a particular coding sequence in a Locus entry. That is, when the coding sequences are extracted from the GenBank record, all entries in the FEATURE TABLE that are listed as "mat_peptide" or "CDS" are extracted and processed. These are numbered sequentially from the start of the record, beginning with 1. "CAI" refers to the "codon adaptation index," a measure of how well the codon usage in a gene matches that found in highly-expressed genes in the same organism (23).

repair. The *vsr* gene product is an endonuclease which nicks one strand at a T:G mismatch, at the T in CTRG or TRGG sequences (26). This mechanism may be the cause of the low frequency of TAG sequences in the genome, and *E. coli* strains that used UAA or UGA signals for the majority of their genes might have had a selective advantage during evolution. In contrast to *E. coli*, UGA is the most common signal (~47%) in human genes, with UAA (~31%) and UAG (~22%), reflecting a more even distribution of the signals in this species. This distribution is similar to the occurrence of these triplets in the

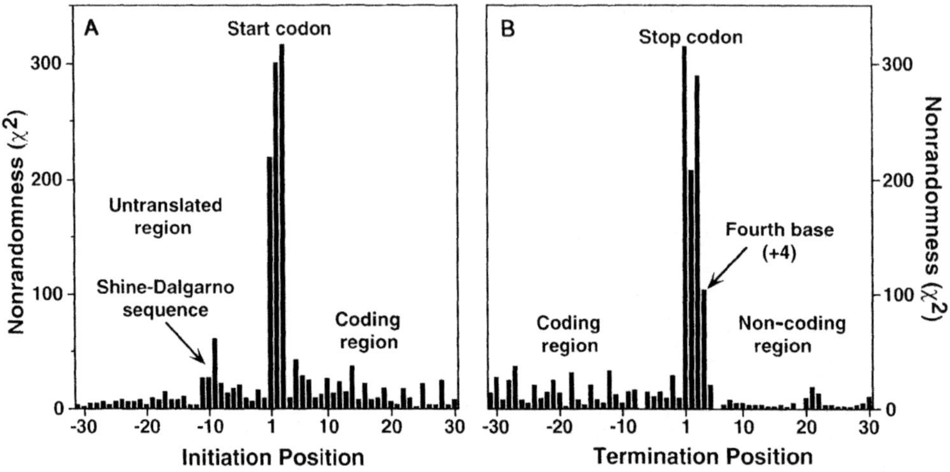

Figure 1. Statistical analysis of the context of *E. coli* initiation and termination codons. The sequences around A) the initiation codons (-31 to +30) and B) the termination codons (-31 to +30) from the highly-expressed *E. coli* genes in TransTerm (1994) with a CAI ≥ 0.6 (~5% of the total genes) were analyzed. The nonrandomness (χ^2) observed at each position was determined.

non-coding region immediately following the stop codon. It seems therefore that from the perspective of the mechanism of termination of protein biosynthesis all three codons are similarly acceptable.

The G+C content of the DNA can have a profound effect on which stop codons are used in a particular organism (27). As shown in Figure 2A, the use of UAA in a number of prokaryotic genomes correlates with the percentage of G+C base pairs in the DNA, varying from ~80% in an organism with only ~30% G+C, to as low as 3% in an organism with a G+C content of 73%. Of the two G-containing stop codons, UAG and UGA, UGA seems to be favored in organisms with higher G+C% DNAs (Figure 2B). It seems that mutation pressure is generally the strongest influence in determining which stop codon is favored, although for individual genes within the group there may be other factors such as efficiency of termination or suppressibility which become important.

AN ANALYSIS OF THE CONTEXTS OF NATURAL STOP CODONS

As indicated in Table 1 and Figure 1, with a selected group of genes from *E. coli*, analysis of the regions surrounding the stop codon hinted that the signal for translational termination might be larger than the triplet codon. Statistical analyses of the sequences around stop codons from large numbers of genes from other organisms showed a significant bias in the nucleotide immediately following the stop codon (22, 24, 28, 29). However, it was not always the same nucleotide that contributed to the 4th base bias in each organism, rather the specificity varied from one organism to another. For example in *E. coli* and *Bacillis subtilis* genes U was the most abundant 4th base, whereas in humans and rats A or G were predominant.

The pattern of nonrandomness surrounding stop codons from the 2,455 *E. coli* genes in our 1994 TransTerm database is shown in Figure 3A (compare with Figure 1B for the highly expressed genes). These genes represent about half of the total genes in this genome. While the most striking finding is the +4 nucleotide bias (the stop codon is +1 to +3), there is also a less striking nonrandomness in the nucleotides following the stop codon as far as position +10, as well as several positions preceding the codon, perhaps

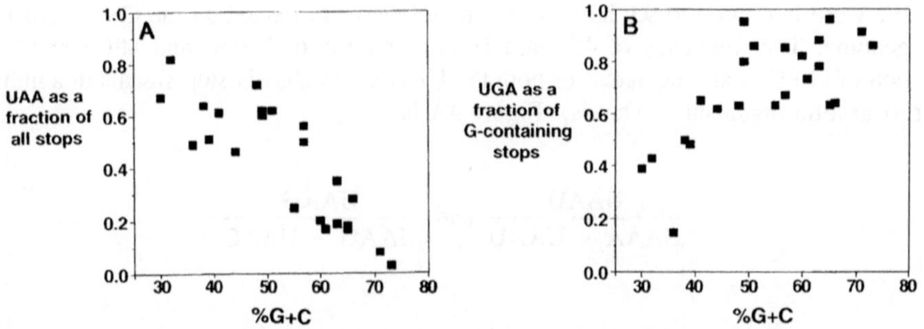

Figure 2. Variation of the stop codon use in genomes of different G+C percentages. A) The fraction of stop signals that are UAA in genes of organisms with genomes of G+C percentages from 30-75%. B) The fraction of G-containing stop signals (UAG and UGA) that are UGA in genes of organisms with genomes of G+C percentages from 30-75%.

indicating that these nucleotides also contribute in some way to the signal. Since the bias in the +4 position was the most significant, an analysis of this large dataset on the basis of a 4-base stop signal (stop codon plus the following nucleotide) was carried out. It revealed strong bias in the frequency of occurrence of particular 4-base stop signals (Figure 3B). Of the 12 possible 4-base signals (UAAN, UAGN, UGAN), UAAU and UAAG were much more frequent at stop signal sites in all genes than expected from their occurrence in non-coding regions, whereas the other 10 occur at the expected frequency or have apparently been selected against. A comparison with non-coding regions is important because the occurrence of these tetranucleotides in such regions is also nonrandom, reflecting a nonrandom distribution of di- tri- and tetranucleotides in the whole genome. Highly-expressed *E. coli* genes have a strong sense codon bias and it is possible to select this subset of genes from the main set for analysis of the 4-base signals. The rationale for considering this set of genes was that the codon bias indicates an evolutionary pressure for the optimization of translation elongation, and therefore they might be expected to use optimal termination signals. In this case the results are quite dramatic; most of the subset (~80%) use UAAU and UAAG, with the hierarchy of usage in this subset being UAAU > UAAG > UGAU > UAAA ~ UAAC and the other 6 are used rarely, if at all (24, 30).

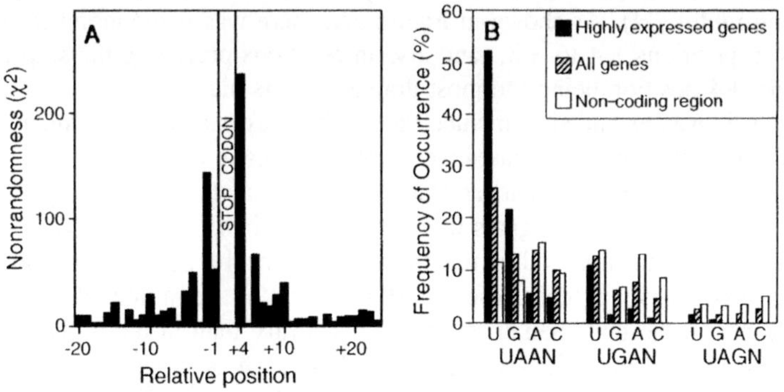

Figure 3. Statistical analysis of *E. coli* termination signals. A) The context of the stop codons. The sequences around the termination codons of 2,455 *E. coli* genes were analyzed and the nonrandomness (χ^2) observed at each position was determined. B) Frequency of occurrence of 4-base stop signals in *E. coli* highly-expressed genes, in all genes and in adjacent non-coding regions. The frequency of occurrence of UAAN, UGAN and UAGN signals is shown with the 4th base indicated for each block of data. Each block contains the frequency in highly-expressed genes (black bars), in all genes (striped bars) and in the 100 nucleotides of non-coding sequence adjacent to the coding region of all genes (white bars).

We were interested to see how the +4 base bias was affected by the G+C content of the genome. The frequency of 4th base U as a fraction of U+A, and 4th base G as a fraction of G+C, was calculated for both the UAAN and UGAN stop signals in a number of prokaryotic organisms. That is, Figure 4A is:

$$\frac{\text{UAAU}}{\text{UAAA} + \text{UAAU}} \text{ and } \frac{\text{UAAG}}{\text{UAAG} + \text{UAAC}}$$

while Figure 4B is:

$$\frac{\text{UGAU}}{\text{UGAA} + \text{UGAU}} \text{ and } \frac{\text{UGAG}}{\text{UGAG} + \text{UGAC}}$$

For both UAAN signals (Figure 4A) and UGAN signals (Figure 4B), the U and G preference for the +4 position generally held (relative number of points above the 0.5 line) irrespective of the organism (and G+C content of the genome). When the data were analyzed to see whether the preference for *either* U *or* G changed with the G+C content of the genome of the organism, perhaps surprisingly, the preferred +4 base of these "top two" was still mainly U for UAAN signals except for the higher G+C genomes (G+C> ~55%) (Figure 4C). For the UGAN signals, U was again preferred in the +4 position in the DNAs up to ~55% G+C, but the preferences were more distinct than for the UAAN signal and the switch from U to G at ~55% G+C can be easily seen in Figure 4D.

What is the situation with the base following the stop codon in eukaryotes? A small initial study (31), and more recent larger studies (29, 32, 33) of eukaryotic stop codon contexts suggested that purines were the most common 3′ nucleotide following the stop codon. The sequences around the stop codons of over 5,200 mammalian genes have now been extracted and compiled as a database representing six species: human, mouse, rat, cow, pig and rabbit (34). As shown in Figure 5A, there was nonrandomness found in the χ^2 analysis in positions +4 to +8, and also in positions preceding the stop codon (-1 to -3), with the +4 position being the most dramatic in its bias.

The analysis was extended to include the +4 base as part of the termination signal as shown in Figure 5B. While the biases were not as dramatic as those for the *E. coli* genes they were consistent with our earlier studies of yeast and *Drosophila* genes (29). The occurrence of the 4-base stop signals in all the mammalian genes reflected the frequency of these sequences in the non-coding regions, but G in the 4th position was more abundant, and U was less abundant than expected ($P<0.001$). A definition of a set of highly-expressed genes in the mammalian set is more arbitrary than in the prokaryotic case but the frequency of the 4-base signals in genes such as globins, immunoglobulins, histones, actins and albumins have been examined and found to be quite different from the complete dataset. Certain signals, UAAG, UAAA and UAGG, are overrepresented and certain signals, UGAU, UGAC, UAGU and UAGC, are not used at all. Individual analyses with the particular species subsets essentially gave the same information.

As seen in Figures 3A and 5A, there is nonrandomness in the nucleotides before the stop codon. This bias may be important for the stop signal and the termination mechanism

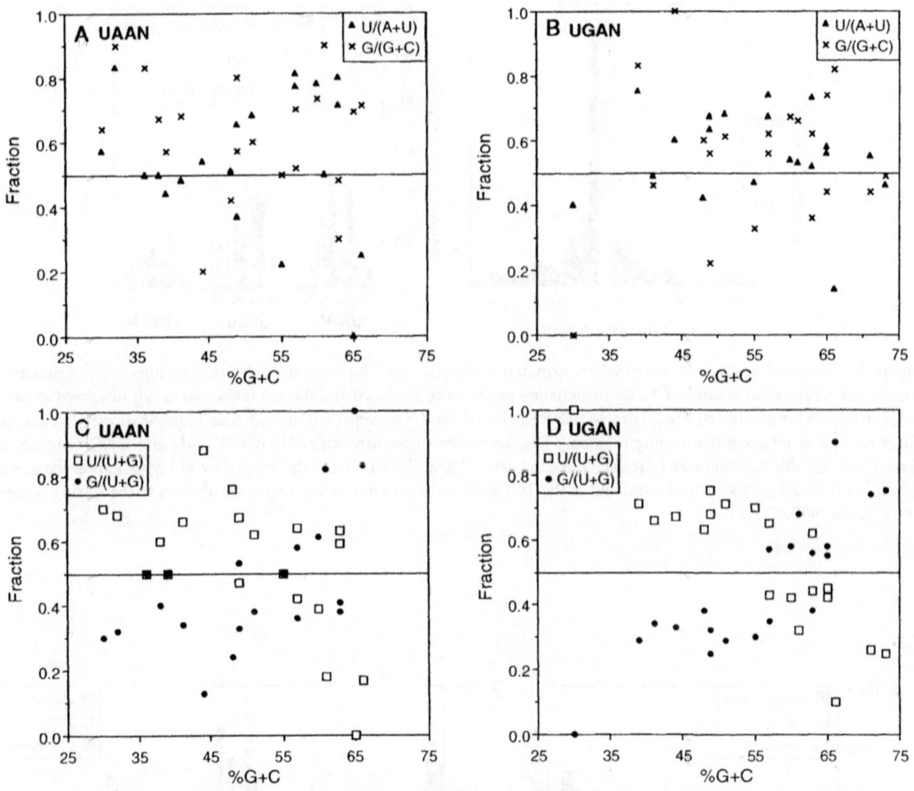

Figure 4. Statistical analysis of the base following the stop codons in prokaryotic genomes with different G+C percentages. A and B) The fraction of 4th base U relative to A+U and 4th base G relative to G+C is shown as a function of the G+C percentages of the genomes for UAAN signals (A) and UGAN signals (B), respectively.
C and D) The fraction of 4th base U relative to U+G and 4th base G relative to U+G is shown as a function of the G+C percentages of the genomes for UAAN signals (C) and UGAN signals (D), respectively.

itself, or it may have more to do with requirements for the end of the polypeptide chain, perhaps for its extrusion from the ribosome. Mottagui-Tabar et al. have shown that the last two amino acids in the nascent chain for *E. coli* are an important context determinant for the decoding of nonsense codons; readthrough of a UGA codon in a model system varied 30-fold (35). Arkov et al. have proposed that the significant bias in the -2 base position may reflect that the factor decoding the stop signal is able to interact with certain tRNAs (36). They have noted similarity between the protein sequence of *E. coli* release factor-2 (a decoding factor) and certain aminoacyl-tRNA synthetases. Figure 6A shows the occurrence of classes of amino acids encoded by the -1 to -6 nucleotides in *E. coli* genes. In both the penultimate and ultimate amino acid positions, there is bias towards a basic amino acid and bias against a large hydrophobic residue. There is also a substantial bias against a polar amino acid in the ultimate position. An analysis at the level of the individual amino acids is shown in Figure 6B, each amino acid being designated by the single letter code. Here it can be seen that lysine (K), arginine (R) and, to a lesser extent, histidine (H), are favored in the final two positions, and several other amino acids are not favored (D, I, L, M, P, V). The most dramatic result however, is the bias against the polar amino acid threonine (T) in the ultimate position. Clearly these biases are interesting

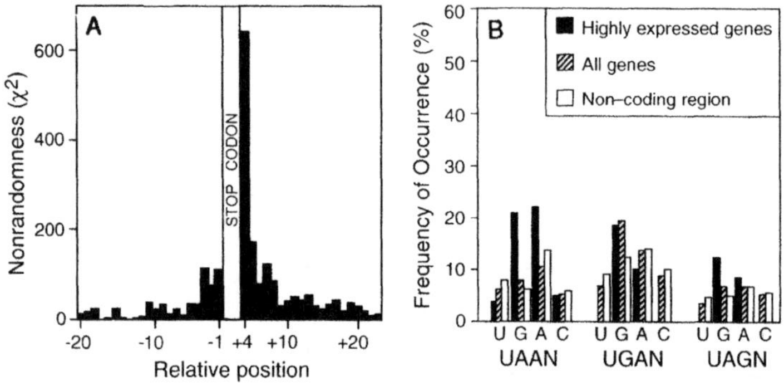

Figure 5. Statistical analysis of mammalian termination signals. A) The context of the stop codons. The sequences around the termination codons of 5,208 mammalian genes were analyzed and the nonrandomness (χ^2) observed at each position was determined. B) Frequency of occurrence of 4-base stop signals in mammalian highly-expressed genes, in all genes and in adjacent non-coding regions. The frequency of occurrence of UAAN, UGAN and UAGN signals is shown with the 4th base indicated for each block of data. Each block contains the frequency in highly-expressed genes (black bars), in all genes (striped bars) and in the 100 nucleotides of non-coding sequence adjacent to the coding region of all genes (white bars).

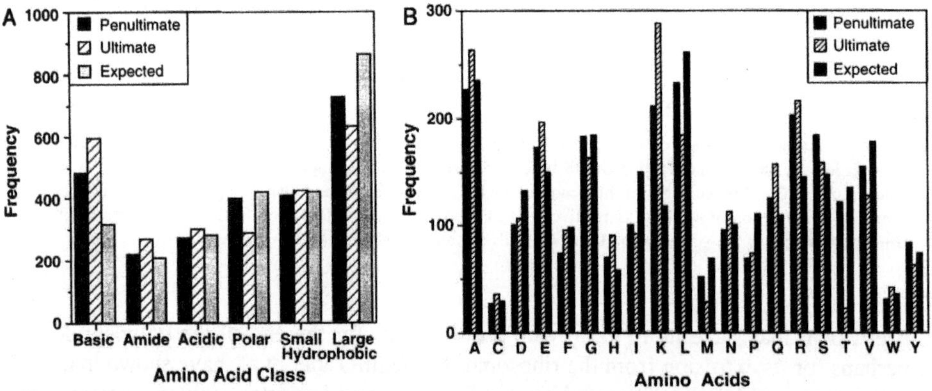

Figure 6. Statistical analysis of the frequency of amino acids encoded by the penultimate and ultimate codon positions of the genes from *E. coli*. The expected frequencies were determined from the average codon usage in *E. coli* genes. A) Frequency of the classes of amino acids encoded by the last two codons compared with that expected. The amino acids are grouped according to Cantor and Schimmel (95). B) Individual amino acids encoded by the last two codons compared with expected.

and it will take some innovative thinking to tease out whether it means anything specific for termination of protein synthesis. Are the -1 to -6 bases part of the termination signal, are the restrictions of the amino acids part of the overall termination mechanism, or do certain amino acids at the end of the chain indirectly cause a longer pause at the termination signal? In the latter case they may have been selected against during evolution because alternative events such as readthrough, frameshifting or peptidyl-tRNA drop off from the ribosome may be permitted if the decoding event occurs with slower kinetics.

IS THE +4 BASE BIAS SIGNIFICANT FOR TRANSLATIONAL TERMINATION?

The analyses of termination signals on a 4-base model suggested that a particular subset of termination signals might be used in special circumstances such as high expression. This led us to test experimentally whether the termination codon context bias had any functional significance for polypeptide chain termination *in vivo* for *E. coli*. The strengths of each of the 12 possible 4-base stop signals were tested in a specific context by the direct competition of termination with frameshifting (25) at the release factor-2 (RF-2) frameshifting sequence motif (37). For frameshifting to occur, the elements 5′ to the stop codon promoting the event must compete with translational termination at the stop codon. If the stop signal is altered in this system then the degree of frameshifting should be influenced by the efficiency of the sequence to signal stop. An immunological assay was used to measure the two products, termination and frameshift, synthesized *in vivo* from constructs varying only in their termination codon or 4th base (Figure 7A). The termination efficiencies of the UGAN, UAAN and UAGN series are shown in Figure 7B. We have calculated rates of release factor selection (38) at each of the signals and there is a 50-fold range, with UAAU the fastest and UGAC the slowest. From the statistical analysis alone UAGN signals might be predicted to be poor. This is not the case, rather they were as efficient as the other two sets. However, as indicated earlier, TAGN occurs with very low frequency throughout the whole genome.

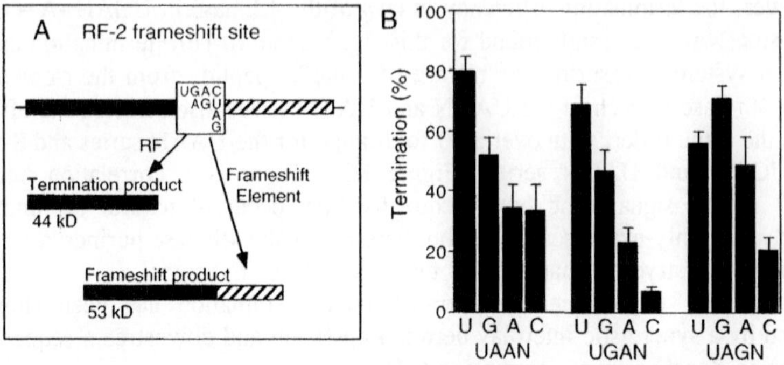

Figure 7. Influence of the 4th base in tetranucleotide stop signals on the efficiency of translational termination in *E. coli*. A) In *E. coli*, the relative strength of each of the UAAN, UGAN and UAGN signals was determined as a competitor of +1 frameshifting at the RF-2 frameshifting site. Each of the 12 possible 4-base stops was cloned into the frameshift site. The ribosome either terminated synthesis at the stop signal to give a 44 kD fusion protein, or frameshifted to give an extended product of 53 kD. B) The strength of each of the 4-base signals was determined from its ability to compete with frameshifting at the RF-2 frameshift site. The particular 4th base is shown under the block for each of the UAAN, UGAN and UAGN series of signals. The percentage termination is shown with each signal, and each bar is derived from several experiments, with the standard deviation indicated on the bar.

As discussed above, bacteria at high growth rates modulate their tRNA populations to match the codons in the highly-expressed mRNAs which predominate under these growth conditions. This is a form of energy minimization for the elongation phase of protein biosynthesis (8). How does the termination phase minimize energy expenditure in fast-growing bacteria? There were at least two attractive possibilities. Bacteria could increase both the synthesis and use of only one of the two protein release factors (the tRNA analogs that decode stop signals), and this would necessitate using termination signals for the highly-expressed genes that were recognized by this factor. This would be analogous to the situation with elongation. Alternatively, the termination signal itself could allow for efficient decoding without the need to significantly raise release factor concentrations in the cell (and thereby saving energy expenditure). We have distinguished these two possibilities; both release factors are coordinately upregulated at fast growth rates but only enough to maintain cellular concentrations, since the cell volume changes under these conditions (39). It is inferred then that the highly-expressed genes maximize their termination kinetics by using what we have determined experimentally to be the most efficient stop signals at selecting the release factors (for example UAAU).

We have tested the influence of the +4 base on mammalian termination signals using constructs containing the rat type I iodothyronine 5′ deiodinase cDNA. This study has been a collaboration with Dr. Marla Berry (Brigham and Woman's Hospital and Harvard Medical School, Boston) who discovered that this gene contains an internal UGA which encodes selenocysteine (40). The recoding event at the internal UGA codon 126 has been used so that the strength of the termination signal can be matched against incorporation of selenocysteine at the codon *in vivo* in human embryonic kidney cells. The base following this UGA has been changed from its natural C to the other three bases (Figure 8A) such that two products could be analyzed depending on whether selenocysteine incorporation had occurred (28 kD product) or whether termination had occurred (14 kD product). The ratio of termination to selenocysteine incorporation at this codon was 1:3 when C or U was in the +4 position and 3:1 when A or G was in this position (Figure 8B). All 12 of the 4-base signals cannot be tested with this system because if UGA is substituted with UAA, for example, the signal no longer encodes selenocysteine and termination occurs with essentially 100% efficiency.

However, the termination efficiency *in vivo* of the 4th base in UGAN (A ~ G > C ~ U) was consistent with the results found for this signal in an *in vitro* termination assay in the mammalian system, measuring the release of a model peptide from the ribosome by the eRF. The 4th base in each of the UAAN and UAGN series also affected the efficiency of release in the same order, with over a 70-fold range for the UAAN series and 8-fold range for the UGAN and UAGN series (Figure 8C). There is a correlation between the "strength" of the signals and how frequently they occur at natural termination sites, particularly in highly-expressed genes, but here it was the 4th base purines A and G rather than U (as in prokaryotes) that boosted the efficiency of the signal.

Very recently the efficiency of translational termination has been shown to be determined by a synergistic interplay between upstream and downstream sequences of the stop codon in *Saccharomyces cerevisiae* (41).

HOW IS THE +4 BASE RELATED TO THE MECHANISM OF TRANSLATIONAL TERMINATION?

The fact that the strength of the termination signal is so affected by the +4 base invokes the question of *how* it affects the mechanism of translational termination. To understand that we first have to understand how the termination signal is decoded. The critical determinant here is the existence of stop signal decoding proteins, the release factors, which mediate the release of a completed polypeptide from the ribosome in response to the termination signal. What are these proteins and how do they recognize the signal?

Termination Involves Two Classes of Extra-Ribosomal Proteins, the Release Factors

Largely through the isolation of genes from a number of organisms that have similarity to a known release factor gene there is a growing database of information on the release factors. This is illustrated in Table 2. Much of the information has been assembled very

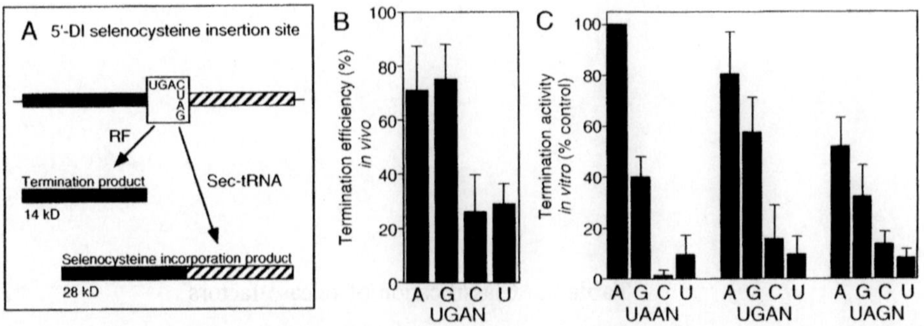

Figure 8. Influence of the 4th base in tetranucleotide stop signals on the efficiency of translational termination in mammalian cells. A) The relative strength of each of the UGAN stop signals was determined as a competitor of the selenocysteine (Sec) incorporation at a UGA in codon position 126 of the type I iodothyronine 5′ deodinase (5′-DI) mRNA, transcribed from a vector transfected into human embryonic kidney cells. B) The *in vivo* termination efficiency of each of the UGAN signals. The base following the UGA codon 126 is shown below the bar. C) The relative termination efficiencies of each of the UAAN, UGAN and UAGN stop signals determined with an *in vitro* termination assay (34). The 4th base is shown below each bar.

recently, in particular with the assignment of the eukaryotic release factors to genes of unknown function already existing in the databases (42). Since some of the release factors are decoding molecules, and some are factors without decoding function but possessing termination enhancing activity, Pel et al. (43) have suggested that decoding molecules be designated as Class I and the enhancing proteins as Class II, discussed below. We have adopted that classification in Table 2.

It had been known for a while before the advent of recombinant DNA techniques that the translational termination process depended on these extra-ribosomal factors. In the late 1960s the development of an *E. coli*-derived *in vitro* translation termination assay had led to the discovery of the three different proteins involved in polypeptide chain termination (44). In this assay the Class I proteins are able to induce hydrolysis of a model peptidyl-tRNA substrate on ribosomes that carry a stop codon at their decoding site. Two Class I factors are found in prokaryotes, RF-1 and RF-2, distinguished by the fact that they each recognize one of the three stop codons uniquely (RF-1, UAG; RF-2, UGA), and they recognize one in common (UAA). The prokaryotic Class II factor, designated RF-3, was recognized as a GTP-binding protein in the early 1970s. Although it enhances the efficiency of termination its presence is not strictly required for *in vitro* termination. RF-3, initially appearing as a rather mysterious unstable *E. coli* protein, is now more firmly established in protein synthesis since the *E. coli* gene was isolated in 1994 and the role of the protein is now clearer. As expected on the basis of the earlier biochemical data, the gene encoding this protein has the characteristic features of a G-protein (45, 46). RF-3 has sequence similarity to both elongation factors, EF-Tu and EF-G.

The organelle release factors have been isolated to date only from mitochondria, and from those mitochondria where UGA is no longer encoding stop but codes for tryptophan. Two factors have been sequenced and display similarity to their prokaryotic counterparts (43, 47). A third factor isolated from rat liver mitochondria resembles prokaryotic RF-1 in its codon recognition (UAA, UAG) (48).

During the last twenty years, a different picture has emerged for the eukaryotic termination process. In the case of the eukaryotic Class I release factors, only one factor (eRF-1) is found and this recognizes signals containing all three stop codons. Studies on a partially-purified rabbit release factor initially led to the idea that a single GTP-binding

Table 2. Classification of release factors*.

Release Factor		Organism	Accession #	Evidence	Reference
Class I Release Factors					
Prokaryotic	RF-1	*Bacillus subtilis*	Z49782	similarity	--
		Coxiella burnetii	X78969	similarity	--
		Escherichia coli	M11519	experimental	(37, 60[a], 81)
		Haemophilus influenzae	U32830	similarity	(82)
		Mycobacterium leprae	U15186	similarity	--
		Mycoplasma genitalium	U01800, partial[b]	similarity	--
		Pseudomonas aeruginosa	X82071, partial	similarity	(83)
		Salmonella typhimurium	J04243	experimental	(84, 85)
		Streptococcus gordonii	L40415, partial	similarity	--
	RF-2	*Bacillus subtilis*	D90218, partial	similarity	(56)
		Escherichia coli	M11520	experimental	(37, 81, 86[a])
		Haemophilus influenzae	U32800	similarity	(82)
		Listeria monocytogenes	L32090, partial	similarity	(43)
		Salmonella typhimurium	M38590	experimental	(87)
Organellar	mRF-1	*Kluyveromyces lactis*	U19586	experimental	(43)
		Rattus norvegicus	--	protein[c]	(48)
		Saccharomyces cerevisiae	X60381	experimental	(47)
Eukaryotic	eRF-1	*Arabidopsis thaliana*	X69375	similarity	--
		Homo sapiens	M75715	experimental	(42[a])
		Oryctolagus cuniculus	--	protein[c]	(50)
		Saccharomyces cerevisiae	Z36012[d]	similarity	(88)
		Xenopus laevis	Z14253	experimental	(89)
Class II Release Factors					
Prokaryotic	RF-3	*Dichelobacter nodosus*	U06471	similarity	(90)
		Escherichia coli	D17724,Z26313	experimental	(45,46)
		Haemophilus influenzae	U32846	similarity	(82)
		Streptococcus mutans	B60328, partial	similarity	(91)
Eukaryotic	eRF-3	*Homo sapiens*	X17644	similarity	(92)
		Pichia pinus	X56910	similarity	(93)
		Saccharomyces cerevisiae	M21129	experimental	(94)
		Xenopus laevis	L37045	experimental	(53)

[a]Contains sequence correction not yet entered in the database; [b]the open reading frame extends beyond the border of the sequence; [c]the release factor has been purified but the corresponding gene has not been cloned; [d]contains unpublished sequence correction.

*The release factors are categorized according to class (see text), factyor type (for example RF-1 or RF-2) and organism or organelle. Where the data come from the gene, the acession number is provided; how it has been designated a release factor is indicated under the title evidence.

protein was involved in termination at all three stop codons (49). Recently a more extensive purification and partial microsequencing of rabbit eRF led to the discovery that this protein had strikingly similar known counterparts in yeast, plants and humans (42). The activity of two of these proteins was tested in the rabbit reticulocyte *in vitro* release assay. Indeed, both proteins were able to direct hydrolysis of a P site bound aminoacyl-tRNA in response to all three stop codons, establishing a family of eukaryotic release factors designated eRF-1. It was quite astonishing, however, to see that none of the members of the family displayed the expected GTP-binding motif. The eukaryotic equivalent of a Class II protein, eRF-3, had been seen as an unstable protein from rabbit reticulocytes in the 1970s (50) although a putative linkage to the well-known yeast gene, *SUP35*, was made only in 1995 (51). The observations suggest that eRF-1s, like their prokaryotic counterparts, act in concert with a separate G-protein. A second indication that the eukaryotic release factor may interact with another extra-ribosomal factor came from the observation that the yeast member of the eRF-1 family, Sup45p, does not cause antisuppression (reduction of the efficiency of nonsense suppression) when overproduced in a tRNA-mediated nonsense suppressor strain. Interestingly, Stansfield and co-workers have shown that Sup45p is able to interact with Sup35p, a yeast G-protein known to be involved in translation but whose function had remained obscure (51). Moreover, the C-terminal part of Sup35p resembles elongation factor EF-1α, the eukaryotic equivalent of bacterial EF-Tu, and overexpression of this part of Sup35p does generate an antisuppressor phenotype (52). It seems that in yeast, as probably in most eukaryotes, a Class I release factor (Sup45p) interacts with a Class II factor (the G-protein Sup35p) to form a heterodimeric release factor complex involved in translational termination. It is interesting that the original studies of the rabbit reticulocyte release factor characterized the eRF as a dimer (Tate, W.P., unpublished data), which was probably a complex of eRF-1 and the eRF-3 (exhibiting GTP-binding activity), analogous to the Sup35p·Sup45p complex. A GTP-binding Sup35p-like protein from *Xenopus laevis* binds to eRF-1 and enhances termination activity, and has been designated eRF-3 (53).

In yeast an extra dimension is added to this picture by the fact that Sup35p is a prion-like protein: the N-terminal part of the protein is required for the maintenance of [*psi*$^+$], a cytoplasmically-inherited enhancer of tRNA-mediated nonsense suppression (54). A current model for prion replication invokes the existence of at least two conformations of the protein, an original form which has full activity (form A) and an alternative form with reduced activity (form B) (55). The essential feature of the model is that form B acts as a template to promote the formation of more form B. Applying this model to Sup35p we speculate that the N-terminal part of Sup35p is responsible for a conformational change involving the C-terminal release factor domain, thus reducing release factor activity and consequently enhancing the efficiency of tRNA-mediated nonsense suppression.

Class I Release Factors: Proteins Mimicking tRNA?

At present, the primary structure of 14 prokaryotic and 2 mitochondrial Class I RFs is either completely or partly known (see Table 2). In addition, several open reading frames whose putative products resemble prokaryotic Class I sequences can be found in the database; however, their relevance remains obscure at present (43). A few years ago, the first comparative study of Class I RFs was published on the basis of essentially four RF sequences (the *Salmonella typhimurium* and *E. coli* factors are over 95% identical and therefore are less useful for a structural comparison) (56). Interestingly, a number of region-specific features suggested on the basis of the first comparison are now more pronounced in an analysis of all currently available sequences. A multiple alignment reveals an overall sequence identity of only 15% (Figure 9). The N-terminal one-third of the proteins (designated Region I) lacks any common residues among all the prokaryotic and mitochondrial factors, representatives of which are displayed in Figure 9. There are small clusters of identity between the mitochondrial factors, or between the RF-1 group or the RF-2 group, but these do not generally extend beyond the subclasses in this region. In contrast, there are two major regions of identity which span all three subclasses of factors (shown by the solid circles): Cluster A is within Region IV and spans residues 230-250 (67% identity), and Cluster B, with 41% identity is within Region III and spans residues 178-199 (*E. coli* RF-1 numbering). Computer-algorithms, developed to indicate the possible secondary structure of a protein, persistently predict extensive alpha-helical structures in Regions I and V of all prokaryotic and mitochondrial Class I RFs. This finding may be significant in the light of the very low levels of primary structure similarity in these regions.

Do we have any clues concerning the function of the various designated regions in the release factor primary structure? Crosslinking experiments have shown that part of *E. coli* RF-2 protein is in contact with a stop codon placed at the ribosome's decoding center (57). The same RF-2 has also been crosslinked to L2, a protein known to be at the heart of the peptidyltransferase center, that could be involved in peptidyl-tRNA hydrolysis (58). Class I RFs thus mimic tRNAs in their two most crucial contact points with the ribosome. The distance between the anticodon loop and the CCA-end of a tRNA is about 75 Å. This distance predicts that distinct release factor domains are involved in the interactions with the decoding and peptidyltransferase centers. Cluster A could be part of one domain and Cluster B part of another. Proteolytic cleavage of *E. coli* RF-2 at position 244 (*E. coli* RF-2 numbering), 2 residues before the start of Cluster A, yields a nicked molecule, though the fragments remain associated, that has lost *in vitro* peptidyl-tRNA hydrolysis activity (59). Interestingly, the nicked factor has an enhanced *in vitro* codon-ribosome binding activity, suggesting that it could be trapped in a pre-hydrolysis binding state. These findings indicate that Cluster A is important for a proper interaction between release factor and the peptidyltransferase center. The updated alignment reveals 15 positions where the mRF-1s (probably recognizing UAA and UAG in view of the fact that UGA encodes tryptophan in *Saccharomyces* and *Kluyveromyces* mitochondria) are identical to prokaryotic RF-1s but differ from RF-2s (open triangles, Figure 9). A group of four such positions coincides with Cluster B, suggesting that this region might play a role in some RF-1 specific trait, such as recognition of UAG (and UAA) codons. It might also simply reflect the slightly different ribosome binding site of RF-1s compared with RF-2s, but the most conserved sites between the mitochondrial and prokaryotic ribosomes are likely to be the decoding site and the peptidyltransferase center, and it seems more likely that conserved primary sequence in the release factors will reflect an interaction with either of these two centers.

Besides the interactions with the decoding and peptidyltransferase centers, crosslinking data have indicated that the RFs also interact with ribosomal proteins L11 and L7/L12 (58). A functional interaction between release factor and L7/L12 was recently confirmed

Figure 9. Multiple alignment of prokaryotic and mitochondrial Class I release factors. The sequences were aligned with the aid of the PileUp program (GCG package). Residues identical between RFs of the same type (mRF-1, RF-1 or RF-2) are boxed. • residues identical between all RFs, ○ residues identical between all RFs except one (shown below the sequence blocks), ▵ residues exclusively identical between mRF-1s and RF-1s (shown between the mRF-1 and RF-1 blocks), ▿ residues exclusively identical between mRF-1s and RF-2s (shown between the RF-1 and RF-2 blocks). Only a subset of the prokaryotic and mitochondrial Class I factors listed in Table 2 is displayed; however, all factors were taken into account for the determination of amino acid identities. Clusters A and B highlight areas of high sequence similarity. The RF-2 frameshift sites are underlined. Vertical lines and roman numerals above the alignment indicate region borders as explained in the text.

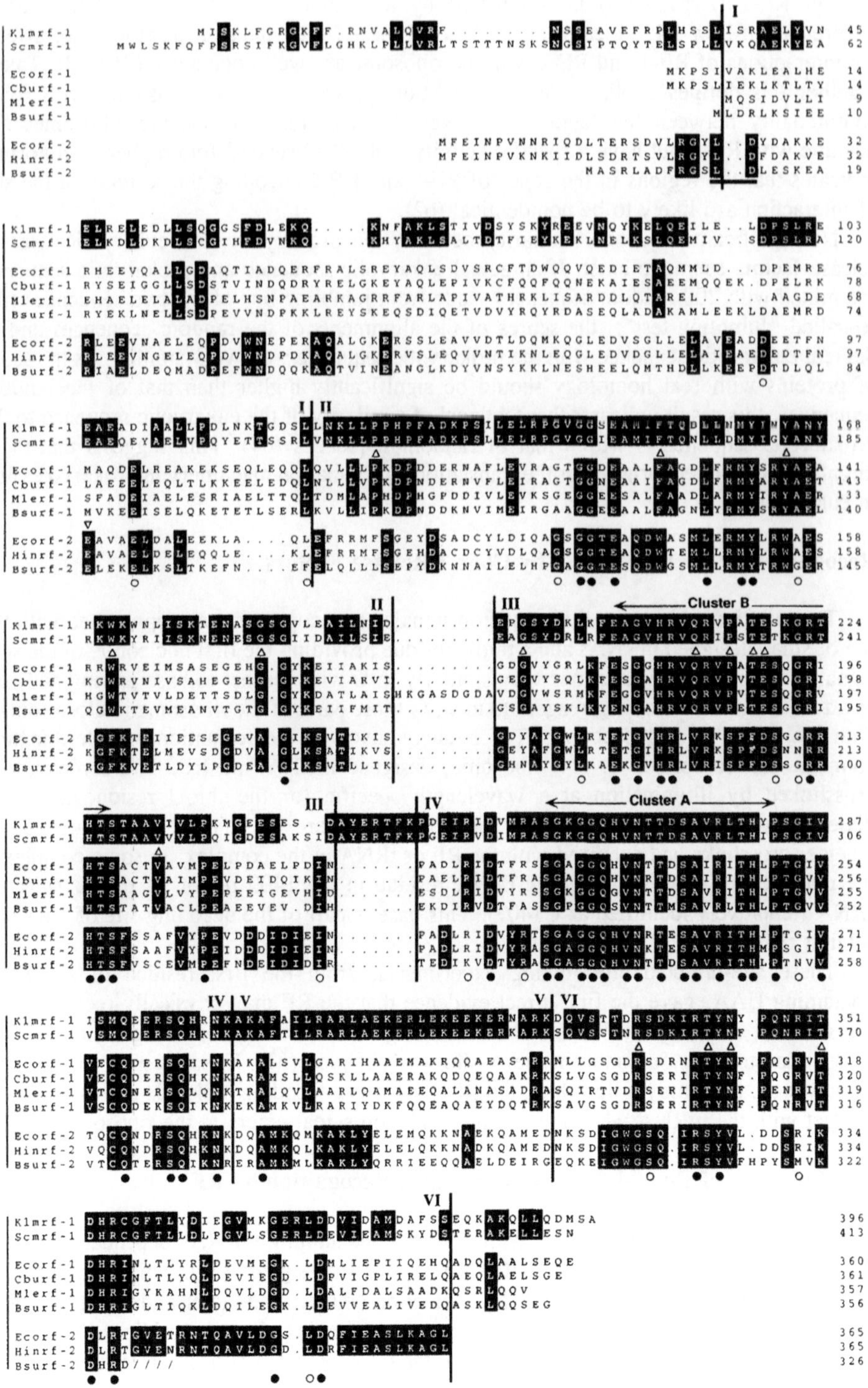
Cluster B
Cluster A

by the finding that the temperature-sensitive phenotype of a strain mutated in region II of *E. coli* RF-1 (Arg137Pro) is suppressed by a mutant form of L7/L12 (60). Subtle differences in how L11, located at the base of the L7/L12 stalk of the ribosome, affects the interactions of RF-1 and RF-2 with the ribosome are well documented (61-63). Taken together this defines another site of interaction between RF and ribosome, with some commonality between the factors. However, the fact that L11-lacking ribosomes are inactive with RF-1 but have a specific activity with RF-2 several-fold higher than normal indicates that the regions in the genes of RF-1 and RF-2 encoding the residues at the site of interaction are likely to be nonidentical (62).

No significant sequence similarity between the prokaryotic and eukaryotic Class I release factors seems to exist. If an actual eukaryotic sequence and prokaryotic sequence are mixed with 200 random sequences made from jumbling one of the two sequences (the so-called "Jumbling test"), the scores of the alignments of the random sequences and of the real eukaryotic sequence with the prokaryotic sequence were very similar. As the score of proteins with real homology should be significantly higher than that of the random sequences, this result indicates that the level of similarity of the eukaryotic sequence to the prokaryotic sequence is within that of random sequences (64). This suggests that there might be independent origins for the translational termination machinery in prokaryotes and in eukaryotes.

Recognition of the Stop Signal by Class I Release Factors

The question of how the different stop signals are recognized was approached with the use of small designed mRNAs and a thioU residue providing the first nucleotide of the stop signal. The thioU base functions very similarly to U with respect to base pairing and acts as a zero-length crosslinking reagent to identify those parts of the ribosome in close contact with the codon during decoding. The strategy involved the formation of a termination complex that comprised *E. coli* ribosome, designed mRNA and RF. The complex was crosslinked by illumination at a wavelength specific for the thioU residue and then dissected to identify the position on the ribosome where the thioU had crosslinked, or whether crosslinks had occurred with the RF or tRNA in the complex. In some complexes the thioU crosslinked to a specific rRNA residue in the 1400 region of the *E. coli* 16S rRNA (tentatively identified as C1407). This base is part of the decoding site of the small subunit rRNA (65).

These site-directed crosslinking experiments, from the first residue of a mRNA containing UAA, gave the first direct evidence that the RF may physically recognize the translational stop signal. With UGA-containing mRNA templates a high yield of RF crosslinked to the stop codon in the A site was obtained (57). These results suggest that whatever the template for decoding, the RF is an active decoding molecule making intimate contact with the mRNA when the translational stop signal enters in the ribosomal A site.

The Influence of the +4 Base on Stop Signal Recognition by RF

The yield of RF-2 complexed with the thioU-containing mRNAs depended upon the identity of the fourth base for the UGAN series of stop signals (57). This suggested that the fourth base of the signal affects the interaction between the factor and the stop codon, with purines at this position promoting more crosslinking than pyrimidines. The likely explanation may be that the conformation of the thioU in the first position is altered by the fourth base purine so as to improve crosslinking. This may imply that the stop signal is in a stacked conformation during decoding. We have tried to determine whether a thioU in the +4 base position crosslinks to the release factor but a very high yield crosslink to rRNA (~70%) occurs from this position, which has prevented us from determining whether the +4 base and the RF are in intimate contact.

The crosslinking of stop signals to the RF and the 1400 region of the rRNA is consistent with either of two orientations for the bases of the stop signal. They could be oriented towards the RF which then makes direct interactions with them. For example, the common keto hydrogen bond acceptor and the imino hydrogen bond donor groups of the fourth base U or G (promoting strong signals in *E. coli*) might be hydrogen bonded to an amino acid in the RF structure. Equally possible, the backbone of the mRNA could make interactions with the RF leaving the bases to pair with the rRNA. The secondary structure of the stop signal in the mRNA is unknown, but RNA is conformationally flexible. If it is modelled as a single-stranded A helix, the crosslinking moiety of the thioU-containing signals would be located immediately over a common N7 of the second base A or G in UAAN and UGAN, the signals recognized by RF-2. If hydrogen bonding were to occur from this N7 to the RF then it would explain why crosslinking from the thioU to the RF is possible, since any part of the RF molecule contacting N7 would be close to the activated (thioU) ring (57). The challenge now is to identify the sites on RF-2 which make contact with the stop signal.

The Template for Release Factor-Stop Signal Recognition

The simplest model of stop-signal decoding by the RF would be that the factor binds directly to the single-stranded mRNA. The rRNA and proteins would provide binding determinants for the factor so it had just the right orientation to the codon in the ribosomal A site. However, there is a provocative alternative, namely that decoding is actually an RNA:RNA event involving the rRNA and the single strand of the mRNA, with the RF recognizing this complex. In this model the RF, when bound to this complex, would trigger the subsequent events of polypeptide chain termination, and in particular the hydrolysis of the completed polypeptide from the terminal tRNA. This RNA:RNA decoding scheme for stop codons is attractive if the modern ribosome evolved from a RNA protoribosome, and a specific, albeit inefficient, mechanism for termination had originated before proteins became part of the ribosome structure. There is circumstantial evidence to support this idea. A thioU crosslink from the first position of the stop codon, when positioned in the A site, occurred not only to the release factor, but also to C1407 in the region linking helix 44 to helix 28 of the 16S rRNA. A mRNA containing UAAU, with thioU in the first and fourth positions, was found to crosslink to both C1407 and C1395 of the rRNA (Figure 10A). The nucleotides in positions 1406-1408, Um^{5}CA, have the potential to form two base pairs with UAA and UAG and three base pairs with UGA, and could form a double-stranded template which stacks on helix 44 that is recognized by the RF (66). Additionally, a third base pair is possible with UAA and UAG through an interaction of m^{5}C with the A in the second position of these codons. Such a m^{5}C-A interaction occurs during the decoding of UGA as selenocysteine by a specific tRNA, where m^{5}C is able to base pair with A in the wobble position (67).

The 1400 region of the 16S rRNA between helix 28 and helix 44 was originally thought to be single stranded (as shown in Figure 10A), but phylogenetic constraints (68), crosslinking data (69), and mutational analyses (70, 71) have established a functionally important base-pairing interaction between the 1400 and the 1500 regions. Recent nuclear magnetic resonance spectroscopy studies of a model oligonucleotide containing residues 1404-1411 and 1489-1497 have shown secondary structure with some unusual base pairing, for example between U1406 and U1495 (Fourmy, D., Recht, M. and Puglisi, J.D., unpublished data), proposed earlier by Gutell (72). This model for the structure of the oligonucleotide is supported by its specific interaction with the antibiotic paromomycin (Fourmy, D., Recht, M. and Puglisi, J.D., unpublished data).

How can a secondary structure for this region be accommodated with a model of a double-stranded template for RF recognition of the stop signal? If the structure of this region were metastable with its unusual base pairing, the appearance of the stop codon in the A site might disrupt the existing structure of the decoding site, forming its own base pairs with residues 1406-1408, and thereby creating a critical binding determinant at the decoding site for the RF. Indeed, several mutations in helix 44 in yeast mitochondrial ribosomes (gray circles with white lettering, Figure 10B) compensate for a release factor mutation *in vivo*, and hence these bases apparently affect RF-ribosome interaction at the site (Pel, H.J. and Bolotin-Fukuhara, M., unpublished data). The footprint made by an A site tRNA (73) (white circles), and the C1407 crosslinked to the stop codon (black circle) are also shown in Figure 10B.

The overall picture that has emerged is that the RF probably recognizes a number of nucleotides of the rRNA in and around the decoding site, perhaps some in a base-paired structure with the mRNA, and perhaps some unpaired. Therefore, codon recognition by RF seems to be controlled by contacts with the termination signal in the mRNA and with adjacent decoding site rRNA.

G-Proteins: Now Also in Termination

The fact that G-proteins are involved in termination indicates a striking parallel with the other phases of translation: initiation and elongation. The three complete RF-3 sequences identified so far (Table 2) all contain a classical GTP/GDP-binding domain as indicated by the presence of a series of short characteristic sequence motifs. G-proteins have an intrinsic GTPase activity that is activated upon interaction with an effector. The conformation of the protein depends on whether GTP or GDP is present in the nucleotide-binding pocket. G-proteins thus transduce the interaction with the effector into a conformational change. Besides RF-3 four G-proteins have been implicated in protein synthesis in prokaryotes. Initiation factor IF-2 complexed with GTP selectively binds N-blocked aminoacylated initiator tRNA to the 30S ribosomal subunit (74). The GTP form of elongation factor EF-Tu complexes with aminoacylated elongator tRNAs and is involved in the correct selection and binding of elongator tRNAs to the ribosome (75). The SelB protein fulfills a similar role as an elongation factor but specifically for selenocysteyl-tRNA (76). Elongation factor EF-G-GTP promotes translocation of mRNA, peptidyl-tRNA and deacylated tRNA from the ribosomal A and P sites to the P and E sites, respectively (77). *E. coli* RF-3 resembles both elongation factors: EF-G more closely than EF-Tu (27% *versus* 22% identity). There is compelling evidence that the translating ribosome undergoes reversible conformational transitions between pre- and post-translocational states promoted by the active GTP conformation of EF-G and EF-Tu, respectively (65, 78). Mesters et al. have shown that the factors induce similar transitions even in an uncoupled system lacking mRNA and tRNAs (79).

As decoding proteins, Class I release factors presumably bind ribosomes that are in a post-translocational state. The greater sequence similarity between RF-3 and EF-G could suggest that RF-3 preferentially binds ribosomes that are in a pre- rather than a post-translocational state, although since the identity of RF-3 with EF-Tu is not too dissimilar to that with EF-G, this may be misleading. Indeed members of the eukaryotic eRF-3 family resemble the eukaryotic counterpart of EF-Tu, EF-1a. We currently envisage that just as

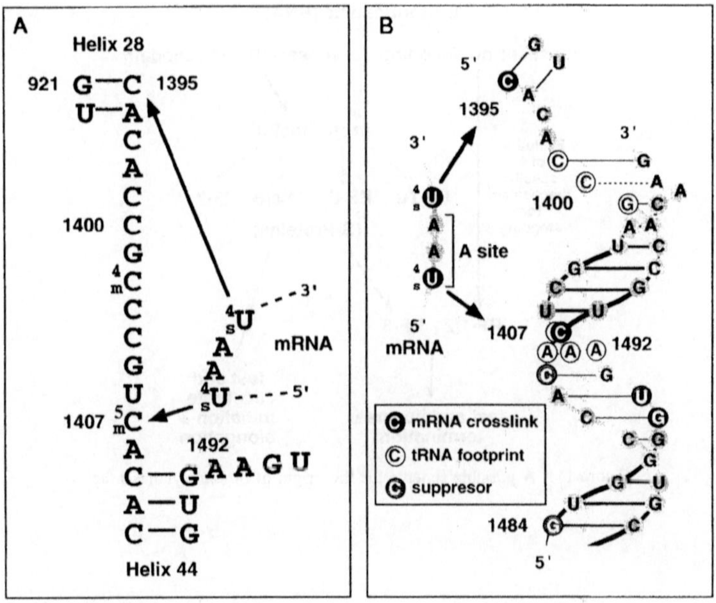

Figure 10. The decoding site of the small subunit rRNA. A) Part of helix 44 and the 1400 region between this helix and helix 28. Two crosslinks (arrows) are shown from thioU (4sU) residues in an artificial mRNA, from the first position of the stop signal to C1407 (66) and from the fourth position to C1395. B) A possible secondary structure for this region. The 1400 region is shown now base-paired with the 1500 region as proposed by Gutell (68) and is drawn as an A helix. U1406 interacts with U1495 as proposed by Gutell (72) and C1403 base-pairs with U1498 as proposed by Cunningham et al. (71). The speculative C1400:A1503 pair is indicated by a broken line and the A1408:A1493 pair by a single point. The nucleotides C1407 and C1395 involved in mRNA crosslinking are represented by black circles; base substitutions compensating for a release factor mutation in yeast mitochondria are shown by the gray circles with white lettering. A site (around A1492) and P site (around C1400) specific tRNA footprints (73) are indicated as white circles.

EF-Tu helps the cognate aminoacyl-tRNA into the ribosomal A site, Class II RFs would help the Class I RFs bind into the decoding site when its specific recognition template was in place, although the experimental evidence for such conclusions is weak at present. The G-protein may provide a directionality to the events of protein synthesis at each step, preventing the potential reverse reactions occurring.

One interesting outcome from this is that it provides a unity for the origin of the protein synthesis factors. A possible evolutionary scheme is shown in Figure 11. It is now believed the original protoribosome was made entirely of RNA and that tRNA-like molecules were involved in decoding. The ribosome acquired proteins which improved efficiency and extra-ribosomal factors developed to aid the efficiency of the RNA molecules interacting with the protoribosome. From an initial factor with GTP-binding activity, today we recognize prokaryotic factors IF-2, EF-Tu, EF-G, SelB and RF-3 as specialized factors that have retained G-protein function. Originally there was probably no specific termination mechanism. Rather when an unassigned codon (a true nonsense codon) was encountered, protein synthesis either could continue through misreading, or frameshifting, or the product could drop off the ribosome (80). Presumably as a later event in evolution a specific mechanism for termination arose with the origin of the Class I release factors, and later still this process would have been made more efficient with the

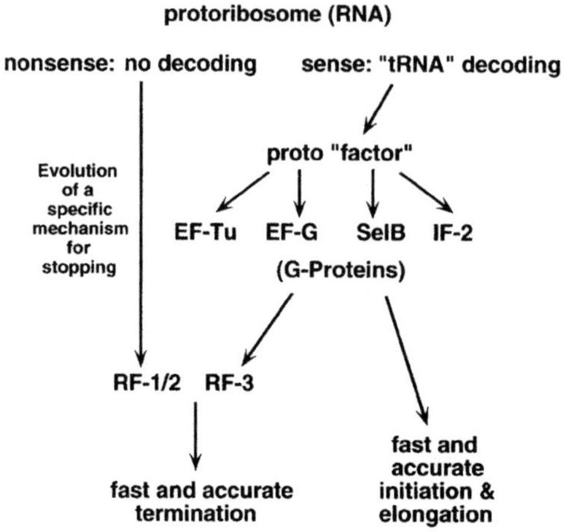

Figure 11. A possible scheme for the origin of protein synthesis factors.

addition of a G-protein, RF-3, to aid the mechanism. With the modern ribosome we have a mechanism of protein synthesis today which is both relatively accurate and efficient for the initiation, elongation and termination stages of the production of a protein.

HOW DOES THE EFFICIENCY OF TRANSLATIONAL TERMINATION RELATE TO THE TERMINATION SIGNAL AND ITS DECODING?

Current understanding is that decoding of stop signals involves a careful positioning of the RF with respect to the stop codon in the ribosomal A site. That this involves a number of interactions with the rRNA and ribosomal proteins seems highly likely. With this scenario any feature of the stop signal in the mRNA which distorts the normal position of the stop codon in the A site, or allows it to form interactions with the rRNA (for example the +4 base and other bases on either side of the stop codon), could perturb the delicate balance of the termination event. If such a balance were disturbed, then it is not too difficult to imagine that the specific decoding mechanism, mediated through the RFs, might become more efficient or might compete less well with other potential events. Such competing events include non-cognate decoding by tRNAs (readthrough), or recoding mediated by specific features of the mRNA that have evolved for a physiologically important alternative translational event at the site of the stop codon. Recoding events such as translational frameshifting or selenocysteine incorporation at UGA stop codons benefit from having a low efficiency stop signal at the site so they have a competitive advantage. Indeed UGAC, a rare signal at natural termination sites and one which we have determined to be weak, is found at several recoding sites. In contrast, highly-expressed genes presumably need strong stop signals which are rapidly decoded for a high translational rate, and not surprisingly, such genes do have the most efficient signals.

ACKNOWLEDGMENTS

The authors are supported by an International Scholar award of the Howard Hughes Medical Institute to W.P.T., a Human Frontier Science Program grant (awarded to

Yoshikazu Nakamura and W.P.T.) and grants from The Health Research Council of New Zealand and the NZ Lotteries Board.

REFERENCES

1 Shine, J. and Dalgarno, L. (1974) Proc. Nat. Acad. Sci. U.S.A. 71, 1342-1346.
2 Gold, L. (1988) Ann. Rev. Biochem. 57, 199-233.
3 Chen, H., Bjerknes, M., Kumar, R. and Jay, E. (1994) Nucl. Acids Res. 22, 4953-4957.
4 McCarthy, J.E.G. and Brimacombe, R. (1994) Trends Genet. 10, 402-407.
5 Brombach, M. and Pon, C.L. (1987) Mol. Gen. Genet. 208, 94-100.
6 Kozak, M. (1995) Proc. Nat. Acad. Sci. U.S.A. 92, 2662-2666.
7 Sharp, P.M., Stenico, M., Peden, J.F. and Lloyd, A.T. (1993) Biochem. Soc. Trans. 21, 835-841.
8 Kurland, G.C. (1993) Biochem. Soc. Trans. 21, 841-846.
9 Collins, R.F., Roberts, M. and Phoenix, D.A. (1995) Biochem. Soc. Trans. 23, 76S.
10 Saier, M.H., Jr. (1995) FEBS Lett. 362, 1-4.
11 Goldman, E. and Zubay, G. (1994) in Molecular Microbiology Techniques, Part A (Adolph, K.W. ed.) Vol. 3, pp. 159-178, Academic Press, Inc., San Diego, CA.
12 Bernardi, G. and Bernardi, G. (1985) J. Mol. Evol. 22, 363-365.
13 Brenner, S., Stretton, A.O.W. and Kaplan, S. (1965) Nature 206, 994-998.
14 Brenner, S., Barnett, L., Katz, E.R. and Crick, F.H.C. (1967) Nature 213, 449-450.
15 Ganoza, M.C. (1966) Cold Spring Harbor Symp. Quant. Biol. 31, 273-278.
16 Caskey, T., Scolnick, E., Tompkins, R., Goldstein, J. and Milman, G. (1969) Cold Spring Harbor Symp. Quant. Biol. 34, 479-491.
17 Capecchi, M.R. and Klein, H.A. (1969) Cold Spring Harbor Symp. Quant. Biol. 34, 469-477.
18 Kopelowitz, J., Hampe, C., Goldman, R., Reches, M. and Engelberg-Kulka, H. (1992) J. Mol. Biol. 225, 261-269.
19 Buckingham, R.H. (1990) Experientia 46, 1126-1133.
20 Buckingham, R.H., Murgola, E.J., Sörensen, P., Pagel, F.T., Hijazi, K.A., Mims, B.H., Figueroa, N., Brechemier-Baey, D. and Coppin-Raynal, E. (1990) in The Ribosome: Structure, Function and Evolution (Hill, W.E., Dahlberg, A.E., Garrett, R.A., Moore, P.B., Schlessinger, D. and Warner, J.R., eds.), pp. 541-545, American Society for Microbiology, Washington, DC.
21 Buckingham, R.H., Sörensen, P., Pagel, F.T., Hijazi, K.A., Mims, B.H., Brechemier-Baey, D. and Murgola, E.J. (1990) Biochim. Biophys. Acta 1050, 259-262.
22 Brown, C.M., Dalphin, M.E., Stockwell, P.A. and Tate, W.P. (1993) Nucl. Acids Res. 21, 3119-3123.
23 Sharp, P.M. and Li, W.-H. (1987) Nucl. Acid. Res. 15, 1281-1295.
24 Brown, C.M., Stockwell, P.A., Dalphin, M.E. and Tate, W.P. (1994) Nucl. Acids Res. 22, 3620-3624.
25 Poole, E.S., Brown, C.M. and Tate, W.P. (1995) EMBO J. 14, 151-158.
26 McClelland, M. and Bhagwat, A.S. (1992) Nature 355, 595-596.
27 Brown, C.M. (1993) Ph.D. Thesis, University of Otago, Dunedin, New Zealand.
28 Brown, C.M., Stockwell, P.A., Trotman, C.N.A. and Tate, W.P. (1990) Nucl. Acids Res. 18, 2079-2086.
29 Brown, C.M., Stockwell, P.A., Trotman, C.N.A. and Tate, W.P. (1990) Nucl. Acids Res. 18, 6339-6345.

30 Tate, W.P. and Brown, C.M. (1992) Biochemistry 31, 2443-2450.
31 Kohli, J. and Grosjean, H. (1981) Mol. Gen. Genet. 182, 430-439.
32 Cavener, D.R. and Ray, S.C. (1991) Nucl. Acids Res. 19, 3185-3192.
33 Sharp, P.M., Burgess, C.J., Cowe, E., Lloyd, A.T. and Mitchell, K.J. (1992) in Transfer RNA in Protein Synthesis (Hatfield, D.L., Lee, B.J. and Pirtle, R.M., eds.), pp. 397-425, CRC Press, Boca Raton, FL.
34 McCaughan, K.K., Brown, C.M., Dalphin, M.E., Berry, M.J. and Tate, W.P. (1995) Proc. Nat. Acad. Sci. U.S.A. 92, 5431-5435.
35 Mottagui-Tabar, S., Björnsson, A. and Isaksson, L.A. (1994) EMBO J. 13, 249-257.
36 Arkov, A.L., Korolev, S.V. and Kisselev, L.L. (1993) Nucl. Acids Res. 21, 2891-2897.
37 Craigen, W.J., Cook, R.G., Tate, W.P. and Caskey, C.T. (1985) Proc. Nat. Acad. Sci. U.S.A. 82, 3616-3620.
38 Pedersen, W.T. and Curran, J.F. (1991) J. Mol. Biol. 219, 231-241.
39 Adamski, F.M., McCaughan, K.K., Jørgensen, F., Kurland, C.G. and Tate, W.P. (1994) J. Mol. Biol. 238, 302-308.
40 Berry, M.J., Banu, L., Chen, Y., Mandel, S.J., Kieffer, J.D., Harney, J.W. and Larsen, P.R. (1991) Nature 353, 273-276.
41 Bonetti, B., Fu, L., Moon, J. and Bedwell, D.M. (1995) J. Mol. Biol. 251, 334-345.
42 Frolova, L., Le Goff, X., Rasmussen, H.H., Cheperegin, S., Drugeon, G., Kress, M., Arman, I., Haenni, A-L., Celis, J.E., Philippe, M., Justesen, J. and Kisselev, L. (1994) Nature 372, 701-703.
43 Pel, H.J., Rozenfeld, S. and Bolotin-Fukuhara, M. Current Genetics (in press).
44 Caskey, C.T. (1980) Trends Biochem. Sci. 5, 234-237.
45 Mikuni, O., Ito, K., Moffat, J., Matsumura, K., McCaughan, K., Nobukuni, T., Tate, W. and Nakamura, Y. (1994) Proc. Nat. Acad. Sci. U.S.A. 91, 5798-5802.
46 Grentzmann, G., Brechemier-Baey, D., Heurgue, V., Mora, L. and Buckingham, R.H. (1994) Proc. Nat. Acad. Sci. U.S.A. 91, 5848-5852.
47 Pel, H.J., Maat, C., Rep, M. and Grivell, L.A. (1992) Nucl. Acids Res. 20, 6339-6346.
48 Lee, C.C., Timms, K.M., Trotman, C.N.A. and Tate, W.P. (1987) J. Biol. Chem. 262, 3548-3552.
49 Beaudet, A.L. and Caskey, C.T. (1971) Proc. Nat. Acad. Sci. U.S.A. 68, 619-624.
50 Konecki, D.S., Aune, K.C., Tate, W.P. and Caskey, C.T. (1977) J. Biol. Chem. 252, 4514-4520.
51 Stansfield, I., Jones, K.M., Kushnirov, V.V., Dagkesamanskaya, A.R., Poznyakovski, A.I., Paushkin, S.V., Nierras, C.R., Cox, B.S., Ter-Avanesyan, M.D. and Tuite, M.F. (1995) EMBO J. 14, 4365-4373.
52 Ter-Avanesyan, M.D., Kushnirov, V.V., Dagkesamanskaya, A.R., Didichenko, S.A., Chernoff, Y.O., Inge-Vechtomov, S.G. and Smirnov, V.N. (1993) Mol. Micro. 7, 683-692.
53 Zhouravleva, G., Frolova, L., Le Goff, X., Le Guellec, R., Inge-Vechtomov, S., Kisselev, L. and Philippe, M. (1995) EMBO J. 14, 4065-4072.
54 Ter-Avanesyan, M.D., Dagkesamanskaya, A.R., Kushnirov, V.V. and Smirnov, V.N. (1994) Genetics 137, 671-676.
55 Cohen, F.E., Pan, K.-M., Huang, Z., Baldwin, M., Fletterick, R.J. and Prusiner, S.B. (1994) Science 264, 530-531.
56 Pel, H.J., Rep, M. and Grivell, L.A. (1992) Nucl. Acids Res. 20, 4423-4428.
57 Brown, C.M. and Tate, W.P. (1994) J. Biol. Chem. 269, 33164-33170.
58 Stöffler, G., Tate, W.P. and Caskey, C.T. (1982) J. Biol. Chem. 257, 4203-4206.

59 Moffat, J.G. and Tate, W.P. (1994) J. Biol. Chem. 269, 18899-18903.
60 Zhang, S., Rydén-Aulin, M., Kirsebom, L.A. and Isaksson, L.A. (1994) J. Mol. Biol. 242, 614-618.
61 Tate, W.P., Dognin, M.J., Noah, M., Stöffler-Meilicke, M. and Stöffler, G. (1984) J. Biol. Chem. 259, 7317-7324.
62 Tate, W.P., Schulze, H. and Nierhaus, K.H. (1983) J. Biol. Chem. 258, 12816-12820.
63 Tate, W.P., McCaughan, K.K., Ward, C.D., Sumpter, V.G., Trotman, C.N.A., Stöffler-Meilicke, M., Maly, P. and Brimacombe, R. (1986) J. Biol. Chem. 261, 2289-2293.
64 Doolittle, R.F. (1986) URFs and ORFs: A Primer on How to Analyze Derived Amino Acid Sequences, University Science Books, California.
65 Noller, H.F. (1991) Annu. Rev. Biochem. 60, 191-227.
66 Tate, W., Greuer, B. and Brimacombe, R. (1990) Nucl. Acids Res. 18, 6537-6544.
67 Hatfield, D.L., Smith, D.W.E., Lee, B.J., Worland, P.J. and Oroszlan, S. (1990) Crit. Rev. Biochem. Mol. Biol. 25, 71-96.
68 Gutell, R.R. (1993) Nucl. Acids Res. 21, 3051-3054.
69 Döring, T., Greuer, B. and Brimacombe, R. (1992) Nucl. Acids Res. 20, 1593-1597.
70 Cunningham, P.R., Nurse, K., Weitzmann, C.J., Nègre, D. and Ofengand, J. (1992) Biochemistry 31, 7629-7637.
71 Cunningham, P.R., Nurse, K., Weitzmann, C.J. and Ofengand, J. (1993) Biochemistry 32, 7172-7180.
72 Gutell, R.R. (1993) in The Translational Apparatus: Structure, Function, Regulation, Evolution (Nierhaus, K.H., Franceschi, F., Subramanian, A.R., Erdmann, V.A. and Wittmann-Liebold, B., eds.), pp. 477-488, Plenum Press, New York and London.
73 Moazed, D. and Noller, H.F. (1990) J. Mol. Biol. 211, 135-145.
74 Ochoa, S. and Mazumder, R. (1974) in The Enzymes (Boyer, P.D., ed.) Vol. 10, Third Edition, pp. 1-51, Academic Press, New York and London.
75 Lucas-Lenard, J. and Beres, L. (1974) in The Enzymes (Boyer, P.D., ed.) Vol. 10, pp. 53-86, Academic Press, New York and London.
76 Böck, A., Forchhammer, K., Heider, J. and Baron, C. (1991) Trends Biochem. Sci. 16, 463-467.
77 Rheinberger, H.-J., Geigenmüller, U., Gnirke, A., Hausner, T.-P., Remme, J., Saruyama, H. and Nierhaus, K.H. (1990) in The Ribosome: Structure, Function and Evolution (Hill, W.E., Dahlberg, A.E., Garrett, R.A., Moore, P.B., Schlessinger, D. and Warner, J.R., eds.), pp. 318-330, American Society for Microbiology, Washington, DC.
78 Nierhaus, K.H. (1993) Mol. Micro. 9, 661-669.
79 Mesters, J.R., Potapov, A.P., de Graaf, J.M. and Kraal, B. (1994) J. Mol. Biol. 242, 644-654.
80 Bain, J.D., Switzer, C., Chamberlin, A.R. and Benner, S.A. (1992) Nature 356, 537-539.
81 Lee, C.C., Kohara, Y., Akiyama, K., Smith, C.L., Craigen, W.J. and Caskey, C.T. (1988) J. Bacteriol. 170, 4537-4541.
82 Fleischmann, R.D., Adams, M.D., White, O., Clayton, R.A., Kirkness, E.F., Kerlavage, A.R., Bult, C.J., Tomb, J.-F., Dougherty, B.A., Merrick, J.M., McKenney, K., Sutton, G., FitzHugh, W., Fields, C.A., Gocayne, J.D., Scott, J.D., Shirley, R., Liu, L.-I., Glodek, A., Kelley, J.M., Weidman, J.F., Phillips, C.A., Spriggs, T., Hedblom, E., Cotton, M.D., Utterback, T.R., Hanna, M.C., Nguyen, D.T., Saudek, D.M., Brandon, R.C., Fine, L.D., Fritchman, J.L., Fuhrmann, J.L., Geoghagen, N.S.M., Gnehm, C.L., McDonald, L.A., Small, K.V., Fraser, C.M.,

Smith, H.O. and Venter, J.C. (1995) Science 269, 496-512.
83 Hungerer, C., Troup, B., Römling, U. and Jahn, D. (1995) J. Bacteriol. 177, 1435-1443.
84 Elliott, T. (1989) J. Bacteriol. 171, 3948-3960.
85 Elliott, T. and Wang, X. (1991) J. Bacteriol. 173, 4144-4154.
86 Mikuni, O., Kawakami, K. and Nakamura, Y. (1991) Biochimie 73, 1509-1516.
87 Kawakami, K. and Nakamura, Y. (1990) Proc. Nat. Acad. Sci. U.S.A. 87, 8432-8436.
88 Breining, P. and Piepersberg, W. (1986) Nucl. Acids Res. 14, 5187-5197.
89 Tassan, J.P., LeGuellec, K., Kress, K., Faure, M., Camonis, J., Jacquet, M. and Philippe, M. (1993) Mol. Cell. Biol. 13, 2815-2821.
90 Billington, S.J., Jost, B.H. and Rood, J.I. (1995) Microbiol. 141, 945-957.
91 Ogier, J.A., Schöller, M., Lepoivre, Y., Gangloff, S., M'zoughi, R. and Klein, J.-P. (1991) Infect. Immun. 59, 1620-1626.
92 Hoshino, S.-i., Miyazawa, H., Enomoto, T., Hanaoka, F., Kikuchi, Y., Kikuchi, A. and Ui, M. (1989) EMBO J. 8, 3807-3814.
93 Kushnirov, V.V., Ter-Avanesyan, M.D., Didichenko, S.A., Smirnov, V.N., Chernoff, Y.O., Derkach, I.L., Novikova, O.N., Inge-Vechtomov, S.G., Neistat, M.A. and Tolstorukov, I.I. (1990) Yeast 6, 461-472.
94 Kushnirov, V.V., Ter-Avanesyan, M.D., Telckov, M.V., Surguchov, A.P., Smirnov, V.N. and Inge-Vechtomov, S.G. (1988) Gene 66, 45-54.
95 Cantor, C.R. and Schimmel, P.R. (1980) Biophysical Chemistry. Part I: The conformation of biological macromolecules, p. 52, W.H. Freeman and Company, San Francisco, CA.

MECHANISM OF REPLICATION AND COPY NUMBER CONTROL OF PLASMIDS IN GRAM-POSITIVE BACTERIA

Saleem A. Khan

Department of Molecular Genetics and Biochemistry
University of Pittsburgh School of Medicine
Pittsburgh, PA 15261

INTRODUCTION

Ever since plasmids were first discovered, they have served as important model systems for the study of various biological processes, including DNA replication, gene expression, recombination and gene transfer. Most of the earlier intensive studies carried out with plasmids of Gram-negative bacteria, such as ColE1, R6K, RK2 and plasmids of the incFII family, revealed that they replicated by either a unidirectional or bidirectional theta-type mechanism. These studies also provided the first evidence of an antisense mechanism for the regulation of DNA replication in the case of the ColE1 plasmid (1). Similarly, initiator-dependent "handcuffing" was shown to be responsible for the regulation of replication of R6K, RK2 and P1 plasmids (2-4). Studies with *E. coli* plasmids also identified directly repeated sequences (iterons) as elements critical in the initiation and regulation of DNA replication (5-8). In recent years, significant progress has also been made in our understanding of the biology of plasmids in Gram-positive bacteria. These studies have revealed some similarities as well as significant differences in the mechanism of replication and copy number control of these plasmids as compared from those of the Gram-negative organisms. In this review, the replication and copy number control of plasmids in Gram-positive bacteria is discussed with emphasis on the molecular aspects.

ROLLING CIRCLE REPLICATION OF PLASMIDS

Evidence has accumulated over the past decade that most small, multicopy plasmids of Gram-positive bacteria replicate by a rolling circle (RC) mechanism. Such plasmids have been found in a large number of Gram-positive organisms, including *Staphylococcus aureus, Bacillus subtilis, Clostridium butyricum, Lactococcus lactis, Lactobacillus plantarum, Streptococcus agalactiae, Streptomyces lividans* and *Bacillus thuringiensis* (for reviews, see ref. 9-11). RC plasmids are not unique to Gram-positive organisms, and more recently, plasmids have been identified in Gram-negative bacteria which also replicate by a RC mechanism; these include pKYM from *Shigella sonnei,* pHPK255 from *Helicobacter pylori* and pCA2.4 from cyanobacteria (12,13). RC plasmids have also been

Genetic Engineering, Vol. 18, Edited by J.K. Setlow
Plenum Press, New York, 1996

reported from *Mycoplasma mucoides* (pAD01 and pKMK1) and the spirochete *Treponema denticola* (pTD1) (14,15). The discovery of the origin-specific nicking-closing activity of the replication initiator protein of the *S. aureus* plasmid pT181 (16) provided the first evidence that this plasmid replicates by a RC mechanism and showed that this mode of replication was not confined to those of single-stranded (SS) DNA bacteriophages of *E. coli* (17,18). Subsequent studies revealed that a large majority of small plasmids of Gram-positive bacteria replicate by a RC mechanism. Most RC plasmids of Gram-positive bacteria have a broad host range and their copy number generally varies between 10 and 50. Plasmids such as pC194, pMV158 and pWV01 have also been found to replicate in *E. coli* (9,10).

Rolling circle plasmids of Gram-positive bacteria can be classified into five groups based on the homology in their initiator proteins and the leading strand origins, namely pT181, pC194/pUB110, pE194/pMV158, pSN2 and pTX14-3 families (Table 1). Plasmids within the individual groups have a highly conserved overall structure, including the location of the replication region and the drug resistance genes (Figure 1, and ref. 9, 11 and 19). The replication origins and initiator proteins (Rep) of the plasmids of the pT181 family share homology with those of the filamentous phages of *E. coli* (17) while those of the pC194 family have homology with ϕX174 (13). The events during the replication of the RC plasmids can be summarized as follows (Figure 2). The plasmid-encoded Rep protein nicks the leading strand origin (termed double-strand origin, *dso*) to generate a free 3′ OH end and becomes covalently attached to the 5′ phosphate of the nick through a phosphotyrosine linkage (16,50,51). A host-encoded helicase displaces the strand and the 3′ OH end serves as a primer for extension synthesis by DNA polymerase III (11,52). After the leading strand is completely synthesized, the Rep protein recognizes a subregion of the origin as the termination signal and carries out a second nicking event, with the concomitant ligation of the released SS DNA corresponding to the leading strand (53,54). This completes the replication of the leading strand of the DNA. The Rep protein is inactivated after one round of replication either due to the attachment of a small oligonucleotide (55) or by another as yet unknown mechanism. The released SS DNA is converted to a double-stranded (DS) form utilizing the lagging or single-strand origin, termed *sso*. The *cis*- and *trans*-acting elements involved in replication of the RC plasmids of Gram-positive bacteria are summarized below.

Initiation of Leading Stand Replication

Replication of RC plasmids proceeds unidirectionally (56) and initiates from their *dso*s which contain the binding sites as well as the nick sites of the initiator proteins (16,51,52). The *dso* of the plasmids of the pT181 family is located within the gene encoding their initiator proteins (19). *In vivo* and *in vitro* studies have localized the *dso* of pT181 to within a 70-bp region (57). A smaller 43-bp region (nt 44 to 86) is sufficient to support replication in the absence of competition from a wild-type origin. The origins of replication of the plasmids of the pT181 family contain three sets of inverted repeat (IR) elements, IRI, IRII and IRIII (Figure 3). The IRII region is conserved in the pT181 family and contains the RepC nick site located within the loop between pT181 nt 70 and 71 (11,16,58). Because of the conservation of IRII, all initiators of the pT181 family are able to nick/close all such plasmids (51,58). The origin contains a static bend which is enhanced by the binding of RepC (59). IRII has also been shown to form a cruciform structure and RepC enhances cruciform extrusion (60). The initiator binding site encompasses the complete IRIII element and the Right (R) arm of IRII (Figure 3 and ref. 52). No function has yet been identified for IRI. The IRIII element is divergent amongst

Table 1. Rolling Circle Plasmids of Gram-Positive Bacteria

Plasmid	Size (kb)	Resistance Gene	Original Host	Reference
pT181	4.4	Tc	*Staphylococcus aureus*	(19)
pC221	4.6	Cm	*Staphylococcus aureus*	(20)
pC223	4.6	Cm	*Staphylococcus aureus*	(20)
pCW7	4.2	Cm	*Staphylococcus aureus*	(11)
pHD2	2.1	cryptic	*Bacillus thuringiensis*	(21)
pS194	4.4	Sm	*Staphylococcus aureus*	(20)
pT127	4.4	Tc	*Staphylococcus aureus*	(11)
pTZ12 (pTZ10)	2.5	Cm	*Corynebacterium xerosis*	(22)
pUB112	4.1	Cm	*Staphylococcus aureus*	(20)
pE194	3.7	Em	*Staphylococcus aureus*	(11)
pA1	2.8	cryptic	*Lactobacillus plantarum*	(23)
pFX2	2.5	cryptic	*Lactococcus lactis*	(24)
pMV158 (pLS1)	5.5	Tc	*Streptococcus agalactiae*	(9)
pSH71	2.1	cryptic	*Lactococcus lactis*	(25)
pWV01	3.3	cryptic	*Lactococcus lactis*	(26)
pC194	2.9	Cm	*Staphylococcus aureus*	(11)
pAMα1	9.6	Tc	*Streptococcus faecalis*	(27)
pBAA1	6.8	cryptic	*Bacillus subtilis*	(28)
pBC1	1.6	cryptic	*Bacillus coagulans*	(29)
pBC16	4.6	Tc	*Bacillus cereus*	(30)
pBS2	2.3	cryptic	*Bacillus subtilis*	(31)
pC30il	2.1	cryptic	*Lactobacillus plantarum*	(32)
pCB101	6.0	cryptic	*Clostridium butyricum*	(33)
pFTB14	8.2	cryptic	*Bacillus liquefaciens*	(34)
pIJ101	8.9	cryptic	*Streptomyces lividans*	(35)
pLAB1000	3.3	cryptic	*Lactobacillus hilgardii*	(36)
pLP1	2.1	cryptic	*Lactobacillus plantarum*	(37)
pOX6	3.2	Cd	*Staphylococcus aureus*	(11)
pRBH1 (pTB19)	1.75	Km	Bacilli[a]	(38)
pSN1981	4.9	Tc	*Bacillus subtilis*	(39)
pTA1060	8.6	cryptic	*Bacillus subtilis*	(40)
pTHT15	4.5	Tc	Bacilli[a]	(41)
pUB110	4.5	Km	*Staphylococcus aureus*	(42)
pUH1	5.7	γ-glut[b]	*Bacillus subtilis*	(43)
pVA380-1	4.2	cryptic	*Streptococcus ferus*	(44)
pWGB32	2.4	Smr	*Staphylococcus aureus*	(45)
p353-2	2.4	cryptic	*Lactobacillus pentosus*	(46)

Table 1 (continued)

pSN2	1.3	cryptic	*Staphylococcus aureus*	(47)
pE5	2.1	Em	*Staphylococcus aureus*	(48)
pE12	2.2	Em	*Staphylococcus aureus*	(48)
pIM13	2.1	Em	*Bacillus subtilis*	(48)
pNE131	2.1	Em	*Staphylococcus epidermidis*	(48)
pT48	2.1	Em	*Staphylococcus aureus*	(11)
pTCS1	1.3	cryptic	*Staphylococcus aureus*	(11)
pTX14-3	7.5	cryptic	*Bacillus thuringiensis*	(49)

[a] thermophillic bacilli
[b] γ-glutamyltranspeptidese; Cm, chloramphenicol; Cd, cadmium; Em, erythromycin; Km, kanamycin; Sm, streptomycin; Smr, staphylococcal multidrug resistance; Tc, tetracycline

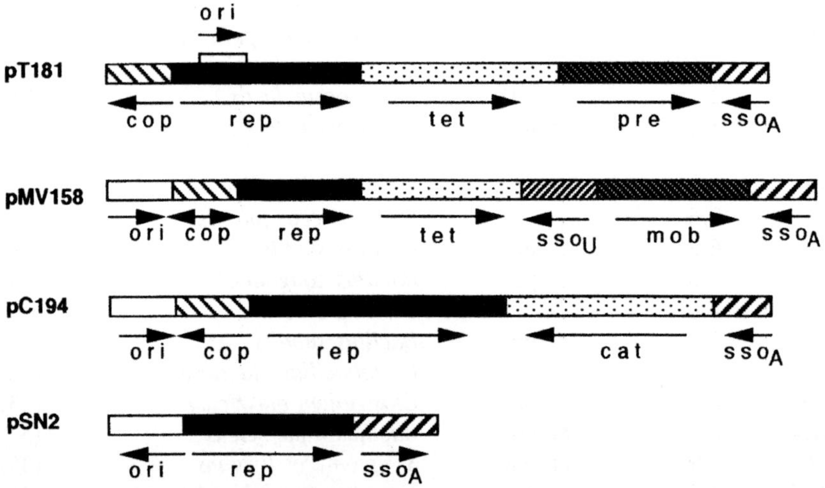

Figure 1. Functional organization of RC plasmids belonging to four different families. The maps are not drawn to scale. The direction of transcription of various genes and the direction of replication from the origins is shown. Plasmid pMV158 encodes two copy-control genes (an antisense RNA and a repressor) which are transcribed in opposite directions. Plasmid pMV158 contains both sso_A and sso_U sequences whereas the other plasmids contain only sso_A. ori, double-strand origin; sso, single-strand origin; rep, initiator gene; cop, copy-control gene(s); pre, recombinase gene; mob, mobilization function; tet, tetracycline-resistance gene; cat, chloramphenicol-resistance gene.

plasmids of the pT181 family and is responsible for specific Rep-origin interactions (51,52). The *dso* of the plasmids of the pC194 family is located immediately upstream of their *rep* genes (53). The Rep nick sites are located in a region with a potential to form a hairpin structure, but the requirement for such a structure in initiation has not been demonstrated. The nick sequence within the *dso* among the other members of the pC194/pUB110 family is conserved and is expected to correspond to the initiator nick site in all such plasmids (13,53). The binding site of the Rep proteins of pC194 and pUB110 within their respective origins has so far not been biochemically identified. The *dso* of the

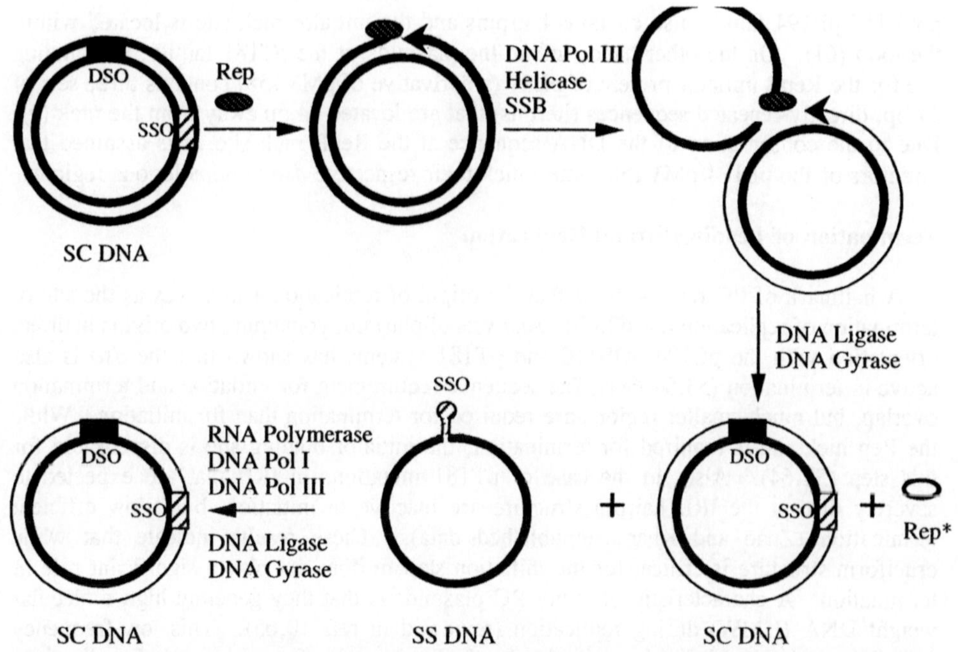

Figure 2. A model for the replication of RC plasmids. The plasmid initiator protein (Rep) binds to the leading-strand origin through sequence-specific interaction and nicks the DNA. This results in the generation of a free 3′ OH group and Rep is covalently attached to the 5′ phosphate of the DNA through its active tyrosine residue. Replication then initiates by extension synthesis at the nick involving DNA polymerase III, helicase and single-stranded DNA binding protein (SSB). After one round of replication, the Rep protein cleaves the regenerated nick site in the DS DNA followed by nicking-closing of the displaced leading-strand DNA. Replication of the leading strand is completed upon sealing of the nick by DNA ligase followed by supercoiling of the DNA by DNA gyrase. The Rep protein is inactivated (Rep*) after supporting one round of replication. Conversion of the SS DNA intermediate to the DS form initiates at the *sso* upon synthesis of RNA primers by RNA polymerase followed by elongation by DNA polymerase III. The primer is presumably removed by DNA polymerase I, and following repair synthesis, supercoiled (SC) plasmid DNA is generated by the actions of DNA ligase and DNA gyrase. The Rep protein shown in the model can function as a dimer or multimer. DSO, double strand origin; SSO, single strand origin.

```
                                                                Specificity Domain
                                                                |---------------------|

                                        Initiator Nick        Initiator Binding
                                                    ↓ |-------------------------------|
pT181     105 5' acGAtttagacAATTTTtcTAAAACCGGcTACTCTAATAGCCGGTTggacgcacAtactgTgtgcAtatCtgAt 3'  32
                     -------> <--------|------->      <--------    ---->     <----

pC221    1239 5' agGAtttagacAATTTTtcTAAAACCGGcTACTCTAATAGCCGGTTaagtggtaAttttTtaccAcccCtcAa 3'  1322
                     ------->  <-------|------->      <--------  ------>     <------

pC223     459 5' acGAttcagacAATTTTtcTAAAACCGGcTACTCTAATAGCCGGTTagacgcacAtttcgTgtgcActtCtaAt 3'  354
                     ------->  <-------|------->      <--------    ---->     <----

pS194    3417 5' agGAaaactcaAATTTTtcTAAAACCGGaTACTCTAATAGCCGGTTaaaccgacAtattaTgtacAcccCcgAa 3'  3490
                   ------>   <-----    -------->      <--------      ----> <----

pUB112   1316 5' gaGAaaattcaAATTTTgtTAAAACCGGaTACTCTAATAGCCGGTTaagtggtcAaacttTgggaAaatCtcAa 3'  1389
                    ----->  <-----     -------->      <--------       ---> <---
                      I        I           II              II          III   III
```

Figure 3. Comparison of the nucleotide sequences of the leading strand origins of plasmids of the pT181 family. Nucleotides which are completely conserved in all the five plasmids are shown in upper case letters while the variable nucleotides are shown in lower case. Inverted repeat sequences (IR) I, II and III are shown by horizontal arrows. The initiator binding sequence and the IRIII region that determines the specificity of initiator-origin interactions are shown by dotted lines. The nick site of the initiators present within the conserved IRII region is indicated by a vertical arrow. Replication of the leading strand proceeds in a rightward direction.

pMV158/pE194 family also contains hairpins and the initiator nick site is located within the loop (61). On the other hand, unlike the plasmids of the pT181 family, the binding site for the RepB initiator protein of pLS1 (a derivative of pMV158) contains three sets of 11 bp, directly-repeated sequences (iterons) that are located 84 bp away from the nick (9). Due to the conservation of the DNA sequence at the RepB nick site, it is assumed that initiators of the pE194/pMV158 family nick their respective *dso* at homologous regions.

Termination of Leading Strand Replication

A hallmark of RC replication is that the origin of replication also serves as the site of termination of replication (54,62,63). Analysis of plasmids containing two origins in direct orientation with the pC194/pUB110 and pT181 systems has shown that the *dso* is also active in termination (53,54,64). The sequence requirement for initiation and termination overlap, but much smaller regions are required for termination than for initiation. While the Rep nick site is required for termination, the initiator binding site is dispensable for this step (53,64). Also, in the case of pT181 mutations in IRII that are expected to severely disrupt the IRII hairpin structure are inactive in initiation, but allow efficient termination (Zhao and Khan, unpublished data). These results indicate that while cruciform structure is critical for the initiation step, it does not play a significant role in termination. A characteristic of many RC plasmids is that they generate high molecular weight DNA (HMW) during replication (reviewed in ref. 10,65). This low-frequency event occurs due to a lack of termination by the Rep protein after one round of replication. This results in the production of a sigma-type structure containing long tails (65).

The Initiator Proteins

The RC plasmids encode initiator proteins which have origin-specific nicking activity similar to those of SS DNA bacteriophages of *E. coli* (16,51,61,66,67). The initiator protein of pT181, RepC, was the first to be purified and shown to have origin-specific nicking-closing activity (16,56). Subsequently, the initiator proteins of pC221, pMV158 and pUB110 were purified and shown to have similar activities (51, 61, 66). Genetic analyses have shown that the initiators of many other RC plasmids also contain similar activities (10,53). The initiator proteins of the pT181 family bind adjacent to their nicking sites and also induce DNA bending and cruciform structure at the origin of replication (52,59,60). The initiator proteins belonging to individual plasmid families have extensive amino-acid sequence homology (10,11,13). An extensive alignment of the RC initiator proteins having nicking-closing activities has previously been published (68). The initiator proteins encoded by the members of the pT181 family initiator proteins have ~80% amino-acid identity (20). The initiator genes of this family also contain their origins of replication (19). When present in low to moderate levels, the initiators of the pT181 family are specific for the replication of their cognate origins *in vivo* (69). However, when present in excess, these initiators can support replication from non-cognate origins in the absence of a competing cognate origin (69). Similar results have also been obtained with *in vitro* replication systems for the pT181 and pC221 plasmids (58,70). RepC, the initiator of pT181, consists of 314 amino acids and has a molecular weight of 38,000 (56). It exists as a dimer and this form of the initiator is active in its various biological activities (55, Ansari and Khan, unpublished data). However, the dimerization domain of RepC and its related initiators has not yet been identified. The initiators of the pT181 family have been shown to contain distinct domains that are involved in their nicking and DNA binding activities. The active tyrosine residue involved in the nicking of the origin corresponds to position 191 in RepC and 188 in the pC221 initiator, RepD (50,71). A 17 amino-acid region surrounding this active tyrosine (residues 186 to 202 in RepC) is completely conserved in most members of this family (Figure 4) and may be important in nicking-closing of the DNA. Because of the conservation of this region and the Rep nicking sequence within their origins, all initiators of the pT181 family are able to nick the origins of all such plasmids (50,72). The variable regions of these initiators are located near their carboxyl terminal ends and contain their DNA binding domain (Figure 5). Amino acids 267 to 270 of RepC have been shown to be critical for its DNA-binding activity (72). These amino acids are also critical in determining the origin recognition and replication specificity of the initiators of the pT181 family. Exchanging amino acids 265 to 270 of RepC with those present in RepD and RepE (initiator encoded by pS194) has been shown to switch the DNA binding and replication specificity of the Rep protein (72). Unlike the RC initiators encoded by SS DNA bacteriophages of *E. coli*, the initiators of the RC plasmids are unable to support multiple rounds of replication. This is based on the observed regulated replication of these plasmids which is primarily regulated at the level of synthesis of their initiator proteins. The initiators of the pT181 and pUB110 families have been found to be inactivated after one round of replication by the attachment of a small oligonucleotide whose sequence corresponds to that located immediately downstream of the initiator nick sites (55,69). The modified initiators (RepC-RepC* dimer in the case of pT181 and RepU-RepU* multimer in the case of pUB110) are inactive in their nicking activity. The inactive RepU-RepU* protein is able to bind to the DNA (69) while the DNA-binding activity of RepC-RepC* has not yet been investigated. The initiators of bacteriophages such as ϕX174 contain two closely juxtaposed tyrosine residues (18). One tyrosine carries out the nicking at the origin during the initiation of replication, while the second tyrosine cleaves at the regenerated nick site after one round of replication and promotes reinitiation. Most RC plasmid initiators lack a second tyrosine near their active tyrosine residue. In the case of pT181, it has been suggested that termination of the first round of leading-strand replication may be carried out by the second monomer of the initiator (55). In this model, the oligonucleotide would be attached to the second monomer which was involved in nicking the DNA during the termination of replication. This model, however, does not explain why a Rep dimer (or multimer) is inactive in nicking since at least one monomer would still contain an active tyrosine residue. One possibility is that only one monomer is active in replication while the other subunit(s) is passively carried to provide a stable structure, but its active sites are not exposed/available for replication. However, this model is inconsistent with our *in vitro* data showing that heterodimers of two inactive RepC mutants (one defective in its nicking-closing and the other in its DNA-binding activity) can support replication (Ansari and Khan, unpublished data). Another possibility to prevent Rep reutilization includes the involvement of an amino acid other than tyrosine in the second nicking event during the termination step as observed with the pC194 Rep protein (see below).

Recent studies with the initiator protein RepA encoded by pC194 have revealed some novel properties of RC initiator proteins (73). At least three amino acids are critical for the nicking activity of this protein. They correspond to Glu-142, Glu-210 and Tyr-214 (73). The data obtained show that Tyr-214 is involved in the nicking of the DNA and covalent attachment to the 5′ phosphate to initiate replication. At the end of one round of replication, Glu-210 terminates replication by activating a water molecule to carry out a transesterification reaction which generates a circular SS DNA and the Rep protein is released. This prevents reinitiation of a new round of DNA replication (73). Recent studies also indicate that conversion of Glu-210 to Tyr (along with modification of Ser-213

```
          185         191                    200   203
           .           .                      .     .
pT181      N R F I R I Y N K K Q E R K D N A D A
pC221      D - - - - - - - - - - - - - - - - - V
pC223      - - - - - - - - - - K - - - E - - - V
pS194      D - - - - - - - - - - - - - - - - - I
pUB112     D - - - - - - - - - - - - - - - - - V
```

Figure 4. Alignment of the amino-acid sequence at the active site of the initiator proteins of the pT181 family. Numbering corresponds to the amino acids of the pT181 initiator protein, RepC. Identity of amino acids with those of RepC is indicated by dashes. The active tyrosine residue involved in the nicking of the DNA is underlined.

```
         250                 260                 270                 280                 290
          .                   .                   .                   .                   .
pT181     I V F M L L S D E E E W G K L H R N S R T K Y K N L I K E I S P V D L T D L M K S
pC221     M - Y L - - N E - G T - - - - E - H A K Y - - - Q - - - - - - - I - - - E - - - -
pC223     M - Y L - - H E - S K - - E - - - - - - R - - - Q I - Q - - - S I - - - - - - - -
pS194     M I Y - - I H E - S T - - - - E - R T K N - - R E M L - S - - E I - - - - - - - L
pUB112    M - Y T - - H E - S M - - - - S K - T K - - F - K M - R - - - - I - - - E - - - -
```

Figure 5. Amino-acid sequence of the DNA-binding domains of the initiator proteins of the pT181 family. Numbering corresponds to the amino acids of the pT181 initiator, RepC. Identity of amino acids with those of RepC is indicated by dashes. Amino acids that determine the specificity of the initiators in origin binding and replication are underlined.

to Ala to maintain local helical structure) results in a Rep protein that can reinitiate replication (Noirot-Gros and Ehrlich, unpublished data). Thus, the Rep protein of pC194 (and possibly other RC initiators) can behave like those of the SS DNA bacteriophages upon the presence of two tyrosines in the active site. These data indicate that since plasmid replication is regulated, their initiators contain nicking domains that do not allow reinitiation to occur and the initiators are presumably inactivated after one round of replication. In the case of pUB110, which is a member of the pC194 family, there is some evidence that its initiator protein (RepU) is inactivated after one round of replication by its conversion to a multimeric RepU-RepU* form that is attached to an oligonucleotide (69). This is surprising since these plasmids are closely related and the utilization of a glutamic acid residue in the termination step by the pC194 Rep protein would appear to rule out its attachment to an oligonucleotide. This issue needs to be further investigated.

The RepB protein of pMV158 has also been isolated and characterized for its biochemical activities (61). In general, its activities are similar to those of the initiators of the pT181 family. RepB exists as a hexamer in solution (61). Covalent DNA complexes with RepB have not been observed, suggesting that it may be similar to the gp II protein of filamentous bacteriophages (17,61). However, it is possible that RepB attaches covalently to the DNA in a transient manner.

The Single-Strand Origin of Replication

Most RC plasmids replicate by an asymmetric RC mechanism involving a SS DNA intermediate (54,63). The leading strand of the DNA released after one round of replication is converted to a DS form utilizing solely the host proteins (74). The single-strand origin (*sso*) is located close to the termination site for the leading-strand replication. Most RC plasmids have been found to contain palindromic sequences that act as *sso*. Plasmids of the pT181, pE194, pSN2 and pC194 families contain a single-strand origin termed sso_A (previously called *palA*) (reviewed in ref. 10,11). The sso_A contained in plasmids of different families have only limited sequence homology, but their folded structures are very similar (Figure 6 and ref. 11). The pUB110 plasmid contains a different *sso*, termed sso_U (75). The *sso* sequences are active in only one orientation. Deletion of sso_A results in plasmid instability and a reduction in copy number *in vivo* (11). However, since plasmids deleted in this region can still replicate to a limited extent, alternate sites could be used for lagging-strand synthesis (10,11). An *in vitro* system that supports SS→DS replication in the presence of various *sso* sequences has been developed (74). Both *in vivo* and *in vitro* experiments have shown that lagging-strand synthesis requires RNA polymerase-dependent synthesis of RNA primers from both sso_A- and sso_U -type origins (63,74-77). The sso_A-type origins contain a conserved central loop in their folded structures and a recombination site, RS_B, at the base of their stem in the palindromic structure (Figure 6). We have recently shown that RNA to DNA transition during lagging-strand synthesis from sso_A occurs at multiple positions (76). The first initiation point from the various sso_A sequences corresponds to initiation within or adjacent to the conserved central loop (Figure 6). This could be due to the synthesis of an RNA primer of a fixed size by RNA polymerase initiating at various positions within the *ssoA*. Alternatively, RNA polymerase may initiate primer synthesis at a unique position upstream of the central loop of palindromic sso_A structure and synthesize primers of various lengths followed by transition from RNA to DNA at multiple positions (76). The sso_A sequences do not appear to contain any promoter sequences. It is likely that the secondary structure of sso_A acts as a recognition signal for RNA polymerase. It is possible that the conserved RS_B sequence may serve as the entry point for the host RNA polymerase. In contrast to the sso_A-type origins, synthesis from sso_U of pUB110 initiates from a predominantly single position (76). The sso_A sequences appear to function well only in their natural hosts and are known to be inactive in *B. subtilis* (10,11). On the other hand, sso_U is active in a number of Gram-positive organisms (10). It is possible that RNA polymerase containing a similar sigma factor in various hosts is able to initiate primer synthesis from a specific site within the sso_U. It is intriguing that although RNA polymerase is involved in initiating lagging-strand synthesis from both the sso_A- and sso_U-type elements, different sequences and structures are used for initiation in these cases. The lagging-strand synthesis initiating from the sso_U- type origin is similar to those of the filamentous phages of *E. coli* such as M13, fd and f1 in which synthesis initiates from a single, unique position present within the origin (reviewed in ref. 17). The streptococcal plasmid, pMV158, contains both sso_A- and sso_U- type origins, and in addition, may contain an additional *sso* (78). The *sso* sequences of pMV158 and pWV01 can function in staphylococcal cell extracts *in vitro*. The *sso* of the lactococcal plasmid pWV01, termed sso_W, contains two inverted repeat sequences, IRI and IRII. Both of these inverted repeats are required for full *sso* activity, although IRI alone has limited *sso* activity (77). Interestingly, sso_W activity is only partly sensitive to rifampicin, suggesting that this *sso* can function in both RNA polymerase-dependent and independent manner (77). The RNA polymerase-independent mechanism may involve the primosome complex as found for ϕX174 (18). It is possible

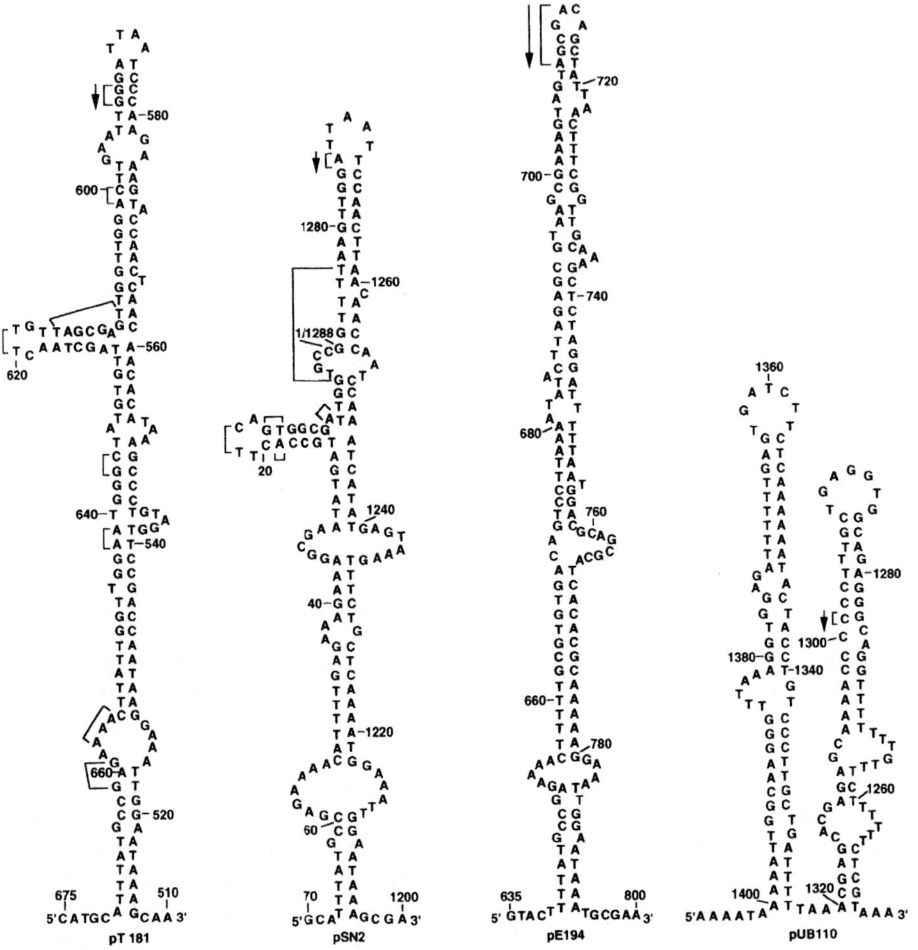

Figure 6. The folded *sso* structures of the various RC plasmid families. The parentheses indicate the major start sites of lagging-strand replication for various plasmids. The arrows indicate the initiation point for replication from the pUB110 sso_U, or the first initiation point from the pT181, pE194 and pSN2 sso_A sequences.

that the promiscuity of plasmids of Gram-positive bacteria that replicate by an RC mechanism may be determined, at least in part, by the efficiency of their *sso*s to function in various hosts. Therefore, the ability of *sso*s to support replication in different hosts may play a significant role in the horizontal spread of these plasmids.

The *cmp* Region

A genetic element termed *cmp* that effects the efficiency of origin utilization has been identified in the pT181 plasmid (79). This element acts as an enhancer of replication and is likely to facilitate the interaction of the initiator protein with the *dso*. The activity of *cmp* is orientation-independent and is inversely proportional to its distance from the origin (79). Deletion of *cmp* does not affect normal plasmid replication, but decreases its ability to compete with a wild-type origin for the initiator protein. A *cmp*-like element has also been found in the pUB110 plasmid (79,80).

REPLICATION OF THETA-TYPE PLASMIDS

In contrast to the small multicopy plasmids which replicate by an RC mechanism, several large (> 10 kb) plasmids have been found in Gram-positive bacteria that replicate by a theta-type mechanism. Two closely-related, low copy number conjugative plasmids, pIP501 and pAMβ1 isolated from *Streptococcus agalactiae* and *Enterococcus faecalis*, respectively, replicate by a unidirectional theta-type mechanism and constitute a large group of plasmids (Table 2). The replication regions of these two plasmids are 98% homologous (reviewed in ref. 65). The pIP501 and pAMβ1 plasmids encode replication initiator proteins of approximately 57 kD which act at their origins of replication. The origin has been localized to within a 52 bp region for pIP501 and does not contain directly-repeated sequences commonly found in many theta-type replicating plasmids of *E. coli* (86). The origin is located downstream of the initiator *repR* gene and replication initiates at a single site located 27 nt downstream of the termination codon of RepR (86,87). Replication then proceeds in the direction of transcription of this gene (88). Lagging-strand synthesis is initiated within 30 bp of the replication terminus at multiple positions. The termination site contains sequences that resemble the binding site of the replication terminator protein (RTP) of *B. subtilis* (87). Replication of pAMβ1 in *B. subtilis* can be arrested by a cloned chromosomal replication terminus sequence downstream of the origin (87). Thus, it is likely that the movement of the replication fork in pIP501, pAMβ1 and related plasmids is arrested during termination in a manner similar to that of the *B. subtilis*

Table 2. Theta-Type Replicating Plasmids in Gram-Positive Bacteria

Plasmid	Size	Resistance Gene	Original Host	Reference
pAMβ1	26.5	MLS	*Enterococcus faecalis*	(81)
pAC1	25.5	MLS	*Streptococcus pyogenes*	(82)
pEL1	26	MLS	*Streptococcus pyogenes*	(83)
pIP501	30.2	MLS,Cm	*Streptococcus agalactiae*	(84)
pIP612	34	MLS,Cm	*Streptococcus agalactiae*	(84)
pIP613	26	MLS,Tc	*Enterococcus faecalis*	(84)
pIP920	30	MLS,Cm	*Enterococcus faecium*	(85)
pMV103	26.5	MLS	*Streptococcus agalactiae*	(84)
pMV141	25	MLS	*Streptococcus agalactiae*	(84)
pRI405	25.5	Em	*Enterococcus faecalis*	(84)
pSM10419	22.5	MLS	*Streptococcus pyogenes*	(84)
pSM19035	27	Em,Lm	*Streptococcus pyogenes*	(84)
pSM22095	27	MLS	*Streptococcus pyogenes*	(84)

MLS, Macrolide, lincosamide, streptogramin B; Lm, lincosamine; Em, erythromycin; Cm, chloramphenicol; Tc, tetracycline

chromosome. In contrast, replication of RC plasmids does not terminate at a cloned termination sequence (89). These results suggest that many of the proteins involved in the replication of RC and theta-type plasmids are expected to be different. The replication properties of the *S. pyogenes* plasmid pSM19035 are also similar to those of pIP501 and pAMβ1 since these plasmids are highly homologous (90). These three plasmids belong to the incompatibility group *inc18* and their initiator proteins can recognize the replication origins of all the three plasmids (65). These plasmids have also been shown to replicate in *B. subtilis* (65). A number of studies have shown that RC plasmids of Gram-positive bacteria are structurally unstable (10). This is likely to be due to the presence of significant levels of SS DNA which are intermediates during RC replication. Additionally, the nick generated by the initiator proteins during the initiation and termination of replication may also be recombinogenic (10,65). On the other hand, plasmids of the pAMβ1 family that replicate by a theta-type mechanism are structurally stable (65). It is likely that differences in the mode of replication play an important role in plasmid stability. It would, therefore, appear that plasmids that replicate by a theta-type mechanism would be more suitable for use as cloning vectors for stable gene expression in Gram-positive organisms.

OTHER PLASMIDS OF GRAM-POSITIVE BACTERIA

A large number of other theta-type replicating plasmids that do not belong to the plasmids of the pAMβ1 family have also been isolated from Gram-positive bacteria (65). A number of linear plasmids have been isolated from *Streptomyces* that appear to replicate by a novel mechanism (91). These plasmids include pSLC and pSL2 and contain proteins bound at their termini (91). Interestingly, their replication does not initiate by protein-priming and strand-displacement mechanism observed for linear viruses such as ϕ29 and adenovirus that also contain a covalently-bound protein at their ends. Replication of these plasmids initiates from an internal origin and proceeds bidirectionally (91). The linear pSLC and pSLA2 plasmids contain telomeres at their ends (91). Upon deletion of their telomeric sequence, replication of these plasmids can still initiate from the internal origin but the plasmids replicate in a circular form (91).

Host Proteins Involved in Plasmid Replication

The information available on host proteins that are required for plasmid replication is limited. The PcrA helicase and DNA polymerase III have been shown to be involved in the RC replication of pT181 (11,92). Recent studies have also shown that DNA polymerase I is required for the normal stable replication of pLS1 in *S. pneumoniae* (93). However, this requirement is not absolute since pLS1 and pUB110 plasmids can be established in *B. subtilis polA* mutants (94). Replication of RC plasmids presumably also requires the SS DNA-binding protein, DNA ligase and DNA topoisomerases. DNA polymerase I has also been shown to be involved in the initiation of replication of pAMβ1-type plasmids and is subsequently replaced by PolIII about 200 bp downstream of the origin (95). Replication of pSM19035 in *B. subtilis* has been shown to require the products of the *dnaB, dnaC, dnaI* and *dnaG* (primase) genes (96). On the other hand, replication of pSM19035 does not require the DnaA protein. As discussed above, RTP is required for the termination of theta-type replicating plasmids but not for RC plasmids (89).

COPY-NUMBER CONTROL

The copy-number control of the RC- and theta-type replicating plasmids will be discussed together due to many similarities in the mechanisms of regulation of their replication. In all the cases where plasmid replication has been studied in detail in Gram-positive bacteria, the initiator protein has been found to be rate-limiting for replication. For most of these plasmids, copy number is regulated at the level of the synthesis of their initiator proteins (Table 3). A common feature in the regulation of most RC as well as theta-type plasmids is the involvement of an antisense RNA (countertranscript) which is complementary to the 5′ leader region of the initiator mRNAs. However, the mechanism of action of the antisense RNAs is variable in different systems. For the pT181, pIP501 and pAMβ1 plasmids, small antisense RNAs induce premature termination of transcription of the initiator mRNA (97,98). In the case of pT181, two small RNAs of 87 and 155 nt (RNAI and RNAII) are involved in transcriptional attenuation of the RepC mRNA (97). The leader region of the initiator mRNA can fold into an extensive secondary structure and contains four sequence elements I, II, III and IV which can generate hairpin structures that are critical for regulation. Region I can either pair with region II or III. Region III can also pair with IV, and the III-IV hairpin is an active rho-independent transcription terminator (97). The antisense RNAs can also fold into hairpin structures containing stem and loops. Interaction between antisense RNAs and the leader of RepC mRNA initiates by the formation of a "kissing" complex originally described for the ColE1 plasmid (1). This interaction involves the single-stranded loops of the antisense RNAs and RepC mRNA. The initial kissing complex presumably leads to the formation of a complete duplex in a zipper-like manner (1). As the RepC mRNA is being synthesized, the loop regions of RNAI can pair with complementary sequences in the loops of the RepC mRNA leader sequence. This would generate a folded structure in which I-II pairing occurs in RepC mRNA. As transcription of the RepC mRNA proceeds further, the leader region paired with RNAI would generate a new hairpin structure in which regions III and IV are paired. This structure causes premature termination of *repC* transcription (97). In the absence of RNAI, a stable I-III stem is formed in the leader region that preempts the formation of the transcription terminator (III-IV hairpin), and complete, functional RepC mRNA is synthesized. Consistent with this model is the observation that many point mutations that increase RepC synthesis and plasmid copy

Table 3. Elements Involved in the Copy-Number Control of Plasmids in Gram-Positive Bacteria

Plasmid	*cop* Product(s)	Mode of Action
pT181	Antisense RNAs; 85 and 150 nt	transcriptional attenuation of Rep mRNA
p353-2	Antisense RNAs; 75 and 250 nt	transcriptional attenuation of Rep mRNA
pLS1	1. Antisense RNA, 50 nt	inhibition of Rep mRNA translation
	2. CopG repressor, 5.1 kD	inhibition of *rep* gene transcription
pUB110	1. Antisense RNA	inhibition of Rep mRNA translation?
	2. RepU-RepU* complex	inhibition of *rep* gene transcription?
pIP501	1. Antisense RNA, 136 nt	transcriptional attenuation of Rep mRNA
	2. CopR repressor, 14 kD	inhibition of *rep* gene transcription
pAMβ1	1. Antisense RNA	transcriptional attenuation of Rep mRNA
	2. CopF repressor, 14 kD	inhibition of *rep* gene transcription

number map in the III-IV stem (97). It has been estimated that this antisense RNA-driven attenuation mechanism allows the synthesis of only about 3% of the full-length RepC mRNA (97). All the plasmids of the pT181 family contain copy control and Rep leader sequences that can form structures similar to those of pT181. It is therefore extremely likely that all such plasmids regulate their initiator synthesis by an identical mechanism. The copy number of the RC plasmid p353-2 of *L. pentosus* is also regulated by two antisense RNAs of 75 and 250 nt in a manner similar to that of pT181 (46). Although for pT181 and related plasmids replication is controlled at the level of RepC, overexpression of RepC has recently been shown to result in a decrease in plasmid copy number (99). Furthermore, RepC overexpression also inhibited chromosome replication and cell division. A number of point mutations in pT181 with increased copy number have previously been described, but these generally produce similar levels of RepC per plasmid copy as the wild-type plasmid (99). However, in the RepC overexpression system, the ratio of RepC per plasmid copy was approximately 15 times higher than in the case of wild-type pT181 (99). Although the mechanism of inhibition of pT181 replication in the presence of high levels of RepC is not understood, it may be due to the formation of inactive RepC-origin complexes (99). The decrease in chromosome replication may be due to the trapping of the PcrA helicase by RepC-pT181 origin complex, thereby sequestering PcrA which is also required for chromosome replication (92,99).

The copy number of theta-type replicating plasmids pIP501 and pAMβ1 is controlled by a small repressor protein in addition to an antisense RNA (98,100,101). These effects are exerted by negative regulation of the levels of the rate-limiting initiator proteins. In the case of the pIP501 plasmid, CopR and RNAIII have been shown to act independently in regulating the levels of the initiator protein, RepR (98,100). The antisense RNAIII as well as its target (leader region of RepR mRNA) can assume secondary structures (98). The secondary structure of RNAIII includes five loops, L1 to L5 (98). L3 and L5 are G-rich and G→A mutations in these loops result in an increase in plasmid copy number. This is probably due to reduced kissing complex formation between RNAIII and the leader region of RepR mRNA. Consistent with this hypothesis is the observation that deletions in the A-T rich loops L1 and L2 do not alter plasmid copy number (98). These results suggest that the interaction between RNAIII and the leader sequence initiates by the formation of a kissing complex. In the presence of RNAIII, folding of the leader region of RepA mRNA generates an inverted repeat, IRII, which acts as a transcription terminator (98). However, in the absence of RNAIII or in mutants carrying mutations in L3 or L5, deletion in the IRII sequence or those which alter the folded structure of RNAIII, the leader region of RepA mRNA folds into an alternate structure which lacks the transcription terminator structure. This, consequently, results in an increase in plasmid copy number. With both pT181 and pIP501 plasmids, mutations affecting the secondary structure of the antisense RNAs result in an increase in plasmid copy number. This demonstrates the importance of secondary structure in the regulatory functions of the antisense RNA molecules. A similar antisense RNA regulatory mechanism is also found in the case of the pAMβ1 plasmid (65). Plasmid pSM19035, which is closely related to pIP501 and pAMβ1, also contains Rep leader RNA and antisense RNA sequences that can fold into extensive secondary structures, and this plasmid is also likely to involve a copy-control mechanism that involves transcriptional attenuation. It is interesting that plasmids that replicate by a RC mechanism (pT181-type) and those by the theta-type mechanism (pIP501-type) have a conserved copy-control mechanism. The copy number of pIP501 and pAMβ1 is also regulated by *trans*-acting repressors CopR and CopF, respectively (100,101). These repressors are small polypeptides having a molecular weight of approximately 14 kD (as compared to 10.8 kD deduced from the DNA sequence). CopR and CopF have been shown to bind to inverted repeat sequences located immediately upstream of the promoter sequences of their respective initiator genes, *repR* and *repE* (100,101). Deletion of the *copR* and *copF* genes results in an approximately 10-fold increase in plasmid copy number (100,101). As stated earlier, the antisense RNAs and the Cop repressors act independently to regulate copy number in these systems. In the case of pAMβ1, these two levels of inhibition seem to be additive whereas this is not the case for pIP501 (100,101). The reason for this discrepancy is not clear. The copy-control repressors and their binding sites are highly conserved in the closely-related plasmids pAMβ1, pIP501 and pSM19035. Also, the CopF repressor of pAMβ1 is able to regulate the copy number of pIP501 and pSM19035. These data show that the CopF, CopR and CopS (repressor of pSM19035 replication) also correspond to the incompatibility function of these plasmids (101).

The copy-control mechanism of the streptococcal plasmid pLS1 (a derivative of pMV158) and belonging to the pE194 family has also been studied in some detail. Interestingly, although pLS1 belongs to the group of small multicopy RC plasmids, its copy number is controlled by both an antisense RNA and a repressor protein (102). The plasmid pLS1 encodes the RepB initiator protein whose synthesis is regulated at the transcriptional level by the 5.1 kD repressor protein, CopG (previously called RepA). The RepB and CopG proteins are part of a single transcription unit. CopG binds to a 13 bp symmetrical sequence which includes the -35 sequence of its promoter and inhibits the transcription of the initiator/repressor mRNA. This results in an inhibition of synthesis of both RepB and CopG (102). A five-fold increase in plasmid copy number is observed with plasmids containing mutations in the *copG* gene (102). CopG resembles several bacterial repressors and has a helix-turn-helix motif (9,102). The second copy control system of pLS1 involves an approximately 50 nt antisense RNA, termed RNAII (102). RNAII is transcribed from a constitutive promoter and its levels are proportional to the gene dosage. Although the mechanism of action of RNAII has not yet been elucidated, it has been postulated that it may pair with the putative ribosome binding site of the RepB mRNA and inhibit its translation (9,102). Several other members of the pLS1 family such as pE194, pADB201 and pLB4 may also potentially encode a repressor protein and an antisense RNA similar in function to those encoded by pLS1 (102). Thus, it is possible that the mechanism of copy-number control of several plasmids of the pE194 family may be similar. The RC plasmids pC194 and pUB110 are also likely to encode antisense RNAs which are complementary to the leader region of the initiator RNA, including the ribosome binding site (103). It has been suggested that these RNAs regulate the synthesis of the initiator proteins by inhibiting the translation of the Rep mRNA (103,104). Recently, it has been proposed that an inactive oligomer of pUB110 initiator protein (RepU-RepU*) which is unable to reinitiate replication can still bind to the *dso* (66). It was suggested that the binding site of the inactive initiator extends into the promoter region of the *repU* gene which could interfere with its transcription (66). This may represent another mechanism for the copy-number control of pUB110-type plasmids.

CONCLUDING REMARKS

In recent years, significant progress has been made in our understanding of the mechanism of replication and regulation of copy number of plasmids in Gram-positive bacteria. The initiator proteins from a large group of small, multicopy plasmids have been shown to contain origin-specific nicking activity and their replication involves a SS DNA intermediate. These and other results have revealed that such plasmids replicate by an RC mechanism previously described only for the SS DNA bacteriophages of *E. coli*. Indeed, there is a strong evolutionary relationship between RC plasmids of Gram-positive bacteria and SS DNA bacteriophages of *E. coli*. More recently, RC plasmids have also been identified in Gram-negative bacteria. Studies on the replication of RC plasmids have also revealed novel regulatory mechanisms. More recently, theta-type replicating plasmids of Gram-positive bacteria have also been studied and their mechanism of replication and its regulation is understood in some detail. The increase in our understanding of the replication properties of plasmids of Gram-positive bacteria has also facilitated the development of broad host-range vectors for these organisms. The pace of research in this field is likely to continue at a rapid pace and details of the biochemical mechanisms involved in replication and copy-number control of these plasmids are expected to be revealed.

ACKNOWLEDGMENTS

I thank Rais Ansari and Adam Zhao for helpful discussions. The author is grateful to M. Noirot-Gros and S. D. Ehrlich for helpful discussions and for providing unpublished results. The work in the author's laboratory is supported by grant GM 31685 from the National Institutes of Health.

REFERENCES

1 Tomizawa, J.-I. and Som, T. (1984) Cell 38, 871-878.
2 Abeles, A.L. and Austin, S.J. (1991) Proc. Nat. Acad. Sci. U.S.A. 88, 9011-9015.
3 Kittell, B.L. and Helinski, D.R. (1991) Proc. Nat. Acad. Sci. U.S.A. 88, 1389-1393.
4 McEachern, M.J., Bott, M.A., Tooker, P.A. and Helinski, D.R. (1989) Proc. Nat. Acad. Sci. U.S.A. 86, 7942-7946.
5 Abeles, A.L., Snyder, K.M. and Chattoraj, D.K. (1984) J. Mol. Biol. 173, 307-324.
6 Filutowicz, M., Dellis, S., Levchenko, I., Urh, M., Wu, F. and York, D. (1994) Progress in Nucl. Acid Res. and Mol. Biol. 46, 239-272.
7 Lin, L.-S. and Meyer, R.J. (1986) Plasmid 15, 35-47.
8 Murotsu, T., Tsutsui, H. and Matsubara, K. (1984) Mol. Gen. Genet. 196, 373-378.
9 del Solar, G.H., Moscoso, M. and Espinosa, M. (1993) Mol. Microbiol. 8, 789-796.
10 Gruss, A. and Ehrlich, S.D. (1989) Microbiol. Rev. 53, 231-241.
11 Novick, R.P. (1989) Annu. Rev. Microbiol. 43, 537-565.
12 Yang, X. and McFadden, B.A. (1993) J. Bacteriol. 175, 3981-3991.
13 Seery, L.T., Nolan, N.C., Sharp, P.M. and Devine, K.M. (1993) Plasmid 30, 185-196.
14 King, K.W. and Dybvig, K. (1992) Plasmid 28, 86-91.
15 MacDougall, J., Margarita, D. and Girons, I.S. (1992) J. Bacteriol. 174, 2724-2728.
16 Koepsel, R.R., Murray, R.W., Rosenblum, W.D. and Khan, S.A. (1985) Proc. Nat. Acad. Sci. U.S.A. 82, 6845-6849.
17 Baas, P.D. and Jansz, H.S. (1988) Curr. Top. Microbiol. Immunol. 136, 31-70.
18 Sims, J., Koths, S. and Dressler, D. (1979) Quant. Biol. 43, 349-365.
19 Khan, S.A. and Novick, R.P. (1983) Plasmid 10, 251-259.
20 Projan, S.J. and Novick, R. (1988) Plasmid 19, 203-221.
21 McDowell, D.G. and Mann, N.H. (1991) Plasmid 25, 113-120.
22 Aoki, T., Noguchi, N., Sasatsu, M. and Kono, M. (1987) Gene 51, 107-110.
23 Vujcic, M. and Topisirovic, L. (1993) Appl. Environ. Microbiol. 59, 274-280.
24 Xu, F., Pearce, L.E. and Yu, P.-L. (1991) Mol. Gen. Genet. 227, 33-39.
25 De Vos, W.M. (1987) FEMS Microbiol. Rev. 46, 281-295.
26 Leenhouts, K.J., Tolner, B., Bron, S., Kok, J., Venema, G. and Seegers, J.F.M.L. (1991) Plasmid 26, 55-66.

27 Perkins, J. and Youngman, P. (1983) J. Bacteriol. 155, 607-615.
28 Devine, K., Hogan, S., Higgins, D. and McConnell, D. (1989) J. Bacteriol. 171, 1166-1172.
29 De Rossi, E., Milano, A., Brigdi, P., Bini, F. and Riccardi, G. (1992) J. Bacteriol. 174, 638-642.
30 Bernhard, K., Schrempft, H. and Goebel, W. (1978) J. Bacteriol. 133, 897-903.
31 Darabi, A., Forough, R., Bhardwaj, G., Watabe, M., Goodarazi, G., Gross, S.S. and Watabe, K. (1989) Plasmid 22, 281-286.
32 Skaugen, M. (1989) Plasmid 22, 175-179.
33 Brehm, J.K., Pennock A., Bullman, H.M.S., Young, M., Oultram, J.D. and Minton, N.P. (1992) Plasmid 28, 1-13.
34 Murai, M., Miyashita, H., Araki, H., Seki, T. and Oshima, Y. (1987) Mol. Gen. Genet. 210, 92-100.
35 Kendall, K.J. and Cohen, S. (1988) J. Bacteriol. 170, 4634-4651.
36 Josson, K., Soetaert, P., Michiels, F., Joos, H. and Mahillon, J. (1990) J. Bacteriol. 172, 3089-3099.
37 Bouia, A., Bringel, F., Frey, L., Kammerer, B., Belarbi, A., Guyonvarch, A. and Hubert, J.-C. (1989) Plasmid 22, 185-192.
38 Muller, R.E., Ano, T., Imanaka, T. and Aiba, S. (1986) Mol. Gen. Genet. 202, 169-171.
39 Shishido, K., Noguchi, N., Kim, C. and Ando, T. (1983) Plasmid 10, 224-234.
40 Bron, S., Bosma, P., Van Belkum, M. and Luxen, E. (1987) Plasmid 18, 8-15.
41 Hoshino, T., Ikeda, T., Furukawa, K. and Tomizuka, N. (1985) Can. J. Microbiol. 31, 614-619.
42 McKenzie, T., Hoshino, T., Tanaka, T. and Sueoka, N. (1986) Plasmid 15, 93-103.
43 Hara, T., Nagatomo, S., Ogata, S. and Ueda, S. (1991) Appl. Environ. Microbiol. 57, 1838-1841.
44 LeBlanc, D.J., Lee, L.N. and Jaibet, A.-A. (1992) Plasmid 28, 130-145.
45 Grinius, L., Dreguniene, G., Goldberg, E.B., Liao, C.-H. and Projan, S.J. (1992) Plasmid 27, 119-129.
46 Pouwels, P.H., Luijk, N.V., Leer, R.J. and Posno, M. (1994) Mol. Gen. Genet. 242, 614-622.
47 Khan, S.A. and Novick, R.P. (1982) J. Bacteriol. 149, 642-649.
48 Projan, S.J., Monod, M., Narayanan, C.S. and Dubnau, D. (1987) J. Bacteriol. 169, 5131-5139.
49 Madsen, S.M., Andrup, L. and Boe, L. (1993) Plasmid 30, 119-130.
50 Thomas, C.D., Balson, D.F. and Shaw, W.V. (1988) Biochem. Soc. Trans. 16, 758-759.
51 Thomas, C.D., Balson, D.F. and Shaw, W.V. (1990) J. Biol. Chem. 265, 5519-5530.
52 Koepsel, R.R., Murray, R.W. and Khan, S.A. (1986) Proc. Nat. Acad. Sci. U.S.A. 83, 5484-5488.
53 Gros, M.F., te Riele, H. and Ehrlich, S.D. (1987) EMBO J. 6, 3863-3869.
54 Murray, R.W., Koepsel, R.R. and Khan, S.A. (1989) J. Biol. Chem. 264, 1051-1057.
55 Rasooly, A. and Novick, R.P. (1993) Science 262, 1048-1050.
56 Koepsel, R.R., Murray, R.W., Rosenblum, W.D. and Khan, S.A. (1985) J. Biol. Chem. 260, 8571-8577.
57 Gennaro, M.L., Iordanescu, S., Novick, R.P., Murray, R.W., Steck, T.R. and Khan, S.A. (1989) J. Mol. Biol. 205, 355-362.

58 Zock, J.M., Birch, P. and Khan, S.A. (1990) J. Biol. Chem. 265, 3484-3488.
59 Koepsel, R.R. and Khan, S.A. (1986) Science 233, 1316-1318.
60 Noirot, P., Bargonetti, J. and Novick, R.P. (1990) Proc. Nat. Acad. Sci. U.S.A. 87, 8560-8564.
61 Moscoso, M., del Solar, G. and Espinosa, M. (1995) J. Biol. Chem. 270, 3772-3779.
62 Khan, S.A., Murray, R.W. and Koepsel, R.R. (1988) Biochim. Biophys. Acta 951, 375-381.
63 te Riele, H., Michel, B. and Ehrlich, S.D. (1986) Proc. Nat. Acad. Sci. U.S.A. 83, 2541-2545.
64 Iordanescu, S. and Projan, S.J. (1988) J. Bacteriol. 170, 3427-3434.
65 Janniere, L., Gruss, A. and Ehrlich, S.D. (1993) in *Bacillus subtilis* and Other Gram-Positive Bacteria: Biochemistry, Physiology, and Molecular Genetics (Sonenshein, A.L., Hoch, J.A. and Losick, R., eds.) pp. 625-644, American Society for Microbiology, Washington, DC.
66 Muller, A.K., Rojo, F. and Alonso, J.C. (1995) Nucl. Acids Res. 23, 1894-1900.
67 Koepsel, R.R. and Khan, S.A. (1987) Nucl. Acids Res. 15, 4085-4097.
68 Ilyina, T.V. and Koonin, E.V. (1992) Nucl. Acids Res. 20, 3279-3285.
69 Iordanescu, S. (1989) Mol. Gen. Genet. 217, 481-487.
70 Khan, S.A., Koepsel, R.R., Murray, R.W. and Zock, J.M. (1990) in Molecular Biology of Staphylococci (Novick, R., ed.) pp. 197-210, VCH Publishers, New York, NY.
71 Dempsey, L.A., Birch, P. and Khan, S.A. (1992) Proc. Nat. Acad. Sci. U.S.A. 89, 3083-3087.
72 Dempsey, L.A., Birch, P. and Khan, S.A. (1992) J. Biol. Chem. 267, 24538-24543.
73 Noirot-Gros, M.F., Bidenko, V. and Ehrlich, S.D. (1994) EMBO J. 13, 4412-4420.
74 Birch, P. and Khan, S.A. (1992) Proc. Nat. Acad. Sci. U.S.A. 89, 290-294.
75 Boe, L., Gros, M.-F., te Riele, H., Ehrlich, S.D. and Gruss, A. (1989) J. Bacteriol. 171, 3366-3372.
76 Dempsey, L.A., Zhao, A.C. and Khan, S.A. (1995) Mol. Microbiol. 15, 679-687.
77 Seegers, J.F.M.L., Zhao, A.C., Khan, S.A., Venema, G. and Bron, S. Mol. Gen. Genet. (in press).
78 Kramer, M.G., del Solar, G. and Espinosa, M. (1995) Microbiology 141, 655-662.
79 Gennaro, M.L. and Novick, R.P. (1988) J. Bacteriol. 170, 5709-5717.
80 Bidnenko, V.E., Gruss, A. and Ehrlich, S.D. (1993) J. Bacteriol. 175, 5611-5616.
81 Clewell, D.B., Yagi, Y., Dunny, G.M. and Schultz, S.K. (1974) J. Bacteriol. 117, 283-289.
82 Clewell, D.B. and Franke, A.E. (1974) J. Antimicrob. Agents Chemother. 5, 534-537.
83 Malke, H., Jacob, H.E. and Storl, K. (1976) Mol. Gen Gen. 144, 333, 338.
84 Horaud, T., Le Bouguenec, C. and Pepper, K. (1985) J. Antimicrob. Agents Chemother. 16, 111-135.
85 Bourgueleret, L., Bieth, G. and Horodniceanu, T. (1981) J. Bacteriol. 145, 1102-1105.
86 Brantl, S. and Behnke, D. (1992) Mol. Microbiol. 6, 3501-3510.
87 Bruand, C., Ehrlich, S.D. and Janniere, L. (1991) EMBO J. 8, 2171-2177.
88 Le Chatelier, E., Ehrlich, S.D. and Janniere, L. (1993) Plasmid 29, 50-56.
89 Kaul, S., Mohanty, B.K., Sahoo, T., Patel, I., Khan, S.A. and Bastia, D. (1994) Proc. Nat. Acad. Sci. U.S.A. 91, 11143-11147.

90 Brantl, S., Behnke, D. and Alonso, J.C. (1990) Nucl. Acids Res. 18, 4783-4790.
91 Chang, P.-C. and Cohen, S.N. (1994) Science 265, 952-954.
92 Iordanescu, S. (1993) Mol. Gen. Genet. 241, 185-192.
93 Diaz, A., Lacks, S.A. and Lopez, P. (1994) Mol. Microbiol. 14, 773-783.
94 Alonso, J.C., Stiege, C.A., Tailor, R.H. and Viret, J.-F. (1988) Mol. Gen. Genet. 214, 482-489.
95 Huberman, J.A. (1995) Cell 82, 535-542.
96 Ceglowski, P., Lurz, R. and Alonso, J.C. (1993) FEMS Microbiol. Lett. 109, 145-150.
97 Novick, R.P., Iordanescu, S., Projan, S.J., Kornblum, J. and Edelman, I. (1989) Cell 59, 395-404.
98 Brantl, S., Birch-Hirschfeld, E. and Behnke, D. (1993) J. Bacteriol. 175, 4052-4061.
99 Iordanescu, S. (1995) Mol. Microbiol. 16, 477-484.
100 Brantl, S. (1994) Mol. Microbiol. 14, 473-483.
101 Le Chatelier, E., Ehrlich, S.D. and Janniere, L. (1994) Mol. Microbiol. 14, 463-471.
102 del Solar, G. and Espinosa, M. (1992) Mol. Microbiol. 6, 83-94.
103 Alonso, J.C. and Tailor, R.H. (1987) Mol. Gen. Genet. 210, 476-484.
104 Maciag, I.E., Viret, J.-F. and Alonso, J.C. (1988) Mol. Gen. Genet. 212, 232-240.

PATHWAYS OF PROTEIN REMODELING BY *ESCHERICHIA COLI* MOLECULAR CHAPERONES

Marie Pak and Sue H. Wickner

Laboratory of Molecular Biology
Bldg. 37, Rm. 2D19
National Institutes of Health
Bethesda, MD 20892-4255

INTRODUCTION

A new concept in molecular biology that has evolved over the past ten years is that proteins fold with assistance from other proteins, collectively referred to as molecular chaperones. All organisms, from bacteria to humans, possess several classes of highly conserved molecular chaperones, defined generally as proteins that bind to non-native conformations of proteins and facilitate correct folding and to native proteins and promote refolding. Extensive work has shown that the function as well as the structure of chaperones are highly conserved and the results obtained from studies of one organism are usually applicable to other organisms (1-3). This review will focus on several ATP-dependent molecular chaperone systems in *Escherichia coli*, including (i) DnaK, and its co-chaperones, DnaJ and GrpE, (ii) Clp proteins and (iii) GroEL and its co-chaperone, GroES (summarized in Table 1).

DnaK/DnaJ/GrpE, THE *E. coli* Hsp70 MOLECULAR CHAPERONE SYSTEM

DnaK is the major 70 kD heat shock protein (Hsp70) of *E. coli* (4) and is required for cell growth at both low and high temperatures (5). It functions *in vivo* with two other heat shock proteins, DnaJ and GrpE. Examples of some of the many cellular processes that are similarly affected by mutations in *dnaJ, dnaK* and *grpE* include DNA replication of *E. coli*, phage λ and plasmids mini-P1 and mini-F, RNA synthesis, cell division, protein transport, proteolysis, protection of enzymes from misfolding or aggregation during heat shock and the regulation of the heat shock response. Not surprisingly, there are DnaJ and GrpE homologues in eukaryotes (6-9) and recently they have been shown to function with Hsp70s (10-19). A DnaK homologue, named Hsc66 or HscA (20-22) and two DnaJ homologues, referred to as CbpA (for curved DNA binding protein A) and HscB (also called YfhE) (22,23), have recently been found in *E. coli*. However, in normal growth conditions they are expressed at lower levels than DnaK and DnaJ. HscA and HscB are co-transcribed and induced by cold shock (22) and CbpA is induced during stationary phase

Table 1. Major Families of *E. coli* Molecular Chaperones

Family	Chaperone	Partners	*In vivo* Functions	*In vitro* Functions
Hsp70	DnaK	DnaJ,GrpE	DNA replication; RNA synthesis; cell division; protein transport; proteolysis; heat-shock response; protein folding and disaggregation.	DnaK is a peptide-stimulated ATPase, DnaJ stimulates V_{max} and GrpE stimulates nucleotide exchange; together, they remodel the λ DNA replication initiation complex, activate the Rep proteins of plasmids mini-P1 and mini-F, reactivate proteins and prevent protein inactivation.
		CbpA,GrpE	CbpA is induced during stationary phase; substitutes for DnaJ when in multicopy.	CbpA binds curved DNA; functions with DnaK and GrpE in P1 RepA activation[a].
	HscA[b]	HscB	Induced by cold shock.	?
Hsp60	GroEL	GroES	Protein assembly; protein folding.	GroEL is an ATPase; GroES forms a complex with GroEL; GroEL/GroES fold proteins, stabilize unfolded, nascent polypeptides.
Clp[c]	ClpA	ClpP	Protein degradation with ClpP.	ClpA is a peptide-stimulated ATPase; ClpA undergoes ATP-dependent hexamerization; ClpA activates P1 RepA, prevents heat-inactivation of RepA and luciferase; ClpP is a serine peptidase; ClpAP degrades casein, RepA and other proteins.
	ClpB	?	Thermotolerance; Hsp104 (ClpB of yeast) functions in thermotolerance, RNA splicing, protein disaggregation, luciferase reactivation and [*psi*$^+$] regulation.	ClpB is a protein-stimulated ATPase; Hsp104 undergoes ATP-dependent hexamerization.
	ClpX	ClpP	Mu DNA replication; protein degradation with ClpP.	ClpX is a λO-stimulated ATPase; disaggregates λO, replicates Mu DNA with host proteins, disassembles MuA-DNA complexes; ClpXP degrades λO and MuA proteins.
	ClpY	?	?	?

[a] S. Wickner, unpublished data - [b] also referred to as Hsc66 - [c] also referred to as Hsp 100 family

under the control of σ^s (24). *In vivo* experiments have shown that a *cbpA, dnaJ* double deletion strain has more severe defects in cell growth than *dnaJ* or *cbpA* strains, suggesting that CbpA and DnaJ carry out some of the same functions (25). Furthermore, overexpressed CbpA substitutes for DnaJ for replication of phage λ and plasmid mini-F (23). CbpA and DnaJ also substitute for each other for plasmid mini-P1 growth, in that P1 is poorly maintained in a *dnaJ* deletion strain but is unable to grow in a *dnaJ, cbpA* double deletion strain (S. Wickner and K. McKenney, unpublished data).

The catalytic component of the DnaK/DnaJ/GrpE chaperone system is DnaK. It has a high affinity for ATP and very weak ATPase activity (26-28). Unfolded polypeptides bind DnaK and stimulate its ATPase activity (29-33). Analysis of the crystal structure of the ATPase domain of bovine Hsc70 has revealed that Hsp70 proteins are homologous to hexokinase and actin (34-36). Like other ATP-dependent enzymes, DnaK undergoes ATP-dependent conformational changes associated with its activity (37-40). DnaJ and GrpE can be thought of functionally as regulatory components of the chaperone activity of DnaK. DnaJ, a dimer of 39 kD subunits, stimulates the hydrolytic step of the DnaK ATPase reaction (32,33,41). GrpE, a dimer of 23 kD subunits, interacts with DnaK through a conserved loop in the ATPase domain (42) and stimulates the release of DnaK bound nucleotides, both ATP and ADP (33,41,43). GrpE is required for DnaK ATPase activity when the Mg^{2+} concentration is low (about 1 μM), implying that it facilitates the utilization of Mg^{2+} for an essential step of ATP hydrolysis (44). The combination of DnaJ and GrpE stimulates DnaK ATPase activity as much as 190-fold (33,41).

Initial biochemical studies of DnaK/DnaJ/GrpE function in phage λ DNA replication from the laboratories of Echols, Georgopoulos, McMacken and Zylicz laid the foundation for the current models of DnaK, DnaJ and GrpE action (45,46). For λ replication DnaK, DnaJ and GrpE are involved in remodeling the protein-DNA complex that assembles at the λ origin. The first step in the assembly is the binding of the initiator protein, λO, to the λ origin. The other λ initiator protein, λP, forms a complex with DnaB helicase, thereby inhibiting the ATPase and helicase activities of DnaB (47,48). Through interactions between λO and λP, a larger DnaB-λP-λO-DNA complex assembles (49). DnaK, DnaJ and GrpE, in an ATP-dependent reaction, rearrange the DnaB-λP-λO-DNA complex, releasing DnaB helicase from inhibition by λP and initiating replication in conjunction with host replication proteins (49-55).

For plasmid mini-P1 DNA replication, DnaK, DnaJ and GrpE activate origin-specific DNA binding by the initiator protein, RepA, *in vivo* and *in vitro* (5, 56-61). The mechanism of activation is the conversion of inactive RepA dimers to active monomers with the monomer, but not the dimer, binding avidly to RepA DNA binding sites in the P1 origin of replication (58,62). RepA dimers form stable tetramers with DnaJ dimers, suggesting that this complex may be the first intermediate in the activation reaction, targeting RepA for DnaK action (57). DnaK, in a reaction dependent on DnaJ and ATP but independent of DNA, converts RepA to active monomers, as judged by gel filtration. Neither DnaJ nor DnaK is part of the RepA-DNA complex. A requirement for GrpE in the *in vitro* reaction emerges when the Mg^{2+} concentration is very low (about 1 μM) (61).

Treatment of RepA dimers with chemical denaturants, like urea, at low protein concentration followed by dialysis converts RepA dimers to monomers and simultaneously activates P1 origin DNA binding, suggesting that DnaK/DnaJ/GrpE need only to monomerize but not modify RepA (62). Also in support of this suggestion, activation is reversed by increasing the RepA protein concentration. Sedimentation equilibrium experiments show that the K_D for the monomer-dimer equilibrium is 2.6 μM, 50-fold higher than the RepA concentration used in activation reactions, implying that DnaK/DnaJ/GrpE activate inactive RepA monomers (63). However, gel filtration experiments show that RepA exists solely as an inactive dimer at 0.8 μM that is capable of being activated by DnaK and DnaJ [Figure 4 in ref. (62)]. These differences in the monomer-dimer equilibrium may be due to the necessary extrapolation of the conditions used for sedimentation to those used for activation.

The situation with plasmid mini-F is very similar to mini-P1 in that it also requires DnaK, DnaJ and GrpE for plasmid maintenance (64,65). The F initiator protein, RepE, is isolated as a dimer but is unable to bind to RepE binding sites in the origin of replication. The importance of the dimer to monomer transition became clear from the characterization of a mutant *repE* that had been selected for its ability to replicate in a *dnaJ* mutant host (66). Interestingly, the purified mutant protein is a monomer and binds with high affinity to the origin DNA (67). Furthermore, treatment of wild-type RepE with urea converts dimers to monomers and activates the origin binding activity.

DnaK disaggregates the *E. coli* replication initiator protein, DnaA, and also activates it for *in vitro* DNA replication (68). This reaction does not require DnaJ or GrpE. Surprisingly, activation of mutant DnaA proteins requires DnaK and GrpE, but not DnaJ (69-71). Perhaps DnaK itself has a high affinity for DnaA mutant proteins, thus bypassing the need for DnaJ; however, GrpE is required to catalyze the nucleotide exchange reaction that is essential for substrate release.

Studies exploiting model protein reactivation systems, including RNA polymerase (72,73), rhodanese (74), luciferase (75,76), and UmuC (77) have demonstrated details of the mechanism of action of DnaK/DnaJ/GrpE in preventing inactivation and in reactivating denatured proteins. Schröder et al. showed *in vivo* that thermal inactivation of firefly luciferase expressed in *E. coli* is reversible and depends on the function of DnaK, DnaJ, GrpE, GroEL and GroES (75). In contrast, with purified proteins, DnaK, DnaJ and GrpE are sufficient to reactivate inactivated luciferase. Furthermore, GroEL and GroES do not reactivate luciferase by themselves. DnaJ suppresses aggregation of luciferase during heat-inactivation and this is an essential step in reactivation. DnaK, although not essential for this step, acts cooperatively with DnaJ to prevent aggregation, suggesting that DnaJ is targeting DnaK to the substrate. Further studies showed that denatured luciferase forms stable complexes with DnaJ and DnaK in a reaction dependent on ATP (76). Very likely this intermediate is stabilized by the hydrolysis of DnaK-bound ATP, since ADP is the only nucleotide found associated with the complex. Reactivation requires GrpE and Mg•ATP, indicating that both ADP release by GrpE and rebinding of ATP are essential for the release of the substrate (76).

Langer et al. used guanidine-denatured rhodanese as a model non-native substrate in a protein reactivation reaction (74). They discovered that DnaJ, in the absence of DnaK, GrpE, and ATP, prevents aggregation of rhodanese when the protein is diluted out of guanidine. Complexes of rhodanese and DnaJ have been observed by gel filtration, suggesting again a role of DnaJ in presenting the substrate to DnaK. Complexes of rhodanese, DnaJ and DnaK have also been observed in the presence or absence of ATP. However, in contrast to luciferase reactivation, DnaK, DnaJ, GrpE and ATP are not sufficient for rhodanese reactivation. Reactivation requires, in addition, GroEL and GroES. It was proposed that GrpE functions in the protein folding pathway to couple the transfer of rhodanese from its complex with DnaJ and DnaK to the GroEL/GroES complex. However, protein interactions between the DnaK chaperone system and the GroEL chaperone system have not been seen. An equally attractive model of rhodanese reactivation is that GrpE, by facilitating the exchange of ADP for ATP, stimulates the release of rhodanese in a form that can then bind to GroEL/GroES.

Ten years ago Pelham (78) proposed that in the heat shock response, heat inactivated and aggregated proteins are identified as targets for Hsp70 through the specific recognition of exposed hydrophobic regions. Then, in an ATP-dependent reaction, Hsp70 releases the bound polypeptide enabling it to refold in its native conformation. Our current model includes the action of DnaK on native as well as denatured proteins and the function of DnaJ and GrpE in the refolding pathway (Figure 1). It incorporates and expands major aspects of recently proposed models (17,33,58,62,74-76). For specific native proteins, including inactive RepA dimers (57), σ^{32} (79) and λP protein (54) and for some denatured proteins like heat-treated luciferase (76) and guanidine-treated rhodanese (74), DnaJ forms high affinity complexes, thus tagging the substrates for recognition by DnaK-ATP. DnaJ, by stimulating hydrolysis of DnaK-bound ATP, stabilizes the substrate-DnaK-ADP intermediate. For certain substrates, like RepA, luciferase and rhodanese, DnaJ remains as part of the complex with DnaK while for others, like σ^{32}, DnaJ transfers the substrate to DnaK (80,81). In the presence of Mg•ATP, GrpE catalyzes the exchange of ATP for DnaK-ADP resulting in dissociation of the complex and release of the substrate. This binding and release cycle is repeated until the protein is fully folded in its active conformation and no longer recognized by DnaJ.

While this model is useful for visualizing the DnaK, DnaJ and GrpE reactions, it is clear that more information regarding the interaction of the DnaK, DnaJ and GrpE with each other, with substrates and with nucleotides is necessary before the details of the refolding pathway will emerge.

Clp PROTEINS, A NEWLY DISCOVERED FAMILY OF MOLECULAR CHAPERONES

The molecular chaperone field took fresh, exciting strides in the past few years with the demonstration that members of the Clp protein family are chaperones (82-84) [reviewed in (85)]. Like other classes of ATP-dependent chaperones, Clp proteins are highly conserved and present in all organisms (86-88). All have ATP binding sites and many are induced by heat shock. *E. coli*, like many organisms, has multiple family members, including ClpA (86), ClpB (89), ClpX (90,91) and ClpY (also referred to as HslU) (92). Proteins in the ClpA and ClpB family contain two highly conserved ATP binding sites separated by a spacer region of different lengths, whereas ClpX and ClpY proteins contain a single ATP binding site (86,90,92). Proteins bearing homology to ClpA have been found in all other organisms tested, including yeast [Hsp104 (93) and Hsp78 (94)], plants (86), mice [SKD3 (95)], hamsters and humans (93).

ClpA

The founding member of the Clp family is ClpA. It was first identified by Katayama et al. (96,97) and by Hwang et al. (98,99) as the ATP component of a two component ATP-dependent protease [referred to as ClpAP or protease Ti; see reviews by Gottesman and Maurizi (100) and by Chung (101)]. ClpP, the proteolytic component, derives from an unrelated protein family and homologous ClpP proteins have been identified in plant chloroplasts (102). ClpP requires ClpA and ATP to degrade large protein substrates; however, it is able to degrade small peptides by itself (103-107). *clpA* and *clpP* mutants are defective in the degradation of abnormal proteins containing canavanine, an arginine analog (97), a ClpA-β-galactosidase fusion protein (108) and some derivatives of the β-galactosidase protein containing unstable, non-methionine N-terminal amino acids (109). Interestingly, mutations in the other major chaperone genes, including *dnaJ, dnaK, grpE* and *groEL*, also cause reduced levels of abnormal protein turnover (88,110,111).

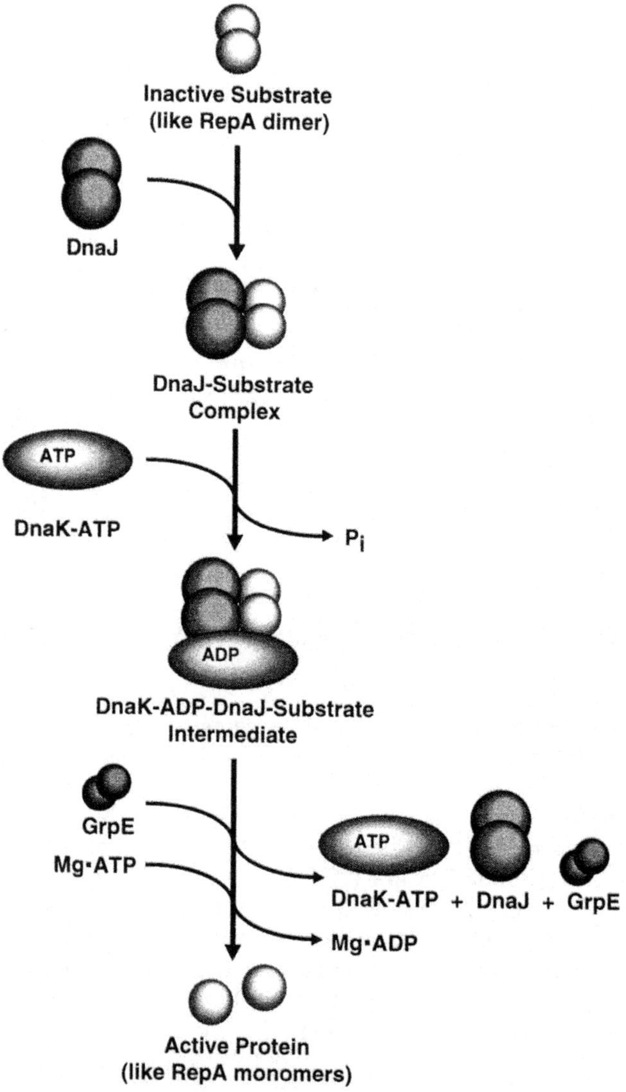

Figure 1. Model of the protein refolding pathway catalyzed by DnaK, DnaJ and GrpE (see text for discussion).

ClpA is isolated as a mixture of monomers and dimers but forms hexamers in the presence of Mg•ATP (112,113). ClpP is a tetradecamer arranged into two rings of seven subunits stacked on the axis of symmetry (113). Electron microscopic studies reveal that the ClpAP complex is strikingly similar in its overall structure to the eukaryotic 26 S ATP-dependent protease, despite little sequence homology between the two protease systems (113,114). Both have in their central domain a bipolar, multi-ring proteolytic component with a 7-fold symmetry. The proteolytic component is flanked by an ATPase component at either or both ends. The overall structural relatedness suggests a common mechanism of proteolysis between the prokaryotic and eukaryotic proteolytic systems. Recent analysis of the three-dimensional structure of the 20 S proteolytic component of the ATP-dependent protease from the archaebacterium *Thermoplasma acidophilum* should enhance current thinking on the structural and functional relationships among proteolytic systems (115).

Recently, Wickner et al. discovered that ClpA, in the absence of ClpP, performs the ATP-dependent chaperone function of DnaJ/DnaK/GrpE in the *in vitro* activation of P1 RepA (83) (discussed above). Studies of the intermediates in the RepA activation reaction have shed light on the mechanism of ClpA action (83) (Figure 2). RepA dimers bind ClpA hexamers in a reaction requiring ATP but not ATP hydrolysis. The stoichiometry of RepA to ClpA in the complex is about one RepA dimer per ClpA hexamer (M. Pak and S. Wickner, unpublished data). Upon ATP hydrolysis, RepA is released from ClpA and simultaneously converted into the active monomer form. ClpA activation of RepA is reversible (M. Pak and S. Wickner, unpublished data), suggesting that ClpA does not modify RepA. RepA is not activated by ClpB, Hsp104, ClpX or GroEL/GroES, in support of the idea that there is substrate specificity in the interaction of polypeptides and chaperones (83).

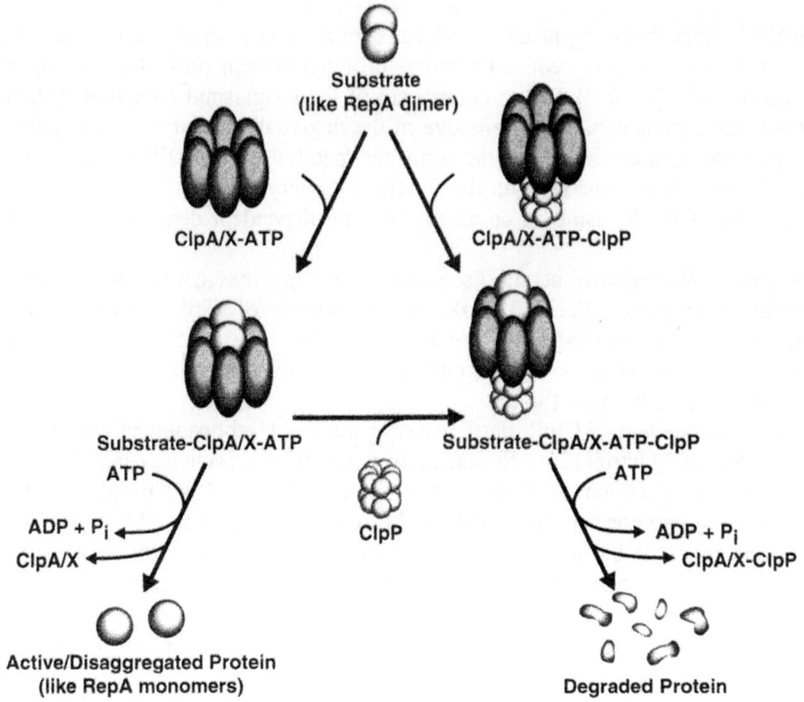

Figure 2. Model of the protein refolding pathway catalyzed by ClpA/X and protein degradation pathway catalyzed by ClpA/X and ClpP (see text for discussion).

The ClpA chaperone activity is further shown by its ability *in vitro* to prevent the irreversible heat-inactivation of firefly luciferase. Unlike DnaK/DnaJ/GrpE, ClpA is unable to reactivate luciferase (83). In another model system, ClpA protects RepA from heat-inactivation in the absence of ATP (D. Skowyra and S. Wickner, unpublished data).

In addition to activation of RepA, ClpA targets RepA for degradation by ClpP *in vitro* in the presence of ATP (83). ClpA is directly required for degradation and not simply generating monomeric substrates for ClpP, since RepA activated by DnaK/DnaJ/GrpE is not degraded by ClpP without ClpA. Interestingly, in the presence of ClpA, ClpP, ATP and DNA, some RepA is activated (and protected from degradation by binding to DNA) while some is degraded (S. Wickner, unpublished data).

Our model for the pathways of substrate activation and/or degradation is shown in Figure 2. First, ClpA interacts with the substrate in a reaction requiring ATP binding, but not ATP hydrolysis. Then, upon ATP hydrolysis, the substrate is released, very likely due to conformational changes in ClpA structure. The substrate has the opportunity to refold in its active form. By analogy with the known functions of chaperones, ClpA may go through cycles of substrate binding and release until the substrate is fully folded and active. Alternatively, if the substrate-ClpA-ATP complex associates with ClpP or if the substrate associates with ClpA-ATP-ClpP, then the substrate is degraded in a reaction also requiring ATP hydrolysis.

ClpX

clpX is a heat shock-regulated gene located in an operon downstream from *clpP* (90). Mutations in *clpX* or *clpP* reduce degradation of λO protein (90), RpoS (116), a mutant Mu repressor (117) and P1 Phd (a component of the P1 plasmid addiction system) (118). However, *clpX* mutants are not defective in the degradation of the ClpA-β-galactosidase fusion protein, demonstrating that the substrate specificity of ClpXP is different than that of ClpAP (90). Simultaneous with the *in vivo* discovery of ClpXP protease, Wojtkowiak et al. purified ClpXP, using as an assay the ATP-dependent degradation of λO protein (91).

Recently, Wawrzynow et al. discovered that ClpX has activities in common with molecular chaperones. Purified ClpX, in the absence of ClpP, prevents heat-induced aggregation of bacteriophage λO protein, a reaction also carried out by DnaJ (84). Furthermore, these workers demonstrated that ClpX disaggregates heat-denatured λO in an ATP-dependent reaction (84).

ClpX, independent of ClpP, participates in the DNA replication of bacteriophage Mu *in vivo* (119) and *in vitro* (120). Recent exciting results of Levchenko et al. have explained the mechanism of action of ClpX in this system (120). They discovered that ClpX functions as a chaperone in disassembling MuA-DNA complexes. MuA is a monomeric protein that forms stable tetramers upon binding to specific Mu DNA sequences during cleavage and joining reactions of Mu DNA transposition and replication. ClpX catalyzes the disassembly of MuA-DNA complexes by releasing MuA tetramers as monomers (120). This reaction, which requires ATP hydrolysis, does not involve ClpP or proteolysis. Once MuA is released, it can undergo further rounds of recombination. Thus in the cases of P1 RepA and MuA, reactions catalyzed by the chaperones are reversible and do not involve covalent modification of the protein substrates (120) (M. Pak and S. Wickner, unpublished data). Deletion analysis reveals that the C-terminal domain of MuA is critical for its disassembly and thus may be the domain that interacts with ClpX. ClpX together with ClpP degrades MuA *in vitro* (120). Interestingly, the C-terminal MuA deletion proteins are not degraded, suggesting that this domain is important in determining substrate specificity of ClpXP.

Thus, the discovery that ClpA and ClpX act alone as molecular chaperones as well as components of protease systems redirects our thinking towards the emerging relationship between pathways of protein folding, protein unfolding and protein degradation (Figure 2).

ClpB and the yeast homologue, *Hsp104*

The ClpB homologue in yeast, a 104 kD heat shock protein (Hsp104) important in thermotolerance (93,121), was one of the first Clp proteins shown to have molecular chaperone activity (82). Purified Hsp104 is a mixture of monomers and dimers but forms hexamers with a ring-like appearance upon addition of ATP resembling ClpA hexamers (122).

Parsell et al. discovered that Hsp104 *in vivo* mediates the reactivation and resolubilization of luciferase aggregates formed during exposure to high temperature (82). Furthermore, while both wild-type and *hsp104* cells accumulate aggregated proteins after heat shock, only in the wild-type cells are the aggregates dissolved during recovery at low temperature. Hsp104 also helps reactivate the RNA splicing reaction in heat-treated yeast cells both *in vivo* and in extracts (123). It has recently been shown that Hsp104 is involved in regulating the propagation of the yeast prion-like factor [psi^+] which causes suppression of nonsense mutations (124). It is proposed that Hsp104 alters the structural conformation of [psi^+] protein so that [psi^+] propagation, i.e., nonsense suppression, is inhibited. All three of the chaperone functions of Hsp104 can be performed at least partially by Hsp70s, indicating a functional relationship between Hsp104 and Hsp70s (82,123-125). Taken together, these results suggest that Hsp104 promotes proper renaturation and reactivation of damaged proteins, very likely by a mechanism similar to that of ClpA and ClpX (Figure 2). Presently, there is no evidence to suggest that Hsp104 participates in the turnover of heat-inactivated proteins.

The *E. coli clpB* gene, first identified by its sequence homology to ClpA, is a heat-shock protein (F84.1) (89). Null mutations in *clpB* reduce the growth rate at 44°C and exacerbate survival at higher temperatures, suggesting that ClpB, like Hsp104, is involved in thermotolerance (89). Also, like its yeast homologue, there is no evidence implicating ClpB in proteolysis *in vivo* or *in vitro*, although the purified protein has ATPase activity (126). Thus far ClpB has not been tested for molecular chaperone activity, with the exception that neither it nor Hsp104 activates P1 RepA *in vitro*.

GroEL/GroES, THE *E. coli* Hsp60 MOLECULAR CHAPERONE SYSTEM

A subset of molecular chaperones, called chaperonins, make up a class of ubiquitous and abundant proteins forming large toroidal complexes that mediate ATP-dependent protein folding. The best characterized chaperonin is GroEL, the *E. coli* 60 kD heat-shock protein (Hsp60) (1-3,127,128). GroEL functions with a co-chaperone, GroES, a 10 kD heat-shock protein. GroEL and GroES are constitutively expressed and essential for growth under all conditions, although they are stress inducible (129). The general role of GroEL in the folding of cytosolic proteins has been demonstrated by the analysis by Horwich et al. of a temperature-sensitive lethal *groEL* mutant (130). They found that after the mutant cells were shifted to the nonpermissive temperature (37°C), a large number of newly-synthesized proteins were found as insoluble aggregates. GroEL also functions during cellular stresses that cause protein denaturation by associating with and stabilizing preexisting proteins (130,131). Similarly, purified GroEL and GroES bind to unfolded proteins, preventing aggregation and promoting refolding and reactivation (132-135).

GroEL is a tetradecamer arranged into two heptameric stacked rings with a central cavity (136-138), whereas GroES is a heptameric protein of a single ring (139). In the presence of Mg•ATP or Mg•ADP, GroES forms a complex with GroEL by binding to either end (138,140,141) or in some non-physiological conditions to both ends of the

GroEL double ring (142,143). The crystal structure of GroEL confirms electron microscopic studies and reveals two stacked doughnut-shaped rings with a hole in the center (144). Chemical crosslinking and electron microscopic studies of substrate-bound GroEL show that substrates are bound in the central cavity (138,145). Furthermore, site-specific mutations in residues in the central cavity abolish polypeptide binding and also prevent the association with GroES (146).

The following general model of GroEL and GroES function has come from the *in vitro* characterization of model protein folding reactions (2,128,134,141,147-152). Briefly, complexes of GroEL and GroES containing tightly bound ADP bind compact molten globule-like substrates. Exchange of ADP with ATP and ATP hydrolysis triggers release of the bound substrate and GroES. The released substrate then has the opportunity to refold properly. Subsequent rounds of GroES and substrate association and dissociation result, after multiple cycles, in the fully folded and released substrate.

SUMMARY AND PERSPECTIVE

Molecular chaperones function in many, if not all, cellular processes that involve folding and remodeling of proteins and assembly and disassembly of protein complexes. Their essential nature is apparent from the many demonstrations *in vivo* and *in vitro* that more than one chaperone system can carry out one reaction. They perform these functions by both protecting nascent polypeptide chains and unfolded proteins from aggregation and proteolysis and by assisting in proper folding. Additionally, an intimate relationship between protein folding and degradation has emerged, suggesting that chaperones not only assist in the proper folding, but direct severely damaged proteins to degradation pathways.

ACKNOWLEDGMENTS

We dedicate this paper to the memory of Nat Sternberg. We are grateful to Keith McKenney, Hae-Yeong Kim and Nancy Trun for helpful discussions and comments.

REFERENCES

1 Gething, M.-J. and Sambrook, J. (1992) Nature 355, 33-45.
2 Hendrick, J.P. and Hartl, F.U. (1993) Annu. Rev. Biochem. 62, 349-384.
3 Morimoto, R.I., Tissieres, A. and Georgopoulos, C., eds. (1994) The Biology of Heat Shock Proteins and Molecular Chaperones. Cold Spring Harbor Laboratory Press, Cold Spring Harbor, NY, 610 pp.
4 Bardwell, J.C. and Craig, E.A. (1984) Proc. Nat. Acad. Sci. U.S.A. 81, 848-852.
5 Bukau, B. and Walker, G.C. (1989) J. Bacteriol. 171, 6030-6038.
6 Silver, P.A. and Way, J.C. (1993) Cell 74, 5-6.
7 Bolliger, L., Deloche, O., Glick, B.S., Georgopoulos, C., Jeno, P., Kronidou, N., Horst, M., Morishima, N. and Schatz, G. (1994) EMBO J. 13, 1998-2006.
8 Ikeda, E., Yoshida, S., Mitsuzawa, H., Uno, I. and Toh-e, A. (1994) FEBS Lett. 339, 265-268.
9 Laloraya, S., Gambill, B.D. and Craig, E.A. (1994) Proc. Nat. Acad. Sci. U.S.A. 91, 6481-6485.
10 Cyr, D.M., Lu, X. and Douglas, M.G. (1992) J. Biol. Chem. 267, 20927-20931.
11 Cheetham, M.E., Jackson, A.P. and Anderton, B.H. (1994) Eur. J. Biochem. 226, 99-107.
12 Cyr, D.M. and Douglas, M.G. (1994) J. Biol. Chem. 269, 9798-9804.

13 Nakai, M., Kato, Y., Ikeda, E., Toh-e, A. and Endo, T. (1994) Biochem. Biophys. Res. Commun. 200, 435-442.
14 Voos, W., Gambill, B.D., Laloraya, S., Ang, D., Craig, E.A. and Pfanner, N. (1994) Mol. Cell. Biol. 14, 6627-6634.
15 Cyr, D.M. (1995) FEBS Lett. 359, 129-132.
16 Freeman, B.C., Myers, M.P., Schumacher, R. and Morimoto, R.I. (1995) EMBO J. 14, 2281-2292.
17 King, C., Eisenberg, E. and Greene, L. (1995) J. Biol. Chem. 270, 22535-22540.
18 Levy, E.J., McCarty, J., Bukau, B. and Chirico, W.J. (1995) FEBS Lett. 368, 435-440.
19 Sugito, K., Yamane, M., Hattori, H., Hayashi, Y., Tohnai, I., Ueda, M., Tsuchida, N. and Ohtsuka, K. (1995) FEBS Lett. 358, 161-164.
20 Kawula, T.H. and Lelivelt, M.J. (1994) J. Bacteriol. 176, 610-619.
21 Seaton, B.L. and Vickery, L.E. (1994) Proc. Nat. Acad. Sci. U.S.A. 91, 2066-2070.
22 Lelivelt, M. and Kawula, T. (1995) J. Bacteriol. 177, 4900-4907.
23 Ueguchi, C., Kakeda, M., Yamada, H. and Mizuno, T. (1994) Proc. Nat. Acad. Sci. U.S.A. 91, 1054-1058.
24 Yamashino, T., Kakeda, M., Ueguchi, C. and Mizuno, T. (1994) Mol. Microbiol. 13, 475-483.
25 Ueguchi, C., Shiozawa, T., Kakeda, M., Yamada, H. and Mizuno, T. (1995) J. Bacteriol. 177, 3894-3896.
26 Zylicz, M., LeBowitz, J.H., McMacken, R. and Georgopoulos, C. (1983) Proc. Nat. Acad. Sci. U.S.A. 80, 6431-6435.
27 McKay, D.B. (1993) Adv. Protein Chem. 44, 67-98.
28 Wilbanks, S.M., Deluca-Flaherty, C. and McKay, D.B. (1994) J. Biol. Chem. 269, 12893-12898.
29 Flynn, G.C., Chappell, T.G. and Rothman, J.E. (1989) Science 245, 385-390.
30 Sadis, S. and Hightower, L.E. (1992) Biochemistry 31, 9406-9412.
31 Blond-Elguindi, S., Fourie, A.M., Sambrook, J.F. and Gething, M.-J.H. (1993) J. Biol. Chem. 268, 12730-12735.
32 Jordan, R. and McMacken, R. (1995) J. Biol. Chem. 270, 4563-4569.
33 McCarty, J.S., Buchberger, A., Reinstein, J. and Bukau, B. (1995) J. Mol. Biol. 249, 126-137.
34 Flaherty, K.M., DeLuca-Flaherty, C. and McKay, D.B. (1990) Nature 346, 623-628.
35 Flaherty, K.M., McKay, D.B., Kabsch, W. and Holmes, K.C. (1991) Proc. Nat. Acad. Sci. U.S.A. 88, 5041-5045.
36 Flaherty, K.M., Wilbanks, S.M., DeLuca-Flaherty, C. and McKay, D.B. (1994) J. Biol. Chem. 269, 12899-12907.
37 Liberek, K., Skowyra, D., Zylicz, M., Johnson, C. and Georgopoulos, C. (1991) J. Biol. Chem. 266, 14491-14496.
38 Banecki, B., Zylicz, M., Bertoli, E. and Tanfani, F. (1992) J. Biol. Chem. 267, 25051-25058.
39 Palleros, D.R., Reid, K.L., Shi, L., Welch, W.J. and Fink, A.L. (1993) Nature 365, 664-666.
40 Buchberger, A., Valencia, A., McMacken, R., Sander, C. and Bukau, B. (1994) EMBO J. 13, 1687-1695.
41 Liberek, K., Marszalek, J., Ang, D., Georgopoulos, C. and Zylicz, M. (1991) Proc. Nat. Acad. Sci. U.S.A. 88, 2874-2878.

42 Buchberger, A., Schroder, H., Buttner, M., Valencia, A. and Bukau, B. (1994) Struct. Biol. 1, 95-101.
43 Schonfeld, H.J., Schmidt, D., Schröder, H. and Bukau, B. (1995) J. Biol. Chem. 270, 2183-2189.
44 Skowyra, D. and Wickner, S. (1995) J. Biol. Chem. 270, 26282-26285.
45 Echols, H. (1990) J. Biol. Chem. 265, 14697-14700.
46 Zylicz, M. (1993) Phil. Trans. Roy. Soc. (Lond.) B339, 271-277.
47 Wickner, S. (1978) Cold Spring Harb. Symp. Quant. Biol. 43, 303-310.
48 Mallory, J.B., Alfano, C. and McMacken, R. (1990) J. Biol. Chem. 265, 13297-13307.
49 Dodson, M., Echols, H., Wickner, S., Alfano, C., Mensa-Wilmot, K., Gomes, B., LeBowitz, J., Roberts, J.D. and McMacken, R. (1986) Proc. Nat. Acad. Sci. U.S.A. 83, 7638-7642.
50 Alfano, C. and McMacken, R. (1989) J. Biol. Chem. 264, 10709-10718.
51 Alfano, C. and McMacken, R. (1989) J. Biol. Chem. 264, 10699-16708.
52 Dodson, M., McMacken, R. and Echols, H. (1989) J. Biol. Chem. 264, 10719-10725.
53 Zylicz, M., Ang, D., Liberek, K. and Georgopoulos, C. (1989) EMBO J. 8, 1601-1608.
54 Liberek, K., Osipiuk, J., Zylicz, M., Ang, D., Skorko, J. and Georgopoulos, C. (1990) J. Biol. Chem. 265, 3022-3029.
55 Osipiuk, J., Georgopoulos, C. and Zylicz, M. (1993) J. Biol. Chem. 268, 4821-4827.
56 Tilly, K., Sozhamannan, S. and Yarmolinsky, M. (1990) New Biol. 2, 812-817.
57 Wickner, S.H. (1990) Proc. Nat. Acad. Sci. U.S.A. 87, 2690-2694.
58 Wickner, S., Hoskins, J. and McKenney, K. (1991) Nature 350, 165-167.
59 Wickner, S., Skowyra, D., Hoskins, J. and McKenney, K. (1992) Proc. Nat. Acad. Sci. U.S.A. 89, 10345-10349.
60 Sozhamannan, S. and Chattoraj, D.K. (1993) J. Bacteriol. 175, 3546-3555.
61 Skowyra, D. and Wickner, S. (1993) J. Biol. Chem. 268, 25296-25301.
62 Wickner, S., Hoskins, J. and McKenney, K. (1991) Proc. Nat. Acad. Sci. U.S.A. 88, 7903-7907.
63 DasGupta, S., Mukhopadhyay, G., Papp, P.P., Lewis, M.S. and Chattoraj, D.K. (1993) J. Mol. Biol. 232, 23-34.
64 Ezaki, B., Ogura, T., Mori, H., Niki, H. and Hiraga, S. (1989) Mol. Gen. Genet. 218, 183-189.
65 Kawasaki, Y., Wada, C. and Yura, T. (1990) Mol. Gen. Genet. 220, 277-282.
66 Ishiai, M., Wada, C., Kawasaki, Y. and Yura, T. (1992) J. Bacteriol. 174, 5597-5603.
67 Ishiai, M., Wada, C., Kawasaki, Y. and Yura, T. (1994) Proc. Nat. Acad. Sci. U.S.A. 91, 3839-3843.
68 Hwang, D.S., Crooke, E. and Kornberg, A. (1990) J. Biol. Chem. 265, 19244-19248.
69 Hwang, D.S. and Kaguni, J.M. (1991) J. Biol. Chem. 266, 7537-7541.
70 Hupp, T.R. and Kaguni, J.M. (1993) J. Biol. Chem. 268, 13143-13150.
71 Hupp, T.R. and Kaguni, J.M. (1993) J. Biol. Chem. 268, 13137-13142.
72 Skowyra, D., Georgopoulos, C. and Zylicz, M. (1990) Cell 62, 939-944.
73 Ziemienowicz, A., Skowyra, D., Zeilstra-Ryalls, J., Fayet, O., Georgopoulos, C. and Zylicz, M. (1993) J. Biol. Chem. 268, 25425-25431.
74 Langer, T., Lu, C., Echols, H., Flanagan, J., Hayer, M.K. and Hartl, F.-U. (1992) Nature 356, 683-689.

75 Schröder, H., Langer, T., Hartl, F.-U. and Bukau, B. (1993) EMBO J. 12, 4137-4144.
76 Szabo, A., Langer, T., Schröder, H., Flanagan, J., Bukau, B. and Hartl, F.-U. (1994) Proc. Nat. Acad. Sci. U.S.A. 91, 10345-10349.
77 Petit, M.-A., Bedale, W., Osipiuk, J., Lu, C., Rajagopalan, M., McInerney, P., Goodman, M.F. and Echols, H. (1994) J. Biol. Chem. 269, 23824-23829.
78 Pelham, H. (1986) Cell 46, 959-961.
79 Gamer, J., Bujard, H. and Bukau, B. (1992) Cell 69, 833-842.
80 Liberek, K., Wall, D. and Georgopoulos, C. (1995) Proc. Nat. Acad. Sci. U.S.A. 92, 6224-6228.
81 Wawrzynow, A., Banecki, B., Wall, D., Liberek, K., Georgopoulos, C. and Zylicz, M. (1995) J. Biol. Chem. 270, 19307-19311.
82 Parsell, D.A., Kowal, A.S., Singer, M.A. and Lindquist, S. (1994) Nature 372, 475-478.
83 Wickner, S., Gottesman, S., Skowyra, D., Hoskins, J., McKenney, K. and Maurizi, M.R. (1994) Proc. Nat. Acad. Sci. U.S.A. 91, 12218-12222.
84 Wawrzynow, A., Wojtkowiak, D., Marszalek, J., Banecki, B., Jonsen, M., Graves, B., Georgopoulos, C. and Zylicz, M. (1995) EMBO J. 14, 1867-1877.
85 Horwich, A.L. (1995) Curr. Biol. 5, 455-458.
86 Gottesman, S., Squires, C., Pichersky, E., Carrington, M., Hobbs, M., Mattick, J.S., Dalrymple, B., Kuramitsu, H., Shiroza, T., Foster, T., Clark, W.P., Ross, B., Squires, C.L. and Maurizi, M.R. (1990) Proc. Nat. Acad. Sci. U.S.A. 87, 3513-3517.
87 Squires, C. and Squires, C.L. (1992) J. Bacteriol. 174, 1081-1085.
88 Gottesman, S., Wickner, S., Jubete, Y., Singh, S.K., Kessel, M. and Maurizi, M. (1995), Cold Spring Harb. Symp. Quant. Biol. (in press).
89 Squires, C.L., Pedersen, S., Ross, B.M. and Squires, C. (1991) J. Bacteriol. 173, 4254-4262.
90 Gottesman, S., Clark, W.P., de Crecy-Lagard, V. and Maurizi, M.R. (1993) J. Biol. Chem. 268, 22618-22626.
91 Wojtkowiak, D., Georgopoulos, C. and Zylicz, M. (1993) J. Biol. Chem. 268, 22609-22617.
92 Chuang, S.-E., Burland, V., Plunkett, G. III, Daniels, D.L. and Blattner, F.R. (1993) Gene 134, 1-6.
93 Parsell, D.A., Sanchez, Y., Stitzel, J.D. and Lindquist, S. (1991) Nature 353, 270-273.
94 Leonhardt, S.A., Fearson, K., Danese, P.N. and Mason, T.L. (1993) Mol. Cell. Biol. 13, 6304-6313.
95 Perier, F., Radeke, C.M., Raab-Graham, K.F. and Vandenberg, C.A. (1995) Gene 152, 157-163.
96 Katayama-Fujimura, Y., Gottesman, S. and Maurizi, M.R. (1987) J. Biol. Chem. 262, 4477-4485.
97 Katayama, Y., Gottesman, S., Pumphrey, J., Rudikoff, S., Clark, W.P. and Maurizi, M.R. (1988) J. Biol. Chem. 263, 15226-15236.
98 Hwang, B.J., Park, W.J., Chung, C.H. and Goldberg, A.L. (1987) Proc. Nat. Acad. Sci. U.S.A. 84, 5550-5554.
99 Hwang, B.J., Woo, K.M., Goldberg, A.L. and Chung, C.H. (1988) J. Biol. Chem. 263, 8727-8734.
100 Gottesman, S. and Maurizi, M.R. (1992) Microbiol. Rev. 56, 592-621.

101 Chung, C.H. (1993) Science 262, 372-373.
102 Maurizi, M.R., Clark, W.P., Kim, S.-H. and Gottesman, S. (1990) J. Biol. Chem. 265, 12546-12552.
103 Woo, K.M., Chung, W.J., Ha, D.B., Goldberg, A.L. and Chung, C.H. (1989) J. Biol. Chem. 264, 2088-2091.
104 Singh, S.K. and Maurizi, M.R. (1994) J. Biol. Chem. 269, 29537-29545.
105 Thompson, M.W. and Maurizi, M.R. (1994) J. Biol. Chem. 269, 18201-18208.
106 Thompson, M.W., Singh, S.K. and Maurizi, M.R. (1994) J. Biol. Chem. 269, 18209-18215.
107 Seol, J.H., Baek, S.H., Kang, M.S., Ha, D.B. and Chung, C.H. (1995) J. Biol. Chem. 270, 8087-8092.
108 Gottesman, S., Clark, W.P. and Maurizi, M.R. (1990) J. Biol. Chem. 265, 7886-7893.
109 Tobias, J.W., Shrader, T.E., Rocap, G. and Varshavsky, A. (1991) Science 254, 1374-1376.
110 Straus, D.B., Walter, W.A. and Gross, C. (1988) Genes Dev. 2, 1851-1858.
111 Keller, J.A. and Simon, L.P. (1988) Med. Microbiol. 2, 31-41.
112 Maurizi, M.R. (1991) Biochem. Soc. Trans. 19, 719-723.
113 Kessel, M., Maurizi, M.R., Kim, B., Kocsis, E., Trus, B.L., Singh, S.K. and Steven, A.C. (1995) J. Mol. Biol. 250, 587-594.
114 Peters, J.-M., Cejka, Z., Harris, J.R., Kleinschmidt, J.A. and Baumeister, W. (1993) J. Mol. Biol. 234, 932-937.
115 Lowe, J., Stock, D., Jap, B., Zwickl, P., Baumeister, W. and Huber, R. (1995) Science 268, 533-539.
116 Lee, K.-H., Schweder, T., Lomovskaya, O. and Matin, A. (1995) ASM General Meeting, Abstract H100.
117 Geuskens, V., Mhammedi-Alaoui, A., Desmet, L. and Toussaint, A. (1992) EMBO J. 11, 5121-5127.
118 Lehnherr, H. and Yarmolinsky, M.B. (1995) Proc. Nat. Acad. Sci. U.S.A. 92, 3274-3277.
119 Mhammedi-Alaoui, A., Pato, M., Gama, M.J. and Toussaint, A. (1994) Mol. Microbiol. 11, 1109-1116.
120 Levchenko, I., Luo, L. and Baker, T.A. (1995) Genes Dev. 9, 2399-2408.
121 Sanchez, L., Taulien, J., Borkovich, K.A. and Lindquist, S. (1992) EMBO J. 11, 2357-2364.
122 Parsell, D.A., Kowal, A.S. and Lindquist, S. (1994) J. Biol. Chem. 269, 4480-4487.
123 Vogel, J.L., Parsell, D.A. and Lindquist, S. (1995) Curr. Biol. 5, 306-317.
124 Chernoff, Y.O., Lindquist, S.L., Ono, B., Inge-Vechtomov, S.G. and Liebman, S.W. (1995) Science 268, 880-884.
125 Sanchez, Y., Parsell, D.A., Taulien, J., Vogel, J.L., Craig, E.A. and Lindquist, S. (1993) J. Bacteriol. 175, 6484-6491.
126 Woo, K.M., Kim, K.I., Goldberg, A.L., Ha, D.B. and Chung, C.H. (1992) J. Biol. Chem. 267, 20429-20434.
127 Landry, S.J. and Gierasch, L.M. (1994) Annu. Rev. Biophys. Biomol. Struct. 23, 645-669.
128 Todd, M.J., Viitanen, P.V. and Lorimer, G.H. (1994) Science 265, 659-666.
129 Fayet, O., Ziegelhoffer, T. and Georgopoulos, C. (1989) J. Bacteriol. 171, 1379-1385.
130 Horwich, A.L., Low, K.B., Fenton, W.A., Hirshfield, I.N. and Furtak, K. (1993) Cell 74, 909-917.

131 Martin, J., Horwich, A.L. and Hartl, F.-U. (1992) Science 258, 995-998.
132 Goloubinoff, P., Christeller, J.T., Gatenby, A.A. and Lorimer, G.H. (1989) Nature 342, 884-889.
133 Buchner, J., Schmidt, M., Fuchs, M., Jaenicke, R., Rudolph, R., Schmid, F.X. and Kiefhaber, T. (1991) Biochemistry 30, 1586-1591.
134 Martin, J., Langer, T., Boteva, R., Schramel, A., Horwich, A.L. and Hartl, F.-U. (1991) Nature 352, 36-42.
135 Mendosa, J.A., Rogers, E., Lorimer, G. and Horowitz, P.M. (1991) J. Biol. Chem. 266, 13044-13049.
136 Hendrix, R.W. (1979) J. Mol. Biol. 129, 375-392.
137 Hohn, T., Hohn, B., Engel, A. and Wurtz, M. (1979) J. Mol. Biol. 129, 359-373.
138 Langer, T., Pfeifer, G., Martin, J., Baumeister, W. and Hartl, F.-U. (1992) EMBO J. 11, 4757-4765.
139 Chandrasekhar, G.N., Tilly, K., Woolford, C., Hendrix, R. and Georgopoulos, C. (1986) J. Biol. Chem. 261, 12414-12419.
140 Engel, A., Hayer-Hartl, M.K., Goldie, K.N., Pfeifer, G., Hegerl, R., Muller, S., da Silva, A.C.R., Baumeister, W. and Hartl, F.-U. (1995) Science 269, 832-836.
141 Hayer-Hartl, M.K., Martin, J. and Hartl, F.-U. (1995) Science 269, 836-841.
142 Azem, A., Kessel, M. and Goloubinoff, P. (1994) Science 265, 653-655.
143 Schmidt, M., Rutkat, K., Rachel, R., Pfeifer, G., Jaenicke, R., Viitanen, P., Lorimer, G. and Buchner, J. (1994) Science 265, 656-659.
144 Braig, K., Otwinowski, Z., Hegde, R., Boisvert, D.C., Joachimiak, A., Horwich, A.L. and Sigler, P.B. (1994) Nature 371, 578-586.
145 Braig, K., Simon, M., Furuya, F., Hainfeld, J.F. and Horwich, A.L. (1993) Proc. Nat. Acad. Sci. U.S.A. 90, 3978-3982.
146 Fenton, W.A., Kashl, Y., Furtak, K. and Horwich, A.L. (1994) Nature 371, 614-619.
147 Martin, J., Mayhew, M., Langer, T. and Hartl, F.-U. (1993) Nature 366, 228-233.
148 Hayer-Hartl, M.K., Ewbank, J.J., Creighton, T.E. and Hartl, F.-U. (1994) EMBO J. 13, 3192-3202.
149 Hartl, F.-U. and Martin, J. (1995) Curr. Opin. Struct. Biol. 5, 92-102.
150 Weissman, J.S., Kashi, Y., Fenton, W.A. and Horwich, A.L. (1994) Cell 78, 693-702.
151 Weissman, J.S., Hohl, C.M., Kovalenko, O., Kashi, Y., Chen, S., Braig, K., Saibil, H.R., Fenton, W.A. and Horwich, A.L. (1995) Cell 83, 577-587.
152 Burston, S.G., Ranson, N.A. and Clarke, A.R. (1995) J. Mol. Biol. 249, 138-152.

PHEROMONES AND PHEROMONE RECEPTORS AS MATING-TYPE DETERMINANTS IN BASIDIOMYCETES

Lisa J. Vaillancourt and Carlene A. Raper

Markey Center for Molecular Genetics
Department of Microbiology and Molecular Genetics
Stafford Building
University of Vermont
Burlington, VT 05405

INTRODUCTION

Sexual variability within the fungi is extreme. For example, some fungi have only two mating types, while others have thousands. Is there a common genetic basis for these different mating systems? If so, how do the genes function to allow recognition of only a few, as contrasted to many thousands of potential mates?

Traditional genetic analyses revealed a unique system of multiple loci with numerous specificities as the basis for thousands of sexes in fungi (reviewed in 1). Recently, a new level of understanding has been achieved through molecular genetic studies. It is now known that different fungal mating systems do have common genetic denominators. Whether the system involves two mating types or thousands, it is based on regulatory genes similar to those found in many other eukaryotic organisms. Two kinds of regulatory proteins are important in fungal mating: transcriptional regulators, some of which belong to the homeodomain-protein family, and membrane-bound receptors and peptide pheromone molecules that initiate a signal-transduction cascade. In this review, we summarize our current understanding of the different ways in which pheromone and pheromone-receptor genes are organized within fungal genomes, and how this affects the spectrum of mating-type systems that exist among the "higher" fungi. Potential functions for these genes in self/non-self recognition and sexual development will also be discussed.

LIFE CYCLES AND MATING SYSTEMS

We will begin by briefly introducing the fungi to be considered in this review. Discussion is limited to members of two major groups: the Ascomycetes, which produce meiotic spores called ascospores in sac-like structures called asci, and the Basidiomycetes, in which meiotic spores called basidiospores are produced by specialized cells called basidia. Ascomycetes are further divided into two subclasses: the Hemiascomycetes, which include forms with naked asci, and the Euascomycetes, in which the asci are contained within a fruiting body. The Hemiascomycetes include the familiar budding yeast *Saccharomyces cereviseae* and the fission yeast *Schizosaccharomyces pombe.* The

Genetic Engineering, Vol. 18, Edited by J.K. Setlow
Plenum Press, New York, 1996

Euascomycetes are represented by *Neurospora crassa*. The Basidiomycetes are also divided into two groups. The Hemibasidiomycetes, which produce naked basidia, include the smuts such as *Ustilago* spp., yeast-like fungi such as *Rhodosporidium toruloides* and *Cryptococcus neoformans*, and jelly fungi including the *Tremella* spp. The Homobasidiomycetes produce basidia on the surfaces of gills or pores within fruiting bodies, and include most of the fungi popularly known as mushrooms. Examples to be discussed here are *Schizophyllum commune* and *Coprinus cinereus*.

Life Cycles

The generalized life cycle of an Ascomycete or Basidiomycete is described simply as follows (see Figure 1). Meiosis is followed by production of haploid ascospores or basidiospores. These propagate either as yeast cells by budding or fission, or as multicellular filaments called hyphae. The sexual part of the fungal life cycle is initiated by plasmogamy, i.e., fusion between two compatible elements; these can be yeast cells, filamentous hyphae, or differentiated sex cells or organs. Plasmogamy is followed by karyogamy to establish a diploid nucleus. Hemiascomycetes differ from the Euascomycetes and Basidiomycetes in that karyogamy occurs directly after plasmogamy. In the Euascomycetes and Basidiomycetes, plasmogamy is followed by production of a

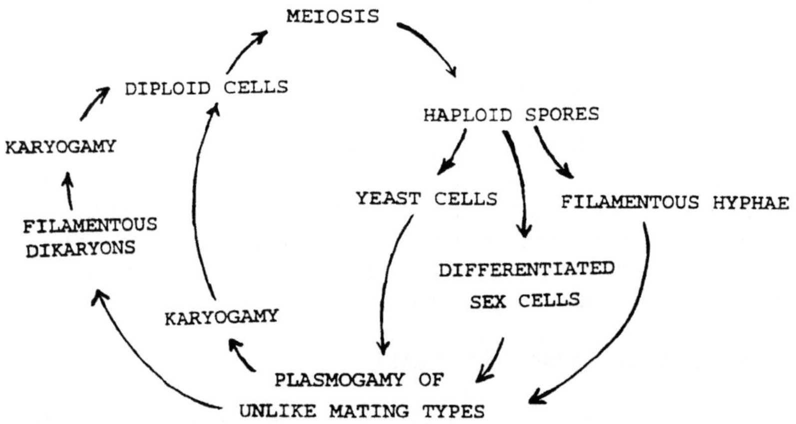

Figure 1. A generalized life cycle of Ascomycetes and Basidiomycetes.

specialized heterokaryotic mycelium known as a dikaryon. Pairs of haploid nuclei, one derived from each mate, are propagated and maintained in each cell. The dikaryon of Basidiomycetes can grow indefinitely before karyogamy finally occurs in basidial cells. The dikaryon of Euascomycetes is more limited and is associated only with the fruiting bodies. Karyogamy is generally followed immediately by meiosis. *S. cerevisiae* is exceptional in that it can propagate indefinitely as a diploid (a more comprehensive discussion of fungal taxa and life cycles can be found in ref. 2).

Mating Systems

Fungi can be divided into two major types: those which are homothallic and can self-fertilize, and those which are heterothallic and can mate only with another compatible individual. Both types are common throughout the fungi. Homothallism would be expected to result in a more genetically uniform population, while heterothallism would tend to increase genetic diversity between individuals. Either may be advantageous in different ecological niches. Among heterothallic fungi, sexual success requires the ability to recognize a compatible mate. This sometimes involves differentiated male and female organs, but sexual dimorphism of this sort is quite rare in fungi. Recognition of compatible mates more often involves genetic mating-type determinants, historically called incompatibility factors. To be sexually compatible, two individuals must have different incompatibility factors.

The number of compatible mating types among different groups of fungi varies. The heterothallic Ascomycetes have only two mating types. Superimposed on this genetic mating-type system, the Euascomycetes also exhibit sexual dimorphism, with one strain acting as a receptive female for a fertilizing conidium or hypha produced by another strain acting as a male. Strains are normally hermaphroditic, having the genetic capacity to act both as male and female. In contrast, the Hemiascomycetes, including *S. cerevisiae* and *Sch. pombe*, accomplish sex through fusion of two morphologically identical haploid cells of opposite mating type.

As a rule, Basidiomycetes do not produce male and female sexual structures. Mating is usually accomplished by fusion between morphologically identical cells. The number of different mating types in various species of Basidiomycetes can be as few as two, or more than 20,000! This fantastic number is not uncommon for the Homobasidiomycetes, some of which are among the most prolific outbreeders in the world (1).

The number of mating types is a product of two genetic features: the number of mating-type loci and the number of alternate versions (or specificities) of each of those loci. Studies of many heterothallic fungi have demonstrated that there are two basic arrangements: one in which mating type is determined by a single, nonrecombinable locus, and one in which it is determined jointly by two or more recombinable loci. Fungi having two mating types always have two alternate mating-type specificities at a single locus. Half the progeny of a cross will have the mating type of one parent and half will have that of the other. Such a bispecific mating system prevails in all of the Ascomycetes and many of the Hemibasidiomycetes (Table 1). Among Basidiomycetes with more than two mating types, many have multiple specificities at a single locus. An individual can mate with any other individual that has a different mating-type specificity. This system increases outbreeding potential. For example, in a one-locus, bispecific system, outbreeding potential is 50%; that is, an individual can mate successfully with only 50% of others within the population. In contrast, if an organism has one locus with ten alternate specificities, outbreeding potential is 90%. The potential for outbreeding increases as the number of alternate specificities increases. If the goal is to maximize outbreeding, then a multispecific system accomplishes that better than a bispecific system.

A small number of the Hemibasidiomycetes and many of the Homobasidiomycetes have two or more mating-type loci. In a two-locus system, four types of progeny result from a cross: one with the mating type of one parent, one with the mating type of the other, and two classes different from either parent resulting from recombination between the factors (e.g. from a cross of A1B1 X A2B2: progeny classes will be A1B1; A2B2; A1B2; A2B1). *Ustilago maydis* has a two-locus mating system. The two unlinked loci control complementary aspects of sexual development, and different specificities of both are

Table 1. Mating Systems of Fungi.

NUMBER OF MATING TYPE LOCI	NUMBER OF SPECIFIC-ITIES	TAXANOMIC GROUPING	PHEROMONES GENES IN MAT LOCUS?	DIMORPHIC? (YEAST-LIKE PHASE)	SPECIES IN THIS CHAPTER
ONE	TWO (BISPECIFIC)	HEMIASCO-MYCETES	NO	YES	*S. cerevisiae* *Sch. pombe*
ONE	TWO (BISPECIFIC)	EUASCO-MYCETES	NO	NO	*N. crassa*
ONE	TWO (BISPECIFIC)	HEMIBASIDIO-MYCETES	YES	YES	*U. hordei* *U. violacea* *R. toruloides* *C. neoformans*
ONE	MANY MULTISPECIFIC	HOMOBASIDIO-MYCETES	?	NO	-
TWO	TWO AT ONE LOCUS, MANY AT SECOND LOCUS	HEMIBASIDIO-MYCETES	YES	YES	*U. maydis* *T. mesenterica* *T.brasiliensis*
MORE THAN TWO	MANY AT ALL LOCI	HOMOBASIDIO-MYCETES	YES	NO	*S. commune* *C. cinereus*

required for mating. Thus, an individual can mate with only 25% of its siblings, *versus* 50% in a one-locus system. A two-locus system decreases inbreeding potential. Interestingly, no two-locus system has been described where there are only two alternate specificities of each of the loci. In the Hemibasidiomycetes one of the loci is bispecific and one is multispecific, and in the Homobasidiomycetes, both are multispecific. Outbreeding potential depends on the number of specificities at each locus. For example, an organism with two unlinked loci and ten specificities at each would have an outbreeding potential of 81%. If both loci have 100 specificities, outbreeding potential rises to 98%.

The best studied example of a multi-locus, multispecific mating system is that of the wood-rotting mushroom *Schizophyllum commune* (reviewed in 1,3,4). Mating is regulated by two unlinked genetic complexes called **A** and **B** which govern separate and complementary pathways of mating and sexual development. Each genetic complex consists of two closely linked loci, called α and β. The α and β loci of each complex function redundantly: strains need to differ at only one of the two to be compatible for the entire complex. For example, an individual with the genotype Aα1β1 is compatible for **A** with a strain having the genotype Aα1β2. In this case, the difference at β is sufficient to allow recognition as a different **A** mating type. In the world-wide population of *S. commune*, there are 9 specificities of Aα, 32 of Aβ (thus 288 different A specificities), 9 of Bα and 9 of Bβ (thus 81 different B specificities) (5-8). A very large number of mating types can be generated (more than 20,000) with relatively few different versions of these four loci.

DIFFUSIBLE PHEROMONES IN FUNGAL SEX

Some fungi produce diffusible substances which are able to affect compatible mates at a distance, causing them to undergo morphological changes in preparation for mating. Such substances were first observed and characterized in the Oomycetes, particularly *Achlya*, in the early part of this century (reviewed in 9). Diffusible sex factors can be divided into two chemically distinct types: isoprenoids, which include sterols such as those produced by *Achlya*, as well as sesquiterpenes and apocarotenoids, and peptides, which are produced predominantly by Ascomycetes and Basidiomycetes. Earlier writers named the diffusible substances produced by *Achlya* hormones, and we suggest retaining that term for the isoprenoid group, while using the term "pheromone" for the peptides which are the subject of this chapter. The use of two terms is justified, we believe, because it is likely that the two are different in their modes of action, as has already been discovered for similar types of signals in plants and animals (a more comprehensive discussion of hormones in fungi can be found in 10 and 11).

Pheromones in *Saccharomyces cerevisiae*

The role of diffusible pheromones in fungal sex is best understood in the budding yeast *S. cerevisiae*. Mating in yeast depends absolutely on the presence of pheromones and the various components of a pathway involved in pheromone response. A fruitful approach to understanding the role of pheromones in yeast mating has been the isolation and characterization of sterile (*STE*) mutants. Many components of the pathway are now identified and their roles have been defined. The yeast system serves as a model for other fungi which are less studied. We will begin, therefore, by summarizing the pheromone response system in yeast. Several recent reviews describe the system in detail (12-14).

In yeast, two morphologically identical haploid cells fuse to initiate the sexual part of the life cycle. Fusion is preceded by morphological changes and various cellular alterations in preparation for the event. *S. cerevisiae* has two mating types, called **a** and α. When mixed together, individual cells arrest in the G1 phase of the cell cycle and then elongate to form pear-shaped cells called "shmoos" (after a L'il Abner cartoon character of the same shape). Shmoos of opposite mating type fuse at the elongated tips, where proteins called agglutinins are deposited, allowing them to stick together. Cell-cycle arrest, shmoo formation, and induction of agglutinin genes and other mating-specific genes all take place prior to contact between compatible cells, in response to diffusible peptide pheromones. The **a** cells produce **a**-factor which induces changes only in α cells, while α cells produce α-factor which has a similar effect only on **a** cells. Active α-factor is a 13-amino-acid peptide (15). Two genes, *MFα1* and *MFα2*, encode large α-factor precursors (16,17). Each precursor contains a leader sequence and several tandem repeats of the mature pheromone sequence separated by protease cleavage recognition sites. The pheromone is glycosylated and processed before being secreted from the cells. The structure of **a**-factor was much harder to determine because of its hydrophobicity. This was eventually shown to be due to a farnesyl group attached to the C-terminal cysteine residue (18). Two genes, *MFa1* and *MFa2*, encode precursors of **a**-factor. Mature **a**-factor is a mixture of two 12-amino-acid lipopeptides, which differ at one amino acid (Table 2) (18,19). The N-terminus of the precursor is cleaved off, the farnesyl group is added to the cysteine four amino acids from the end, the three C-terminal amino acids

Table 2. Structures of Mature Fungal Lipopeptide Pheromones. Farnesyl Moieties Are Indicated by Superscripts; Structure Shown Below.

SPECIES	PHEROMONE	GENE	MATURE PHEROMONE	REFS.
S. cerevisiae	**a**-factor	*MFa1* *MFa2*	Y-I-I-K-G-V/L-F-W-D-P-A-C^1-OCH$_3$	(18,19)
Sch. pombe	M-factor	*mfm1*$^+$ *mfm2*$^+$ *mfm3*$^+$	Y-T-P-K-V-P-Y-M-C^1-OCH$_3$	(28)
R. toruloides	rhodotorucine A	*RHA1* *RHA2* *RHA3*	Y-P-E-I-S-W-T-R-N-G-C^1-OH	(44)
T. mesenterica	tremerogen A-10	?	E-H-D-P-S-A-P-G-N-G-Y-C^2-OCH$_3$	(55)
T. mesenterica	tremerogen A-13	?	E-G-G-G-N-R-G-D-P-S-G-V-C^1-OH	(56)
T. brasiliensis	tremerogen A-9291-1	?	D-S-G-S/G-S-R-D-P-G-A-S-S-G-G-C^2-OCH$_3$	(57)
U. maydis	a1 pheromone	*mfa1*	G-R-D-N-G-S-P-I-G-Y-S-S-C$^{1?}$-OCH4$_{3?}$	(79)
U. maydis	a2 pheromone	*mfa2*	N-R-G-Q-P-G-Y-Y-C$^{1?}$-OCH$_{3?}$	(79)

1. 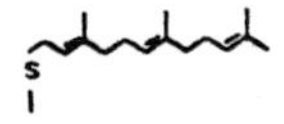

2. 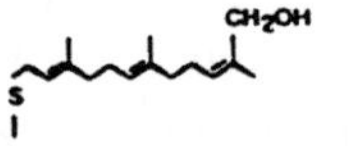

are removed, and the farnesylated cysteine is finally carboxymethylated to produce the active pheromone, which is secreted by a unique pathway (reviewed in 20).

Specific membrane-bound receptors for the two pheromones are also produced by *S. cerevisiae*. Cells of the **a** mating-type produce an α-factor receptor, and α cells produce an **a**-factor receptor. The **a**-factor receptor is encoded by the *STE3* gene, while the α-factor receptor is encoded by *STE2* (21,22). The two receptors are unrelated at the amino-acid sequence level, but their secondary structures are very similar. Each consists of an extracellular N-terminus, seven hydrophobic α-helical regions which probably traverse the cell membrane, and an intracellular C-terminus (22,23). This structure is characteristic of the rhodopsin-like superfamily of membrane-bound receptors which commonly interact with heterotrimeric G proteins. In *S. cerevisiae*, either receptor binds to the same G protein to activate a common signaling pathway (24). Pheromone binding to the receptor causes GTP to replace GDP attached to the Gα subunit, leading to dissociation of the α subunit from the $\beta\gamma$ subunits. G$\beta\gamma$ then activates the downstream components of the signaling pathway, including several serine-theronine protein kinases (reviewed in 14). Ultimately, activation of the pathway results in activation of a transcription factor, encoded by the *STE12* gene, which elicits the transcription of numerous genes involved in the differentiation of a mating-competent cell (25,26). These include agglutinins, cyclins involved in G1 arrest, and the pheromones and receptors themselves, which are upregulated in activated cells. Although this description of yeast mating is simplified, it will serve as a useful baseline for comparing and contrasting pheromone systems in other fungi.

Pheromones in *Schizosaccharomyces pombe*

Other Hemiascomycetes, including the fission yeast *Sch. pombe*, have pheromone response systems which in many ways are similar to that of *S. cerevisiae*. *Sch. pombe* also has two mating types. Plus cells produce a pheromone called P-factor which causes a response specifically in minus cells, and minus cells produce M-factor, active only in plus cells. Active P-factor is a 23-amino-acid, unmodified peptide (27). The *map2*$^+$ gene encodes a large precursor containing a signal leader sequence and several tandem repeats of the mature P-factor sequence, interspersed with protease cleavage recognition sequences. Thus, P-factor is comparable to *S. cerevisiae* α-factor. M-factor is a nine-amino-acid peptide which is modified by addition of a farnesyl residue on the C-terminal cysteine, which is also carboxymethylated (Table 2) (28). M-factor is encoded by three genes, *mfm1*$^+$, *mfm2*$^+$ and *mfm3*$^+$ (28,29). Each gene codes for a precursor which is processed and modified to yield the mature pheromone. All three genes are used for pheromone production by *Sch. pombe*. M-factor is comparable to *S. cereviseae* **a**-factor. *Sch. pombe* also produces two specific membrane-bound pheromone receptors; the P-factor receptor is encoded by the *mam2*$^+$ gene, and is similar to the *S. cerevisiae* α-factor receptor encoded by *STE2*, while the M-factor receptor is encoded by the *map3*$^+$ gene and is similar to the **a**-factor receptor encoded by *STE3* (30,31). Both receptors belong to the rhodopsin-like superfamily, and are bound to a heterotrimeric G protein. The Gα subunit in *Sch. pombe* is encoded by the *gpa1*$^+$ gene (32). One major difference between *S. cerevisiae* and *Sch. pombe* is illustrated by the phenotype in each case of a Gα deletion mutant. In *S. cerevisiae*, these mutants are constitutively activated for mating, which led to the hypothesis that Gβγ was the activating component. In *Sch. pombe*, in contrast, Gα deletion mutants are sterile, indicating that it is the Gα subunit which interacts with downstream components of the pathway (32). Another major difference between the two species is that in *S. cerevisiae,* haploid cells constitutively express all of the components of the pheromone response pathway, while in *Sch. pombe*, much of the pathway is not expressed except under conditions of nitrogen starvation. *Sch. pombe* will mate only under starvation conditions, and remains unresponsive to pheromone as long as nitrogen levels are high. In *S. cerevisiae*, a drop of purified pheromone placed on a lawn of haploid cells of the opposite mating type results in a clear "halo", caused by cells arresting in G1 in response to the pheromone. This halo effect is not seen in *Sch. pombe*. In fact, it was not known until recently that *Sch. pombe* pheromones can actually cause G1 arrest, because cells become responsive to pheromone only when they are starved and the mitotic cycle is arrested anyway. G1 arrest in response to pheromone was demonstrated by testing cells containing two mutations, one resulting in the inability to synthesize cAMP (high levels of cAMP normally repress components of the pheromone response pathway under high nitrogen conditions) and another resulting in the inability to degrade the pheromone from the opposite mating type, thus causing hypersensitivity to the pheromone. G1 arrest in response to pheromone can be observed only in these double mutants (27). Downstream components of the pheromone response pathways in *Sch. pombe* and *S. cerevisiae* are quite similar, including several serine-threonine protein kinases (33,34). Another difference between the two species is that the pheromone response pathway appears to play an important role in meiosis in *Sch. pombe*, but not in *S. cerevisiae* (35). This may be related to the fact that meiosis proceeds quickly from the diploid stage in *Sch. pombe*, whereas the diploid stage can be quite long in *S. cerevisiae*, and meiosis does not occur until after nutrients become limiting. In fact, certain genes involved in the pheromone response pathway are repressed in diploid *S. cerevisiae* cells, and it is therefore impossible for pheromone response to play a role in regulating meiosis (36,37).

Organization of the Mating-Type Loci and Mating-Type Switching in Ascomycete Yeasts

In both *S. cerevisiae* and *Sch. pombe*, mating type is ultimately determined by a single locus with two alternative specificities. The genes of each specificity encode regulatory proteins which either activate or inhibit a subset of unlinked mating-type specific genes, including those for the pheromones and receptors (reviewed in 38,39). In each species, the DNA sequences of the two alternative specificity loci are so different as to suggest that they may have arisen independently of one another. Most strains of *S. cerevisiae* and *Sch. pombe* frequently switch mating types by a mechanism involving translocation of the portion of the chromosome containing the mating-type locus (reviewed in 38,39). Each haploid genome contains genes of both mating-type specificities at different chromosomal locations; one set is active, and the other is "silent". Each haploid genome also contains unlinked genes for both types of pheromones and pheromone receptors. The presence of genes encoding both types of these molecules in each haploid genome is necessary to allow mating-type switching. An important difference between these two Ascomycete yeasts and the Basidiomycetes, which we will discuss next, is that each haploid genome in the Basidiomycetes contains only the pheromone and receptor genes of its own mating type, so mating-type switching cannot occur. Another major difference is that the pheromone and receptor genes are contained within the mating-type loci of Basidiomycetes where they serve as primary mating-type determinants.

Pheromones in the Hemibasidiomycetes

Several of the dimorphic Hemibasidiomycetes, which have yeast-like phases in their life-cycles, have been shown to produce diffusible lipopeptide pheromones. This is perhaps not surprising, given their morphological similarity to the Hemiascomycetes. Dimorphic Hemibasidiomycetes include the plant-pathogenic smuts as well as the saprophytic jelly fungi and the Basidiomycete yeasts. The presence of diffusible mating factors was suspected many years ago based on microscopic observations of some of these Hemibasidiomycetes (e.g., 40-42). The sexual phase begins with fusion between two compatible haploid yeast cells in these fungi, as in the Ascomycete yeasts. Rather than shmoos, however, fusion occurs between distinct conjugation hyphae which are formed in response to a diffusible pheromone produced by the opposite mating type. After plasmogamy, a dikaryotic mycelium is established. Eventually, given the appropriate environmental cues, basidial cells develop and karyogamy and meiosis are followed by production of basidiospores. These haploid cells reproduce vegetatively by budding until they fuse with compatible mates to begin the cycle again.

Rhodotorucine A

The first pheromone to be isolated and characterized from a Basidiomycete fungus was from *Rhodosporidium toruloides*, a saprophytic Basidiomycete yeast. *R. toruloides* has two mating types, **A** and **a**. When haploid cells of the two are mixed, they arrest in G1, and develop conjugation hyphae which eventually fuse and give rise to a dikaryotic mycelium (42,43). The **a** cells require induction by the **A** pheromone before they also secrete a specific **a** pheromone (43). The **A** pheromone, rhodotorucine A, was purified from culture filtrates of type **A** cells, and is an 11-amino-acid peptide with an S-farnesyl cysteine at the C-terminus (Table 2) (44). Rhodotorucine A is encoded by three different genes, *RHA1*, *RHA2* and *RHA3* (45,46). Each gene contains several copies (three to five) of the mature pheromone sequence. Precursors are processed and modified prior to being secreted from the cells. Only type **A** cells contain these three genes, which therefore probably occur within the mating-type locus itself (45). The pheromone produced by **a** cells has not been identified. Purified rhodotorucine A binds specifically to type **a** cells, probably to a site on the outer membrane since the bound pheromone can be inactivated for a limited time (between 10 and 15 minutes) by protease treatment (47). A receptor for rhodotorucine A has not yet been identified, but we would expect it to be similar to other fungal pheromone receptors.

A unique feature of *R. toruloides* is that rhodotorucine A appears to be processed to an active form by a specific protease secreted by **a** cells. Mutants deficient in this protease are insensitive to rhodotorucine A, and inhibitors of cell-surface protein sulfhydryls inhibit the protease as well as the response to rhodotorucine A (48-50). *S. cerevisiae* and *Sch. pombe* also produce proteases which specifically degrade the pheromones of the opposite mating type, but these seem to be involved in pheromone desensitization, since mutations in these proteases cause a pheromone-hypersensitive phenotype, and not a pheromone-insensitive phenotype as seen in *R. toruloides* (51-53).

Tremerogens

Several lipopeptide pheromones have been isolated from the jelly fungi *Tremella mesenterica* and *Tremella brasiliensis*. Unlike the fungi previously discussed, these species have two mating-type loci. One locus is bispecific, either **A** or **a**; the other, called **B**, is multispecific. When cells of different **A** types are mixed, they arrest in G1 and produce conjugation hyphae which eventually fuse (41,54). The **B** mating type is irrelevant in this reaction; however, different **B** mating types are required for establishment and maintenance of the dikaryotic mycelium after fusion (41). It is typical of two-locus systems that the two mating-type loci regulate different, but complementary, stages in dikaryon development. The pheromones produced by *Tremella* spp. are called tremerogens. They are typical fungal lipopeptide pheromones: short peptides with isoprenylated C-terminal cysteines. Some also contain other modifications. Tremerogen A-10, purified from an **A** strain of *T. mesenterica*, is hydroxylated at one of the terminal methyl groups of the farnesyl moiety (Table 2) (55). The farnesyl cysteine is carboxymethylated. Tremerogen a-13 is a 13-amino-acid peptide purified from an **a** strain of *T. mesenterica*. It also contains a C-terminal S-farnesyl cysteine, but the farnesyl group is unmodified, and the C terminus is free (Table 2) (56). Several compounds with pheromone activity were isolated from an **A** strain of *T. brasiliensis* (57). They are all similar in structure, and some appear to be derived from the others by hydrolysis of a C-terminal methyl group. The most active component of the mixture, called tremerogen A-9291-I, contains two almost identical 15-amino-acid peptides in which the fourth amino acid is either serine or glycine (Table 2). Each contains a farnesylated cysteine at the C-terminus, which is carboxymethylated.

Unlike *S. cerevisiae* **a**-factor, tremerogen A-10 appears to be secreted by the usual secretory pathway. Secretion is blocked by tunicamycin, which inhibits attachment to the endoplasmic reticulum, and also by monensin, which inhibits transport through the golgi complex (58).

Pheromones and Receptors are Primary Determinants of Mating Type in Basidiomycetes

In *Tremella*, pheromone-elicited responses (i.e., cell-cycle arrest, conjugation-hyphae formation, cell fusion) are under the control of only one of the two mating-type loci, the bispecific **A** locus. Although genes for the *Tremella* pheromones and their receptors have not been identified, it is likely that they are encoded within the **A** locus itself. This is true in another Hemibasidiomycete, *Ustilago maydis*, which causes corn smut disease. Like *Tremella*, *U. maydis* has two unlinked loci which regulate mating type: the *a* locus with two specificities and the multispecific *b* locus. The *a* locus alone regulates formation of conjugation hyphae and cell fusion, regardless of the *b* type (40,59-61). The *b* locus is involved in proper formation of dikaryotic hyphae and pathogenicity to corn (59,62,63). Specificities at both loci must be different for a completely successful mating to occur. Both specificities of the *a* locus of *U. maydis* have now been cloned and characterized. Each in fact contains two genes, one encoding a mating-type-specific lipopeptide pheromone, and one encoding a receptor for the pheromone of the opposite mating type (65). The pairs of genes of each specificity are embedded within large regions of heterologous DNA, flanked by homologous sequences (64,65). The heterology prevents recombination between homospecific pheromone and receptor gene pairs. This is important to prevent self-activation; recombination could result in a haploid individual containing a receptor responsive to its own pheromone. This possibility has been eliminated by the clustering of genes within large regions of unique DNA. In the *a1* mating type, this region is 4.5 kb, while for *a2* it is about 8 kb (64,65).

Basidiomycete Mating-Type Loci Share an Unusual Structure

Large, heterologous regions containing clusters of two or more genes which interact in heterospecific pairs are a feature of all Basidiomycete mating-type loci which have been analyzed so far. The multispecific *b* locus of *U. maydis* also consists of a large heterologous region containing a pair of genes (66-69). These genes, known as *bE* and *bW*, do not encode components of a signaling pathway, however. They encode homeodomain proteins which interact only in heterospecific combinations (i.e., *bE* of one mating-type with *bW* of another) to transcriptionally activate genes involved in the *b* mating type pathway (69,70). Similar pairs of homeodomain proteins are also encoded by the **A** mating type loci of *S. commune* and *C. cinereus* (71-75). A full discussion of these very interesting genes is unfortunately not possible in this chapter, but the reader may refer to some recent reviews for more information (4,76). Pheromones and receptors also interact only in heterospecific pairs (pheromone of one mating type with receptor of another) to activate the pheromone response part of the mating process. **Thus, we see that the key to self *versus* non-self recognition in Basidiomycetes is specific interactions between molecules encoded by heterospecific gene pairs.** Gene pairs are prevented from recombining by being embedded within regions of unique DNA.

Although they occupy a single map position on a chromosome, fungal mating-type loci are in fact complex, containing two, and often more, different genes. In addition, the DNA sequences of the various specificities are different enough to prevent recombination between them. Because of this complexity, use of the term "allele" to describe a particular mating-type specificity is inappropriate. The alternative term "idiomorph" was proposed for the mating-type loci of Ascomycetes (77). Its purpose was to denote sequences which occupy the same locus, but have so little sequence homology as to suggest that they never had a common ancestor. In the Basidiomycetes, however, the mating-type genes of a particular type (e.g., homeobox genes, pheromone-receptor genes) have clear functional and structural relationships to one another. Furthermore, heterology in DNA sequence is probably an adaptation designed to keep pairs of genes together, rather than a reflection of disparate origins. Neither the term "allele", nor "idiomorph", as originally defined, seem appropriate to describe the mating-type loci of Basidiomycetes. Although we are reluctant to add to the already unwieldy body of terminology in the field, we feel strongly that there is a need for a special term to describe these loci. We suggest **"altermorph"**. The derivation of the word is as follows: "alter", from Latin, meaning different in some particular without a change in basic character; and "morph" meaning a genotypic variant. "Altermorphs" will be used henceforth in this review to denote alternate versions of a particular genetic locus containing two or more genes which interact in heterospecific pairs to determine sexual identity. Included in the definition is the concept that DNA sequences of altermorphs are sufficiently different to prevent separation by recombination of genes encoded within the same altermorph.

Fungal mating-type loci have been compared with heterosomal regions of mammalian sex chromosomes (78). Recent information on the structure and function of fungal mating-type loci leads us to speculate that similar heterosomal gene clusters may have been the origin of true sex chromosomes. As in the sex chromosomes of higher organisms, fungal mating-type genes, which co-evolved to perform related functions in sex determination, are clustered within heterologous regions of DNA to prevent their being separated by recombination.

Pheromones in *Ustilago maydis*

The *a*1 and *a2* mating-type altermorphs of *U. maydis* encode pairs of lipopeptide pheromones and pheromone receptors. The pheromone genes are *mfa1* and *mfa2*, and those encoding the pheromone receptors are *pra1* and *pra2*. The pheromones are generally similar to those of other fungi (Table 2) (79). One interesting difference in *U. maydis* is that each of the pheromones is encoded by only one gene, rather than several related genes (65). Both pheromone genes contain several upstream repeats which are required for the normal pheromone response; one possibility is that *U. maydis* can produce larger quantities of pheromone from its single gene because of more efficient transcriptional control, and therefore it doesn't need extra genes.

Active pheromones have been purified from filtrates of haploid *U. maydis* cultures and have been chemically characterized (79). The purified pheromones induce formation of conjugation hyphae in cells of the opposite mating type in water. As in *Sch. pombe*, spotting the purified pheromone onto a lawn of *U. maydis* cells does not cause a halo, indicating that the pheromone does not induce cell-cycle arrest under these conditions (79). The *Sch. pombe* pheromones were shown to induce cell-cycle arrest only under certain conditions, and a similar effect might also be demonstrated in *U. maydis* given the appropriate environmental parameters. Of relevance is the fact that conjugation hyphae are not formed by *U. maydis* except when the cells are starved for nitrogen (60,61,63). Matings can be made on a relatively rich medium, but only if it contains charcoal. The reason for this is unknown. It has been suggested that the charcoal removes an inhibitor (63). In fact, an inhibitor of the pheromone response was isolated from culture filtrates of both haploid strains of *U.maydis* in the process of purifying the pheromones, although the nature of this inhibitor is unknown (79). Related to this might be the fact that transcript levels of the *mfa1* pheromone gene were significantly higher in haploid strains grown in the presence of charcoal (79). A secreted inhibitor of the pheromone signaling pathway is novel, and its potential role is interesting to contemplate. Might it have something to do with the fact that *U. maydis* is a plant pathogen? Perhaps the inhibitor is inactivated by the host plant, allowing *U. maydis* to mate and form a pathogenic mycelium only when a suitable host is available. Some unknown factors from the host plant are certainly required to allow full development of the dikaryon. Compatible matings on charcoal plates produce filaments, called a fuz+ reaction, but these are not normal dikaryons; most of the filament is devoid of cytoplasm, and only the tip cell is dikaryotic (60,63). Development can continue only within the host plant. Much is still mysterious, therefore, about this very interesting fungus during the dikaryotic part of its life cycle.

The two pheromone receptors of *U. maydis* belong to the rhodopsin-like superfamily of seven transmembrane receptors, and are similar to the receptor of *S. cerevisiae* encoded by *STE3* (see Figure 4) (65). Both pheromones of *U. maydis* are lipopeptides, as are those of *Tremella mesenterica*. In contrast, only one of the two pheromones in each of the Ascomycete yeasts is isoprenylated. It has been suggested that lipopeptide pheromones interact specifically with receptors of the Ste3p type, while unmodified peptide pheromones are likely to interact with receptors similar to Ste2p (65). This leads us to predict that receptors for the tremerogens and for rhodotorucine A are also similar to Ste3p. We might also expect that the pheromone produced by **a** cells of *R. toruloides* is isoprenylated. As yet, unmodified peptide pheromones have not been found among the Basidiomycetes.

Mating tests utilizing various strains of *U. maydis* transformed with cloned pheromone or receptor genes have been informative in elucidating the mode of action of these proteins (65). Transformation of a haploid strain with a pheromone or receptor gene of the opposite *a* type has no obvious morphological effect on that cell. However, it is possible that an effect would be observed only under conditions appropriate for conjugation hypha formation, i.e., nitrogen starvation, and that particular experiment has not been reported in the literature. Transformation of a haploid strain with the pheromone of the opposite *a* type converts it into a double mater; it can mate with cells containing either *a* mating type as long as the *b* mating type is different (successful mating is indicated by a fuz+ reaction on charcoal plates). For example, an *a1b1* cell, transformed with the *a2* pheromone gene *mfa2*, can mate with *a2b2* cells and also with *a1b2* cells. Interestingly, transformation with the pheromone-receptor gene of the opposite mating type does not result in a double mater phenotype. Mating two cells, both of which have been transformed with the receptor of the opposite *a* type, does result in a positive fuz+ reaction, however. Mutants in which the pheromone gene has been deleted are sterile. These mutants can mate only with a transformed or diploid strain that is heterozygous for *a*, and is therefore producing both pheromones. Mutants in which the pheromone receptor gene has been deleted are also sterile, but these mutants cannot mate with strains heterozygous for *a*. An interpretation of these observations is illustrated in Figure 2 and is described as follows (65). To become competent to mate, both partners must be activated by the pheromone of the opposite *a* type (Figure 2A). If the pheromone response pathway of only one partner is activated, mating does not occur. It is likely that both cells must be activated so that both can make conjugation hyphae and become competent for cell fusion. To make the point, we could say that in *U. maydis*, sex must be consensual! Because pheromone molecules are diffusible, it makes no difference if the cell itself produces the pheromone or if the mate produces it. Thus, pheromone-deficient mutants can mate as long as their partners supply the needed pheromone and activate their own pheromone response pathways (Figure 2B). Similarly, strains transformed with the pheromone of the opposite *a* type are able to self-activate, and can also activate and mate with both *a* types, because they produce both pheromones (Figure 2C). The pheromone receptors, on the other hand, are not diffusible. A pheromone-receptor deletion mutant is sterile even if paired with a strain heterozygous for *a* (Figure 2D). A strain transformed with the receptor of the opposite type can self-activate, but it cannot mate with a strain of its own *a* type because it does not produce the appropriate pheromone (Figure 2E). It can mate with a strain of its own type, however, if that strain has also been transformed with a pheromone-receptor gene of the opposite type, and can therefore self-activate (Figure 2F).

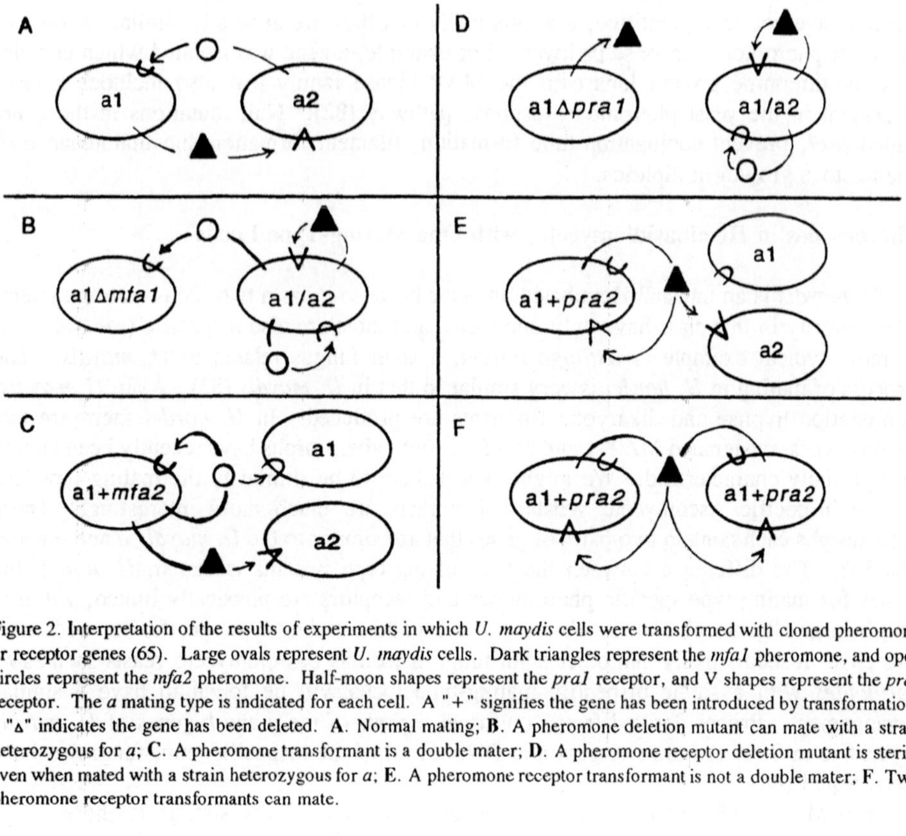

Figure 2. Interpretation of the results of experiments in which *U. maydis* cells were transformed with cloned pheromone or receptor genes (65). Large ovals represent *U. maydis* cells. Dark triangles represent the *mfa1* pheromone, and open circles represent the *mfa2* pheromone. Half-moon shapes represent the *pra1* receptor, and V shapes represent the *pra2* receptor. The *a* mating type is indicated for each cell. A "+" signifies the gene has been introduced by transformation, a "Δ" indicates the gene has been deleted. **A**. Normal mating; **B**. A pheromone deletion mutant can mate with a strain heterozygous for *a*; **C**. A pheromone transformant is a double mater; **D**. A pheromone receptor deletion mutant is sterile even when mated with a strain heterozygous for *a*; **E**. A pheromone receptor transformant is not a double mater; **F**. Two pheromone receptor transformants can mate.

In *U. maydis* the pheromones and receptors play important roles in events subsequent to induction of conjugation hyphae and cell fusion. Heterozygosity at *a* is required, along with heterozygosity at *b*, for maintenance of filamentous growth (80). This was discovered through the study of diploid cells, heterozygous at both *a* and *b*. These cells mimic a positive mating interaction on charcoal medium, and exhibit a fuz+ phenotype. Diploids were subjected to UV mutagenesis and fuz- mutants were selected. These were found to be of two types: homozygous at *b*, or homozygous at *a*. Thus, loss of heterozygosity at either locus results in cessation of filamentous growth. The fuz+ phenotype was restored if the mutant diploid homozygous for *a* was co-cultured with a haploid cell containing the missing *a* altermorph. The fuz+ phenotype also results if the strain is treated with the purified pheromone of the opposite type (79). Thus, continued activation of the pheromone response pathway is required to maintain filamentous growth after cell fusion has occurred. If the stimulation is not maintained, the filaments revert to budding cells. What is the role of the pheromone response pathway in maintenance of filamentous growth? It seems likely that targets of the pheromone signaling pathway, perhaps in combination with the *b* mating-type pathway, include genes that are important in controlling the morphological switch between filamentous and budding growth. Switching between the alternate morphological states is determined by levels of cAMP, known to be a secondary messenger in many signaling pathways (81). The pheromone response pathway may also be important for other aspects of dikaryotic growth, but this will be rather difficult to investigate in *U. maydis*, because a normal dikaryon develops only within the tissues of a host plant.

Although few of the downstream elements of the pheromone response pathway in *U. maydis* have yet been identified, it seems likely that they are generally similar to those of the yeast pheromone response pathway. For example, a gene was isolated which encodes a serine-threonine protein kinase in the MAP kinase family that also includes kinases important in the yeast pheromone response pathway (82). Null mutations in the gene, called *fuz7*, prevent conjugation-tube formation, filament formation and maintenance of filamentous growth in diploids.

Pheromones in Hemibasidiomycetes with One Mating-Type Locus

U. maydis is an unusual Hemibasidiomycete because it has a two-locus mating system. The majority in this class have only one locus, and most are also bispecific (see Table 1). A more typical example is *Ustilago hordei,* a smut fungus related to *U. maydis*. The process of mating in *U. hordei* is very similar to that in *U. maydis* (83). As in *U. maydis*, conjugation hyphae and dikaryotic filaments are produced. In *U. hordei* there are two mating types, designated MAT-1 and MAT-2. Both altermorphs have recently been cloned and partially characterized. We might expect them to be similar to the mating-type loci of the bispecific Ascomycete yeasts. The facts are much more interesting! These altermorphs each contain two pairs of genes that are similar to the *U. maydis a* and *b* genes (84,85). The difference between the two mating-type systems is that in *U. hordei* the genes for mating-type specific pheromones and receptors are physically linked, within a single large altermorph (> 150 kb), to genes encoding mating-type specific homeodomain proteins. It seems likely that other smut fungi as well as Basidiomycete yeasts such as *R. toruloides* with a single bispecific mating-type locus will be found to have a similar arrangement. Probes derived from conserved regions of the *a* and *b* genes of *U. maydis* hybridize in Southern blots to genomic DNA of all of the bispecific smut fungi that have been tested (86).

The MAT-1 pheromone-receptor gene of *U. hordei* is very similar to those of *U. maydis* (85). Furthermore, inserting the MAT-1 *a* genes into *U. maydis* by transformation allows it to mate with the *U. hordei* MAT-2 strain, as indicated by a fuz+ reaction on charcoal plates (85; Bakkeren and Kronstad, personal communication). This is quite remarkable! Normally, *U. maydis* and *U. hordei* do not mate. The result of this experiment implies two things. First, the components of the pheromone response pathway downstream of the pheromone receptor are highly conserved in these two species. And second, the *b* gene products of these two organisms recognize one another as different and interact to produce at least one active *bE/bW* complex. This has been demonstrated in other experiments in which *b* genes from *U. hordei* have been introduced *via* transformation into *U. maydis* (84). The haploid transformants exhibit very weak filamentous growth on charcoal plates, and they are also somewhat pathogenic to corn plants. This is similar to what happens when a non-self *U. maydis b* gene is introduced into a *U. maydis* strain, except that those transformants are strongly pathogenic to corn (69). *U. hordei*, in contrast to *U. maydis*, has only two versions of *b* (85). One is always associated with MAT-1, and the other with MAT-2. The implication of these data is that the mating systems of *U. maydis* and *U. hordei* arose relatively recently from a common origin. We can only speculate as to whether a one-locus or a two-locus system was ancestral. However, given the fact that the majority of smut fungi are bispecific like *U. hordei*, we would guess that the one-locus system was original.

Ustilago violaceae

This smut seems to behave somewhat differently from the others we have discussed. *U. violaceae* has two mating types, **A** and **a**. Although mating-type genes have not been cloned, we would expect them to be similar to those of *U. maydis*, and *U. hordei*, to which they hybridize in Southern blot experiments (86). If they are touching one another, cells of both types can fuse without requiring differentiation of a conjugation hypha, a shmoo-like protuberance, or any other recognizable mating structure (87,88). If the cells are not touching, the **A** type produces a conjugation hypha that grows toward and fuses with the **a** cells, which remain undifferentiated (88,89). It has been proposed that orientation of conjugation hyphae and fusion of *U. violaceae* cells are mediated by contacts between fimbriae, long polymers of a 76 kD subunit extending from the surface of the cells (89,90). Conjugation hyphae seem to grow toward the **a** cells by following a path closely associated with the fimbriae of those cells. Contacts between fimbriae of cells of opposite mating type have been observed under the electron microscope (88,89). There is no evidence that fimbriae of the two mating types are themselves different; they are thought instead to be channels for transport of mating-type specific substances, e.g., pheromones.

Cryptococcus neoformans

This dimorphic fungus is an opportunistic pathogen of mammals and has become very important in recent years as the cause of a serious, often fatal, meningitis in AIDS patients (91). *C. neoformans* has two mating types, MATa and MATα. Interestingly, it has been observed that the MATα strains are more aggressive pathogens of mice than MATa strains (92). The MATα locus was cloned recently as part of an effort to understand the genetic basis for this difference (93). The MATα locus is a large altermorph of about 45 kb. The only gene that has been identified so far within the altermorph appears to encode a typical lipopeptide pheromone (Table 3). When inserted by transformation into MATa strains of *C. neoformans*, the gene induced formation of conjugation hyphae under starvation conditions. Although other genes have not yet been identified, we expect that the altermorph also contains genes encoding a pheromone receptor for the MATa pheromone, and a pair of homeodomain proteins analogous to *bE* and *bW* of *U. maydis* and *U. hordei*. It will be interesting to discover whether the increased virulence of MATα strains is attributable to the mating-type genes themselves, or to other genes linked within the altermorph. It is possible that the signaling and developmental pathways controlled by the mating-type genes interact with cellular processes important to pathogenicity, as they do in *U. maydis* where the *b* genes regulate pathogenicity to corn.

PHEROMONES IN FILAMENTOUS FUNGI

Pheromones in the Euascomycetes

We know that pheromones and receptors are important in mediating recognition and cell fusion among the Ascomycete and Basidiomycete yeasts, but do they also play a role

Table 3. Putative Precursors of Pheromones from *C. neoformans* and *S. commune*. Gaps have been introduced to align identical and similar amino acids in putative pheromone precursors encoded by *bbp1(1)*, *bbp1(3)* and *bap1(3)*.

SPECIES	GENE	AMINO ACID SEQUENCE OF PUTATIVE PRECURSOR PROTEINS	REF.
C. neoformans	*MFα*	MDAFTAIFTTFFSAATSSSEAPRNQEAHPGGMTLCVIA	(93)
S. commune	*bap1(1)*	MDGEGHDINIWGARMSPSPAAAPVSATRGAPWSGCEGCPSRAADRRCVCH	(97)
S. commune	*bap1(2)*	MRSRASAEGIAVLGLRRRGESPVCRRRNVVVCFWGDRSCVEREGCVRG	(97)
S. commune	*bbp1(2)*	MDAFTDFSILADGLASLFDESSHTILAEFSPSILDGPVVADSAPLTEAPCNHDQIADYGSYCVVA	(a)
S. commune	*bbp1(1)*	MDAFTAMFPELFPIEEGLEDALVGSLSDTSAASASATHTSPASTDTFDDADILAILADADDWRGG-NTTAHGWCVVA	(a)
S. commune	*bbp1(3)*	MASSVLARPGPSTVLPAMTRPPPPMAHRAAATPSFARSAQPQTDDAVLALLANAEHTEAGEETTARGWCVVA	(b)
S. commune	*bap1(3)*	MDDFAEFFPTLVLDEPEVARRPARDAEVLAILADAERP-GGSNCTA-WCVVA	(97)

[a] Raudaskoski and Specht, personal communication
[b] Vaillancourt and Raper (unpublished data)

in filamentous fungi? There is convincing, though circumstantial, evidence that pheromones function during sexual development in *Neurospora crassa*, a Euascomycete. As yet, no genes have been cloned, and the pheromones themselves have not been purified. *N. crassa*, like all Ascomycetes, has a one-locus, bispecific mating-type system (Table 1). The two mating types are called **A** and **a**. In addition, *N. crassa* produces specialized male and female organs. Each strain is hermaphroditic, so sexual dimorphism does not normally play a role in determining the compatibility to mate. The female structures are multicellular and are called protoperithecia. They produce long filaments, called trichogynes, which grow until they contact a conidium produced by a strain of the opposite mating type. The two fuse, and the nucleus from the conidium travels back along the trichogyne to the base of the protoperithecium, where a dikaryotic cell is established. This proliferates to produce a limited dikaryotic mycelium, and then karyogamy and meiosis occur. The ascospores and asci are formed inside the protoperithecium, which develops into a flask-shaped fruiting body called a perithecium. It is very likely that the interaction between the trichogyne and the conidium is regulated by diffusible pheromones (94). Trichogynes are attracted and grow toward conidia of the opposite mating type. They also grow toward fragments of charcoal which have been exposed for a time to germinating conidia of the opposite mating type (94). Interestingly, trichogynes of the **a** type are attracted to conidia, or charcoal exposed to conidia, of the **A** type more strongly than in the reverse case. Possibly the **A** type produces more pheromone, or one with a higher activity. The **A** and **a** idiomorphs have both been cloned and characterized (95,96). (We use the term idiomorph because it was originally defined to apply to the mating-type loci of *N. crassa*). The **A** idiomorph is 5.3 kb and the **a** idiomorph is 3.2 kb long. The **A** idiomorph is found only in **A** strains and the **a** idiomorph likewise is found only in **a** strains. *N. crassa* is more like the Basidiomycetes and less like the Ascomycete yeasts in this respect. As would be expected, *N. crassa* does not switch mating types. A single gene has been identified in each idiomorph which is able to restore **A**- or **a**-specific mating-type function to sterile mutants (95,96). The genes share homology with the mating-type genes of *S. cerevisiae* and *Sch. pombe* and are probably transcriptional activators of the same type. Thus, it is likely that in *N. crassa*, as in the Ascomycete yeasts, the pheromones and receptors are under the control of the mating-type genes, and are not themselves primary determinants of mating type. The location of the putative pheromone and receptor genes is, as yet, unknown. We can suggest two possibilities. Either the genes are encoded somewhere within the **A** and **a** idiomorphs (this seems unlikely, since both have been well characterized) or else each haploid strain contains genes for pheromones and receptors specific to both mating types, and these are differentially regulated by the **A** and **a** transcription factors. The latter would be comparable to regulation of pheromones and receptors in the Ascomycete yeasts.

Pheromones in the Homobasidiomycetes

The Homobasidiomycetes are by far the most determined outbreeders among the fungi. All known heterothallic members of this class have multiple mating types, sometimes numbering in the tens of thousands! One might suppose that the genetic basis for so many mating types would be quite different from that in bispecific systems. Recent research has revealed, however, that the mating-type genes of the Homobasidiomycetes are similar to those found in the bispecific Hemibasidiomycete *U. hordei*. The larger number of mating types among Homobasidiomycetes is a result of their having more loci, more altermorphs of those loci, and a different arrangement of the loci on the chromosomes.

To date, the only members of the Homobasidiomycetes that have been examined at the molecular level are *Schizophyllum commune* and *Coprinus cinereus*. Both of these species have two unlinked mating-type complexes called **A** and **B**. In *S. commune* each complex consists of two, closely linked, redundantly functioning loci, each of which occurs as multiple altermorphs. The arrangement in *C. cinereus* is similar, except that recombination occurs only rarely between the two **A** loci, and has never been observed within the **B** complex (98). In both species, the altermorphs of Aα and Aβ each contain one or more pairs of homeobox genes analogous to the *b* genes of *U. maydis* and *U. hordei* (71-76). The Bα and Bβ altermorphs each contain pheromone and receptor genes similar to those of other fungi (97; Vaillancourt and Raper, unpublished data; Raudaskoski and Specht, personal communication; Casselton, personal communication). This latest discovery was very surprising, because there is no obvious indication that diffusible factors play a role in mating in Homobasidiomycetes. No effects of the mating-type genes are observed prior to hyphal fusion, and fusion itself occurs regardless of mating type.

The bipartite nature of the **A** and **B** mating-type complexes in *S. commune* was discovered when it was noticed that new mating types arose occasionally among the progeny of crosses (reviewed in 1). The "new" mating types were actually the result of recombinations between the two loci of the A factor, or of the B factor. The fact that recombination never resulted in activation of the mating-type pathway suggested that the α and β loci are completely independent, and do not interact. At some point in the evolutionary past of this organism, it seems likely that an ancestral locus of each type duplicated, and a copy was inserted into the DNA a short distance away on the same chromosome; subsequent divergence of the two copies resulted in the arrangement that exists today. The relative recombinability of these loci in *S. commune,* as contrasted with *C. cinereus*, is advantageous for genetic analysis of the multitudes of different mating types. It is possible in *S. commune* to identify all of the altermorphs at each locus, and to estimate with some accuracy the number that exist in the world population (1,8). It is also possible to know the precise genetic make-up of individual strains used in the laboratory. The importance of this for experiments on the function of the mating-type genes cannot be overestimated.

The Molecular Characterization of the Bα and Bβ loci of *S. commune*

Two **B** mating type altermorphs, Bα1 and Bβ1, have been cloned (99). Both are about 10 kb in size, and both contain a single pheromone-receptor gene and three putative pheromone genes (97; Vaillancourt and Raper, unpublished data; Raudaskoski and Specht, personal communication). Preliminary analysis of two of the **B** altermorphs of *C. cinereus* has revealed that they also contain putative pheromone and pheromone-receptor genes (Casselton, personal communication). Pheromone receptors of *S. commune*, encoded by *bar1*, *bar2* and *bbr1*, are similar to Ste3p (Figure 4). The first two receptor genes were isolated from two Bα altermorphs, Bα1 and Bα2, and the last was isolated from the Bβ1 altermorph. The six *S. commune* pheromone genes which have been sequenced to date all appear to encode typical lipopeptide pheromones, except that some of the putative precursors are somewhat longer than average (Table 3). Pheromone genes *bap1(1)*, *bap1(2)* and *bap1(3)* were all isolated from the Bα1 altermorph, while *bbp1(1)*, *bbp1(2)* and *bbp1(3)* were isolated from the Bβ1 altermorph. Interestingly, each of the putative pheromone genes within a single altermorph encodes a different pheromone. The Hemiascomycetes and some of the Hemibasidiomycetes also have several pheromone genes encoded within the same specificity, but all of those genes encode the same pheromone, or closely related pheromones differing in only a single amino acid. The complexity of the Homobasidiomycete altermorphs is perhaps not surprising in view of the fact that the

Hemiascomycetes and Hemibasidiomycetes have only two pheromone/receptor specificities, while Bα and Bβ have nine each. The **B** loci of *C. cinereus* have a comparable number of specificities, although the exact number is unknown.

The functional relationships among receptors and pheromones in *S. commune* were examined through a series of transformation experiments (97; Vaillancourt and Raper, unpublished data). Each cloned receptor and pheromone gene of Bα1 and Bβ1 was introduced into various recipient strains of *S. commune*. The recipient strains represented each of the eight non-self Bα or Bβ types, as well as the self type. Activation of the pheromone response pathway in the transformants was indicated by several morphological changes that are normally the result of heterozygosity for the B complex. These changes include growth of the mycelium primarily beneath the agar surface, hyphae which branch in a knobby, irregular pattern, degradation of septa and continuous migration of nuclei between cells. The results of these transformation experiments indicated the following. Each non-self Bα or Bβ altermorph encodes at least one pheromone that can activate the pheromone receptors of Bα1 and Bβ1, respectively. Each non-self Bα or Bβ altermorph also encodes at least one receptor that can be activated by one or more of the pheromones of Bα1 and Bβ1, respectively. Significantly, pheromones are unable to activate receptors that are encoded within their own altermorph. In *S. commune*, some receptors can be activated by more than one pheromone, and some pheromones can activate more than one receptor. This seems logical when one considers that each of the altermorphs analyzed so far contains only three pheromone genes; if each pheromone was unable to activate more than one receptor, one would expect to find at least eight different pheromone genes contained within each altermorph. There is not a one-to-one relationship between pheromones and receptors in Homobasidiomycetes. This is different from the Hemiascomycetes and Hemibasidiomycetes, where each receptor is activated by only one pheromone and each pheromone activates only one receptor.

In matings involving strains transformed either with pheromone genes or receptor genes, qualitative differences were observed (97). Normally, a successful mating results in the reciprocal migration of nuclei from each mate into and throughout the mycelium of its partner (Figure 3). In matings involving transformants in which the pheromone response pathway is activated, nuclei do not migrate into the transformed mate. This might be because the pathway is already activated and an organized process of nuclear migration cannot be superimposed. Strains transformed with non-self pheromone genes were paired with untransformed tester strains of the same **B** type as the transformation recipient. In these matings, nuclei migrated normally from the transformed mate into the untransformed tester. If the two strains differed for the **A** complex, a morphologically and functionally normal dikaryon was subsequently formed. In a similar set of experiments, strains transformed with non-self pheromone-receptor genes were paired with untransformed tester strains of the same **B** type as the transformation recipients. In these matings, however, nuclei did not migrate into the untransformed tester and a normal dikaryon did not result. Although some dikaryotic hyphae were eventually produced, these apparently arose only by proliferation of the initial dikaryotic cells formed by fusion of hyphae from the two mates. An interpretation of these interesting observations is given in the next section.

What is the Role of Pheromones in *S. commune*?

In *S. commune*, a completely successful mating results only if the mates differ in at least one of the **A** loci, and at least one of the **B** loci. It is possible to study heterokaryons

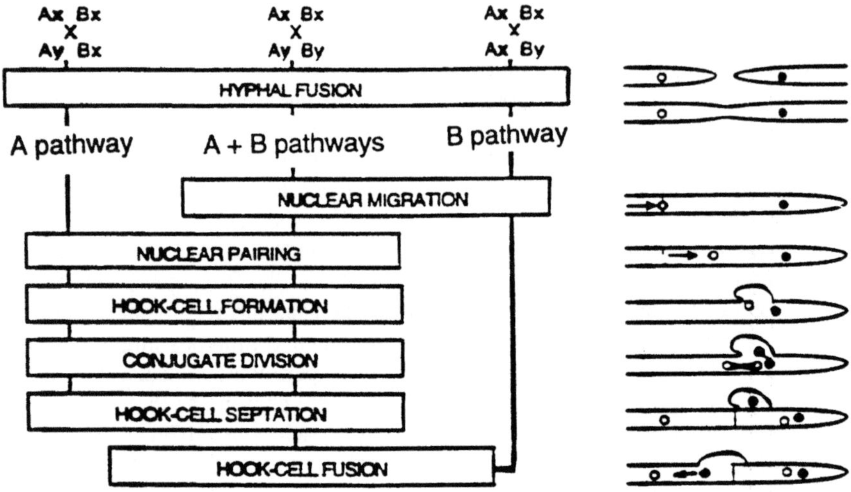

Figure 3. Regulation of mating and dikaryon development by the **A** and **B** genes of *S. commune*. The "x" or "y" designations indicate a difference at α and/or β.

made between two strains that differ only at **A**, or that differ only at **B**, however. The precise regulatory functions of the two mating-type complexes can thus be determined. Mating in *S. commune* consists of a defined sequence of morphological events leading to the formation and maintenance of dikaryotic cells. Morphogenesis of the dikaryon has been described many times (see e.g., 1,3). *S. commune* and other Homobasidiomycetes differ from the other fungi we have discussed in that fusion between hyphae to initiate mating occurs regardless of mating type. Hyphae of one mating type do not appear to be attracted toward hyphae of another mating type. Recognition of self *versus* non-self takes place only after fusion. The first step in a fully compatible mating, reciprocal nuclear migration, is regulated by the B mating-type genes (Figure 3). This is a remarkable process in which fertilizing haploid nuclei migrate rapidly from each mate into and throughout the other. Septa, which normally block nuclear passage, are disrupted in advance of the migrating nuclei. The fertilizing nuclei pass through established cells, divide occasionally, and ultimately reach the hyphal apices where dikaryotic tip cells are established (100,101). It is at this point that the **A** mating-type genes become important. They regulate production and maintenance of a multicellular dikaryotic mycelium once the nuclei of the two mates occur together in the hyphal tip cells. Maintenance of the dikaryon requires precise partitioning of the nuclei as the cells divide so that each new cell has the correct complement of two nuclei, one from each parent. This sequence takes place only when the **A** genes are different. Nuclei in the apical cells pair and divide synchronously while a structure called a hook cell begins to form and bend away from the hyphal tip. One of the nuclei divides into the hook cell; the other divides along the hyphal axis. Septal formation across both planes of division isolates one daughter nucleus in the hook cell and another in the penultimate cell, and leaves a pair of nuclei in the new tip cell. Finally the hook cell fuses to the penultimate cell and discharges its nucleus to establish another pair of nuclei in that cell. Hook-cell fusion occurs only when the two nuclei are derived from mates of different **B** mating types (Figure 3, and reviewed in 1 and 3). Similar functions are attributed to the **A** and **B** mating-type genes of *Coprinus cinereus* and other heterothallic Homobasidiomycetes.

In the sequence of events controlled by the **B** genes, a familiar pheromone response function is seen in the fusion of the hook cells. Activation of the pheromone response pathway is required for cell fusion in the Hemiascomycetes and Hemibasidiomycetes. In *S. cerevisiae*, this is due to induction of cell-fusion proteins encoded by *FUS* genes. One of these genes, called *FUS1*, encodes a protein that is probably membrane-bound and glycosylated with both N- and O-linked sugars (102). The mechanism by which the *FUS* gene products mediate cell fusion is poorly understood, but it is possible that the pheromone response pathway in *S. commune* induces similar proteins that are involved in the fusion of hook cells. A different mechanism, which is not dependent on specific interactions between pheromones and receptors, must be involved in fusions between vegetative hyphae. Why doesn't *S. commune* use the pheromone response pathway to prevent fusion except between compatible mates? We can only speculate. Free fusion of vegetative cells in the absence of mating-type compatibility must be advantageous to the organism, or at least not disadvantageous. This would be a very interesting area to investigate more closely in *S. commune*.

The other process regulated by the **B** genes in *S. commune* is nuclear migration. There is nothing like it among the other fungi we have discussed, with the possible exception of *N. crassa* in which the nucleus of the fertilizing conidium must migrate through the trichogyne to the base of the protoperithecium. Do pheromones and receptors in *N. crassa* regulate this nuclear migration process? This would be an interesting question to investigate, once the pheromone and receptor genes are identified in this organism. The purpose of nuclear migration in the Homobasidiomycetes appears to be to produce a dikaryotic mycelium as rapidly as possible after mating. Although dikaryotic hyphae can be produced by proliferation of the initial fusion cells, nuclear migration dramatically accelerates the process. There must be an advantage to producing a dikaryon quickly; perhaps dikaryons have a survival advantage in the natural ecosystem. Nuclear migration has been described previously in detail (e.g., 1,3). An early event is dissolution of septa in advance of the fertilizing nucleus. To facilitate this process, the **B** genes induce the production of several wall-degrading enzymes, including an R-glucanase (103). These enzymes attack the septa, break them down, and allow the nuclei to pass from cell to cell. The rate at which the nuclei move through the mycelium is very rapid, far exceeding the rate of cytoplasmic flow or hyphal growth. Estimates range between 500 and 3000 μ/hr (100)! To allow the nuclei to move so quickly, septa must start breaking down in advance of their arrival. Here is a potential role for diffusible pheromones! Perhaps the pheromones diffuse through the substrate and activate receptors in cells ahead of the advancing nuclei. The activated pheromone response pathway could initiate the production of wall-degrading enzymes and other proteins, which prepare the cell for the arrival of the nucleus and also play a role in the polarized movement of the nuclei into the cells (97).

This model is supported by the mating experiments involving strains of *S. commune* transformed with cloned pheromone and receptor genes. Strains transformed with non-self pheromone genes were able to donate fertilizing nuclei to test mates. According to the model, diffusible non-self pheromone molecules encoded by the transformed nuclei activate the receptors of the test mate, and thus allow nuclear migration into the mate (97). In contrast, strains transformed with non-self pheromone receptor genes are unable to donate fertilizing nuclei to test mates. According to the model, these transformed strains produce non-diffusible, membrane-bound non-self receptors. While these non-self receptors can be activated by the transformant's endogenous pheromones, the receptors cannot diffuse and so are unable to activate the pheromone response pathway in the test mate. Nuclear migration into the test mate does not occur.

ISOPRENYLATION OF PHEROMONES

Fungal mating pheromones are typically modified post-translationally by addition of an isoprenyl group, and often a methyl ester, onto a C-terminal cysteine. Only two fungal pheromones are known to be unmodified: α-factor of *S. cerevesiae*, and P-factor of *Sch. pombe*. Rhodotorucine A was the first of the lipopeptide pheromones to be chemically characterized (44). Soon after, the tremerogens were also shown to be isoprenylated, but it really was not until Ras proteins were found to contain the same modification that isoprenylation of proteins became a subject of general interest (104). It was realized that this was not just a peculiarity of the fungi, but might also be relevant to other organisms, including humans. Increased interest in the subject has resulted in the publication recently of a number of reviews (105-108). Several classes of proteins have now been identified which are routinely modified by addition of isoprenyl groups *via* a thioether linkage: the fungal mating pheromones, proteins in the Ras superfamily, nuclear lamins and GTP binding proteins. Many of these proteins function in transducing signals across the cytoplasmic membrane. Isoprenylation increases the hydrophobicity of the proteins, and may play a role in directing them or anchoring them to membranes; isoprenylation is also important in the interactions of some of the proteins with membrane-bound receptors (107).

Precursors of isoprenylated proteins all have a similar primary sequence motif at the C-terminus, the so-called CaaX box. This motif consists of a cysteine four amino acids from the end, followed (usually but not always) by two aliphatic amino acids, followed by any amino acid (see Table 3). The CaaX box is a substrate for enzymes involved in protein isoprenylation, primarily farnesyl transferases and geranyl-geranyl transferases. Nearly all of the known isoprenylated proteins are modified by addition of one of these two groups. The fungal pheromones in which the prenyl group is known are all farnesylated. Two genes encoding subunits of a farnesyl transferase, called *RAM1* and *RAM2*, have been identified in *S. cerevisiae* (109). These catalyze farnesylation of the **a**-factor pheromone, and also the product of the yeast *ras* homologue. One of the *STE* genes, *STE14*, encodes an enzyme which carboxymethylates the C-terminal cysteine of **a**-factor (110). The mature prenylated proteins all have a modified cysteine at their C-terminus, hence the three terminal amino acids must be cleaved off during the process of post-translational modification. The steps in the reaction for yeast **a**-factor are farnesylation, followed by cleavage of the terminal amino acids, followed by carboxymethylation (reviewed in 20,108). A special case is the set of *RHA* genes of *R. toruloides*, which encode rhodotorucine A. These genes encode precursors containing several tandem repeats of the mature pheromone sequence. This is the only known case where internal CaaX sequences are isoprenylated (108). However, it is not known whether the protein precursor is cleaved before, or after, isoprenylation.

The function of post-translational modification of lipopeptide pheromones in fungi is unknown. It makes the proteins hydrophobic, which should decrease their diffusion in aqueous substrates. We might expect this would be disadvantageous, but in most cases the modifications are essential for full activity and may play a role in binding to the receptor. Since all of the proteins are similarly modified, such modifications are probably not important in determining specificity: a more likely candidate for that role is the extremely variable peptide portion of the pheromones. A comparison of the peptide moieties of fungal pheromones and pheromone precursors reveals little overall similarity in primary sequence, except for a preponderance of serine, threonine and alanine residues (Tables 2 and 3). Three-dimensional models of the various pheromones are generally similar, however, with the peptide portions forming variable, flexible structures and the lipid portions projecting at an angle as rigid hydrophobic rods (111). Several pheromones with different modifying groups, or without modifying groups, have been tested for changes in activity (reviewed in 111). In most cases, activity was reduced or destroyed if the prenyl group or the carboxymethyl group was removed or changed. There are a few exceptions. For example, the activity of Tremerogen A-10 was increased if farnesyl analogs containing more isoprenyl units than the normal three were substituted (112). The opposite was true for tremerogen a-13, however (113).

RECEPTORS OF FUNGAL LIPOPEPTIDE PHEROMONES

The receptors of fungal pheromones are members of the rhodopsin-like superfamily of seven transmembrane (7tm) receptors. This is a large and diverse family of proteins whose members transduce signals across the cytoplasmic membrane in response to various ligands (reviewed in 114). All are predicted to have an extracellular N-terminus, seven transmembrane regions that give the protein a serpentine shape with three extracellular loops and two cytoplasmic loops, and a cytoplasmic C-terminus. The fungal lipopeptide pheromone receptors are a distinct subgroup of this family, and all are similar at the N-terminus to the yeast Ste3p receptor (Figure 4). The predicted topology of Ste3p is supported by experimental evidence (115). The fungal lipopeptide pheromone receptors are unique among the 7tm receptors in having almost no N-terminal region prior to the first transmembrane helix (116 and Figure 4). This means relatively little of the protein extends into the extracellular matrix. The fungal lipopeptide pheromones must bind either to the helices of the receptors within the plasma membrane, or to extracellular loops, or to both. We would like to suggest that the rigid hydrophobic rod formed by the isoprenyl group of the pheromones could insert and bind in a pocket inside the membrane formed by the hydrophobic transmembrane helices. This could position the variable peptide portion of the molecule correctly so it can interact specifically with the extracellular loops and other helices. This is reminiscent of the "address and message" model of peptide/receptor interactions (117). According to this model, the peptide is divided into two discreet domains, one of which (the "address") directs the peptide to the appropriate place on the receptor, while the other (the "message") contains the specific information necessary to activate the receptor. In the case of lipopeptide pheromones, the lipid portion is the "address", and the peptide portion the "message". Very little is known about binding of lipopeptide pheromones to receptors, but this will certainly change in the next few years. It is interesting to note that there is a conserved region among the fungal lipopeptide pheromone receptors that includes part of the fourth transmembrane region, and extends into the second extracellular loop (Figure 4). The function of this region of the receptor is unknown. Might it bind to the hydrophobic portion of the lipopeptide pheromones? This area certainly deserves further investigation.

A pair of altermorphic pheromone receptors have been analyzed both from *U. maydis* and from the *S. commune* Bα locus (Figure 4). The two receptors of *S. commune*, encoded by *bar1* in the Bα1 altermorph and *bar2* in the Bα2 altermorph, are very similar at the amino acid level (97 and Figure 4). This is in marked contrast to the two *U. maydis* receptors, encoded by *pra1* and *pra2*, which are far less similar (65 and Figure 4). It is reasonable to assume that one or more of the regions of dissimilarity in each pair of altermorphic receptors determines their specific identity. Because the two *S. commune* receptors are more similar to one another, sequences determining specificity should be easier to identify in the *S. commune* receptors than in *U. maydis* receptors.

```
Ste3p        MSYKSAIIGLCLLAVILLAPPLAWHSHTKNIPAIILITWLLTMNLT
Map3        MLPIGIFYQFYAYFALVLSIPILYMQLRARNIPCLLLLFWLTLTTLI
Pra1          MLDHITPFFALVAFFLVLMPFAWHIKSKNVGLIMLSIWLMLGNLD
Pra2      MFSGKENVSFGVLCLLAGCISTSSCLIHLQAKNIGVLLMMFWCFTGLVN
Bar1       MLDPLYPLFPIFAFLGFVLAILPLPWHLQAWNSGTCFFMMWTALGCLN
Bar2       ****T**A********I*CCLV*******S*******L*I***VA***
Bbr1           MHPEFAPVAFLSAASLALPLHLALAAGNVATLSIIAWLFIMNMI

Ste3p  CIVDAAIWSDD-DFLTRWDGKGWCDIVIKLQVGANIGISCAVTNIIYNL
Map3   YVVESAIWSNPYAETIRWMGYGLCDITSRIVTCSSIGIPASAFTLVLYL
Pra1   NFVNSMVWWKT-TAD---LAPAYCELSVRLRHLLFIAIPASNLAIARKL
Pra2   KGINALAFNNS-LRL---AWTLGCDLSAIIERTWQFGICCSALCVLQRL
Bar1   QFINSVAWADD-AMN---KAPVWCEISIRILMGASVGIPASSLCIIRRL
Bar2   N*V**MV*WK*-*Q*---V*******T***TL*************V***
Bbr1   YGINAVIWAGS-ARI---TAVVYCDIGAIGGNFAL---PAACLCLCIHL

Ste3p  HTILKADSVLPDLSSWTKIVKDLVIS-LFTPVMVMGFSYLLQVFRYGIA
Map3   DTVIRRDHPLKRYENW---IWHVCLS-ILLPLIIMAMMVPLESNRYVVI
Pra1   ESIASTRQVRAGPGDHRRAVIIDLLICLGIPIIYTSLMIVNQSNRYGIL
Pra2   EGIASLRQAHSTVWDRKRRLLIDFGVGLGLPALQIPMFFIVQPYRLNVI
Bar1   YYIAKVRAVSKTRAEKMRAILVDALICVLFPLVYIALQYIVQGHRFNIL
Bar2   *S********T**S**F**VM**********IL**I**IV******I**
Bbr1   ERVASVRAAQTTAADKRRRTIFELAMCWLLPIIFMALHYVVQGHRFDIV

Ste3p  RYNGCQNLLSPTWITPVLYTMWMLIWSFVGAVYATLVLFVFYKKRKDVR
Map3   CMNGCYSSFYQTWYTLLFFYIPPCLLSFGGLFFVSRIVVLYWRRQRELQ
Pra1   EEAGCWPMMVFSWLWVLLVAAPVIVVSLCSAVYSALAFRWFWVRRRQFQ
Pra2   ENIGCSAPLYASVPALFIYHLWRLLVSLVCAVYAVLVLRWFMLRRRQFT
Bar1   ENIGCYPAVTNT AVNYVWPVLLGLISATYGVMALLQFNKHRLQFSQFL
Bar2   *****F*GD*** **K*********************************
Bbr1   EDFGCRPATYYSIPAIFIVWVPPLTMAAASLVYASLAIRHFMHRRLSFA

Ste3p  DILHCTNSGLNLTRFARLLIFCFIIILVMFPFSVYTFVQDLQQVEGHYT
Map3   QFFQ-RDSQLTSKRFLRLLCLAAVFFLGYFPLTIFMVVAN GKLQQFLP
Pra1   AVLASSASTINRSHYVRLLLLTAIDMLLFFPIYVGTIAAQIKSSIS-IP
Pra2   AALSAQHSGLSQKKYFRLFALATCERVLVSAGQFYVIIQSLQIGGL-LP
Bar1   ----HTHSTLSASRYLRLMALALTEMMCTMPLGIFVIVLNSKTENI-QP
Bar2   ----***************************I*V***I**A*AKAV-S*
Bbr1   MHLQARSSALTTSRYLR LILMAIVQLVWLVVTTAYTLWFSSMSSTS--P

Ste3p  FKNTHSS-TIW-NTIIKFDPGRPI-----------YNIWLYVLMSYLVF
Map3   FNHEL-VE-AWHQESITYYPTTKV----------GLNDWVPPTVLYLMS
Pra1   YGSWSSVHTGF-N-QIPQYPASLVLMENTFQRNLILARLVCPLSAYIFF
Pra2   YTSWAEVHTNF-N-RILFVPVDTIAHSSLL-SLSIL-RWFSLTPAMALF
Bar1   WVSLAVTHYGF-G-RIDQVPAIVWLSQHLIVVCNELTRWCAPVSAFIFF
Bar2   Y**W******Y-*-******I***R*NR*L*ASY*****S**AI*****
Bbr1   LDHLGGAHSNF-G-RSRHGPPLLPPPSSSAARAHCGGMV--PASTWIFV

Ste3p  LI--FGLGSDALHMYSKFLRSIKLGFVLDMWKRFIDKNKEKRVGILLNK>
Map3   LF--FSTSGGWTEKVALILWSLLVWLPFTKNTALGRHAQFKLDCCKSIE>
Pra1   AM--FGLGLEVRQGYKEAFHRALLFCRLRKEPKASALQHVVADIEVVTF>
Pra2   VF--FGLTEEAQSVYK-ARWKALINLCSSKGKKQTDG-RESLDLEA--F>
Bar1   FY--FGFAEEARRNYAAAWRKVCRALGLPERVPSLPTTKKPFSSSDNKG>
Bar2   **--*****Q**********GW***RV********************VR>
Bbr1   AFFAF GNDAVEE--YKRVLNVVLSGARRVFPRASSRRRSATSRASPSRA>
```

Figure 4. Comparison of N-terminal regions of amino acid sequences of fungal pheromone receptors. Ste3p, *S. cerevisiae* (21,22); Map3, *Sch. pombe* (31); Pra1, Pra2, *U. maydis* (65); Bar1, Bar2, Bbr1, *S. commune* (97; Vaillancourt and Raper, unpublished data). Amino acid identities are indicated in bold type. Amino acid identities between Bar1 and Bar2 are indicated by asterisks. Potential transmembrane regions are indicated by lines below the sequence.

The third intracellular loop of the 7tm receptors is generally thought to be important for binding to the Gα subunits of the G proteins (reviewed in 118,119). There is a conserved region in the third intracellular loop of the fungal lipopeptide pheromone receptors (Figure 4). In the yeast Ste2p receptor, if the entire C-terminus is removed the receptor is still functional, although the cells become hypersensitive to pheromone (120). The C-termini of the fungal pheromone receptors are all highly heterologous. Perhaps they have been less constrained by evolution because they are not directly involved in the transduction of the pheromone signal.

EVOLUTIONARY CONSIDERATIONS AND CONCLUDING COMMENTS

As we discover more about the molecular mechanisms different fungi use to regulate a variety of sexual behaviors, we find common themes. This is certainly true for the fungal pheromones and pheromone receptors. Furthermore, molecules of these basic types are not confined to the fungi. Similar signaling mechanisms are involved in regulation of development in many organisms, including humans. This implies that these mechanisms began evolving eons ago, in a distant eukaryotic ancestor. It seems likely that pheromones and receptors were in use for mating-type recognition by a common ancestor of the Ascomycetes and Basidiomycetes. It is impossible to say whether that ancestor used a system like the Basidiomycetes, in which the pheromones and receptors themselves determined mating type, or like the Ascomycetes, where they were under the control of other regulatory genes. Either could be ancestral. The development of one from the other probably took place about the time the two lines diverged. The early Basidiomycetes might have had either a two-locus system, in which pheromones and receptor genes are separated from transcriptional activators, or a one-locus system, in which both types of genes are contained in a single locus. Again, it is difficult to say which came first, or whether both arose separately several times in different lineages. Given the similarities between the smuts with a one-locus mating system *versus* a two-locus system, it is logical to assume that these systems diverged relatively recently. It is interesting that only the Homobasidiomycetes have evolved multiple pheromone-receptor altermorphs. Why did they do it, and how did they do it? We have come a bit closer to understanding "how", but we are no closer to understanding "why". We can only point out that an increase in the number of **B** altermorphs has had the effect of substantially increasing outbreeding potential in the Homobasidiomycetes. For example, the outbreeding potential of *U. maydis*, with two *a* types and 33 *b* types, is not quite 50%. For *S.commune*, with 288 **A** types and 81 **B** types, it is better than 98%. In fact, we might ask why *U. maydis* has not evolved more that two *a* types, because it cannot take advantage of the full outbreeding potential of its 33 *b* types as long as the *a* types are limited to two. The capacity to outbreed may not be the only selective force at work on these genes. Obviously, many mysteries remain.

Fungi are ideal experimental subjects for studies of molecular aspects of self *versus* non-self recognition. The mechanisms used to achieve mating-type recognition have been highly conserved among the higher fungi. Furthermore, these mechanisms involve molecules similar to those used by many other organisms, including humans, to regulate development. All the components of the relevant pathways are uniquely accessible for manipulation and characterization in fungi. It may be said, therefore, that the more we understand fungi, the more we understand ourselves!

ACKNOWLEDGMENTS

The authors thank Guus Bakkeren, Lorna Casselton, Jim Kronstad, Marjatta Raudaskoski and Charles Specht for sharing unpublished research results with us.

REFERENCES

1 Raper, J.R. (1966) Genetics of Sexuality in Higher Fungi. Ronald Press, New York, NY.
2 Griffin, D.H. (1994) Fungal Physiology, 2nd ed. Wiley Liss, Inc., New York, NY.
3 Raper, C.A. (1983) in Secondary Metabolism and Differentiation in Fungi (Bennett, J.W. and Ciegler, A., eds.) pp. 195-238, Marcel Dekker, New York, NY.
4 Novotny, C.P., Stankis, M.M., Specht, C.A., Yang, H., Giasson, L. and Ullrich, R.C. (1991) in More Gene Manupulations in Fungi (Bennett, J.W. and Lasure, L.L., eds.) pp. 235-257, Academic Press, New York, NY.
5 Raper, J.R., Baxter, M.G. and Middleton, R.B. (1958) Proc. Nat. Acad. Sci. U.S.A. 44, 889-900.
6 Raper, J.R., Krongelb, G.S. and Baxter, M.G. (1958) Amer. Nat. 92, 221-232.
7 Koltin, Y., Raper, J.R. and Simchen, G. (1967) Proc. Nat. Acad. Sci. U.S.A. 57, 55-62.
8 Stamberg, J. and Koltin, Y. (1973) Heredity 30, 15-26.
9 Raper, J.R. (1952) Bot. Rev. 18, 447-545.
10 Gooday, G.W. and Adams, D.J. (1993) in Advances in Microbial Physiology, Vol. 34 (Rose, A.H., ed.), pp. 69-134. Academic Press, San Diego, CA.
11 Gooday, G.W. (1994) in The Mycota I: Growth, Differentiation, and Sexuality (Wessels, J.G.H. and Meinhardt, F., eds.) pp. 401-411, Springer-Verlag, Berlin.
12 Marsh, L., Neiman, A.M. and Herskowitz, I. (1991) Annu. Rev. Cell Biol. 7, 699-728.
13 Kurjan, J. (1993) Annu. Rev. Genet. 27, 147-179.
14 Whiteway, M. and Errede, B. (1993) in Signal Transduction, Prokaryotic and Simple Eukaryotic Systems (Kurjan, J. and Taylor, B., eds.) pp. 189-237, Academic Press, San Diego, CA.
15 Stotzler, D., Klitz, H.H. and Duntze, W. (1976) Eur. J. Biochem. 69, 397-400.
16 Kurjan, J. and Herskowitz, I. (1982) Cell 30, 933-943.
17 Singh, A., Chen, E.Y., Lugovoy, J.M., Chang, C.N., Hitzman, R.A. and Seeburg, P.H. (1983) Nucl. Acids. Res. 11, 4049-4063.
18 Anderegg, R.J., Betz, R., Carr, S.A., Crabb, J.W. and Duntze, W. (1988) J. Biol. Chem. 263, 18236-18240.
19 Betz, R., Crabb, J.W., Meyer, H.E., Wittig, R. and Duntze, W. (1987) J. Biol. Chem. 262, 546-548.
20 Michaelis, S., Chen, P., Berkower, C., Sapperstein, S. and Kistler, A. (1992) Antonie von Leeuwenhoek 61, 115-117.
21 Nakayama, N., Miyajima, A. and Arai, K. (1985) EMBO J. 4, 2643-2648.
22 Hagen, D.C., McCaffrey, G. and Sprague, G.F. (1986) Proc. Nat. Acad. Sci. U.S.A. 83, 1418-1422.
23 Burkholder, A.C. and Hartwell, L.H. (1985) Nucl. Acids Res. 13, 8463-8475.
24 Bender, A. and Sprague, G.F. (1986) Cell 47, 929-937.
25 Dolan, J.W., Kirkman, C. and Fields, S. (1989) Proc. Nat. Acad. Sci. U.S.A. 86, 5703-5707.
26 Errede, B. and Ammerer, G. (1989) Genes Dev. 3, 1349-1361.

27 Imai, Y. and Yamamoto, M. (1994) Genes Dev. 8, 328-338.
28 Davey, J. (1992) EMBO J. 11, 951-960.
29 Kjaerulff, S., Davey, J. and Nielsen, O. (1994) Mol. Cell. Biol. 14, 3895-3905.
30 Kitamura, K. and Shimoda, C. (1991) EMBO J. 12, 3743-3751.
31 Tanaka, K., Davey, J., Imai, Y. and Yamamoto, M. (1993) Mol. Cell. Biol. 13, 80-88.
32 Obara, T., Nakafuku, M., Yamamoto, M. and Kaziro, Y. (1991) Proc. Nat. Acad. Sci. U.S.A. 88, 5877-5881.
33 Nadin-Davis, S.A. and Nasim, A. (1988) EMBO J. 7, 985-993.
34 Toda, T., Shimanuki, M. and Yanagida, M. (1991) Genes Dev. 5, 60-73.
35 Leupold, U., Nielson, O. and Egel, R. (1989) Curr. Genet. 15, 403-405.
36 Dietzel, C. and Kurjan, J. (1987) Cell 50, 1000-1010.
37 Miyajima, I., Nakafuku, M., Nakayama, N., Brenner, C. and Miyajima, A. (1987) Cell 50, 1011-1019.
38 Sprague, G.F., Blair, L.C. and Thorner, J. (1983) Annu. Rev. Microbiol. 37, 623-660.
39 Egel, R., Nielsen, O. and Weilguny, D. (1990) Trends Genet. 6, 369-373.
40 Bowman, D.H. (1946) J. Agric. Res. 72, 233-243.
41 Bandoni, R.J. (1963) Can. J. Bot. 41, 467-474.
42 Banno, I. (1967) J. Gen. Appl. Microbiol. 13, 167-196.
43 Abe, K., Kusaka, I. and Fukui, S. (1975) J. Bacteriol. 122, 710-718.
44 Kamiya, Y., Sakurai, A., Tamura, S., Takahashi, N., Tsuchiya, E., Abe, K. and Fukui, S. (1979) Agric. Biol. Chem. 43, 363-369.
45 Akada, R., Minomi, K., Yamashita, I., Miyakawa, T. and Fukui, S. (1987) Agric. Biol. Chem. 51, 1211-1215.
46 Akada, R., Minomi, K., Kai, J., Yamashita, I., Miyakawa, T. and Fukui, S. (1989) Mol. Cell. Biol. 9, 3491-3498.
47 Tsuchiya, E. and Fukui, S. (1978) Biochem. Biophys. Res. Commun. 85, 473-479.
48 Miyakawa, T., Nishihara, M., Tsuchiya, E. and Fukui, S. (1982) J. Bacteriol. 151, 1184-1194.
49 Miyakawa, T., Kaji, M., Yasutake, T., Jeong, Y.K., Tsuchiya, E. and Fukui, S. (1985) J. Bacteriol. 162, 294-299.
50 Jeong, Y.K., Miyakawa, T., Imabayashi, A., Tsuchiya, E. and Fukui, S. (1987) Eur. J. Biochem. 169, 511-515.
51 MacKay, V.L., Welch, S.K., Insley, M.Y., Manney, T.R., Holly, J., Saari, G.C. and Parker, M.L. (1988) Proc. Nat. Acad. Sci. U.S.A. 85, 55-59.
52 Manney, T. (1983) J. Bacteriol. 155, 291-301.
53 Marcus, S., Xue, C.-B., Naider, F. and Becker, J.M. (1991) Mol. Cell. Biol. 11, 1030-1039.
54 Bandoni, R.J. (1965) Can. J. Bot. 43, 627-630.
55 Sakagami, Y., Isogai, A., Suzuki, A., Tamura, S., Kitada, C. and Fujino, M. (1979) Agric. Biol. Chem. 43, 2643-2645.
56 Sakagami, Y., Yoshida, M., Isogai, A. and Suzuki, A. (1981) Science 212, 1525-1526.
57 Ishibashi, Y., Sakagami, Y., Isogai, A. and Suzuki, A. (1984) Biochemistry 23, 1399-1404.
58 Miyakawa, T., Tabata, M., Tsuchiya, E. and Fukui, S. (1985) J. Biochem. 147, 489-493.
59 Rowell, J.B. (1955) Phytopathology 45, 370-374.
60 Snetselaar, K.M. (1993) Exp. Mycol. 17, 345-355.

61 Banuett, F. and Herskowitz, I. (1994) Exp. Mycol. 18, 247-266.
62 Holliday, R. (1961) Genet. Res. 2, 231-248.
63 Day, P. and Anagnostikis, S. (1971) Nature 231, 19-20.
64 Froeliger, E.H. and Leong, S.A. (1991) Gene 100, 113-122.
65 Bölker, M., Urban, M. and Kahmann, R. (1992) Cell 68, 441-450.
66 Kronstad, J.W. and Leong, S.A. (1989) Proc. Nat. Acad. Sci. U.S.A. 86, 978-982.
67 Kronstad, J.W. and Leong, S.A. (1990) Genes Dev. 4, 1384-1395.
68 Schulz, I., Banuett, F., Dahl, M., Schlesinger, R., Schäfer, W., Martin, T., Herskowitz, I. and Kahmann, R. (1990) Cell 60, 295-306.
69 Gillissen, B., Bergemann, J., Sandmann, C., Schroeer, B., Bolker, M. and Kahmann, R. (1992) Cell 68, 647-657.
70 Kamper, J., Reichmann, M., Romeis, T., Bolker, M. and Kahmann, R. (1995) Cell 81, 73-83.
71 Giasson, L., Specht, C.A., Milgrim, C., Novotny, C.P. and Ullrich, R.C. (1989) Mol. Gen. Genet. 218, 72-77.
72 Stankis, M.M., Specht, C.A., Yang, H., Giasson, L., Ullrich, R.C. and Novotny, C.P. (1992) Proc. Nat. Acad. Sci. U.S.A. 89, 7169-7173.
73 Specht, C.A., Stankis, M.M., Giasson, L., Novotny, C.P. and Ullrich, R.C. (1992) Proc. Nat. Acad. Sci. U.S.A. 89, 7174-7178.
74 Shen, G.P., Park, D.C., Ullrich, R.C. and Novotny, C.P. (1996) Curr. Genet. (in press).
75 Kues, U., Richardson, W.V.J., Tymon, A.M., Mutasa, E.S., Gottgens, B., Gaubatz, S., Gregoriades, A. and Casselton, L.A. (1992) Genes Dev. 6, 568-577.
76 Kues, U. and Casselton, L.A. (1993) J. Cell. Sci. 104, 227-230.
77 Metzenberg, R.L. and Glass, N.L. (1990) BioEssays 12, 53-59.
78 Egel, R. (1992) Nature 357, 23-24.
79 Spellig, T., Bolker, M., Lottspeich, F., Frank, R.W. and Kahmann, R. (1994) EMBO J. 13, 1620-1627.
80 Banuett, F. and Herskowitz, I. (1989) Proc. Nat. Acad. Sci. U.S.A. 86, 5878-5882.
81 Gold, S., Duncan, G., Barrett, K. and Kronstad, J. (1994) Genes Dev. 8, 2805-2816.
82 Banuett, F. and Herskowitz, I. (1994) Genes Dev. 8, 1367-1378.
83 Martinez-Espinoza, A.D., Gerhardt, S. and Sherwood, J.E. (1993) Exp. Mycol. 17, 200-214.
84 Bakkeren, G. and Kronstad, J.W. (1993) Plant Cell 5, 123-136.
85 Bakkeren, G. and Kronstad, J.W. (1994) Proc. Nat. Acad. Sci. U.S.A. 91, 7085-7089.
86 Bakkeren, G., Gibbard, B., Yee, A., Froeliger, E., Leong, S. and Kronstad, J. (1992) Molec. Plant-Microbe Interact. 5, 347-355.
87 Poon, N.H. and Day, A.W. (1975) Can. J. Microbiol. 21, 537-546.
88 Day, A.W. and Garber, E.D. (1988) in Advances in Plant Pathology, Vol. 6 (Ingram, D.S. and Williams, P.H., eds.) pp. 457-482, Academic Press, San Diego, CA.
89 Day, A.W. (1976) Nature 262, 583-584.
90 Day, A.W. and Poon, N.H. (1975) Can. J. Microbiol. 21, 547-557.
91 Rippon, J.W. (1988) Medical Mycology, W.B. Saunders, Philadelphia, PA.
92 Kwon-Chung, K.J., Edman, J.C. and Wickes, B.L. (1992) Infect. Immun. 60, 602-605.
93 Moore, T.D.E. and Edman, J.C. (1993) Mol. Cell. Biol. 13, 1962-1970.
94 Bistis, G.N. (1983) Exp. Mycol. 7, 292-295.
95 Glass, N.L., Grotelueschen, J. and Metzenberg, R.L. (1990) Proc. Nat. Acad. Sci. U.S.A. 87, 4912-4916.

96 Staben, C. and Yanofsky, C. (1990) Proc. Nat. Acad. Sci. U.S.A. 87, 4917-4921.
97 Wendland, J., Vaillancourt, L.J., Hegner, J., Lengeler, K.B., Laddison, K.J., Specht, C.A., Raper, C.A. and Kothe, E. (1995) EMBO J. 14 (21), 5271-5278.
98 Casselton, L.A. (1978) in The Filamentous Fungi, Vol. 3 (Smith, J.E. and Berry, D.R., eds.) pp. 275-297, Edward Arnold, London.
99 Specht, C.A. (1995) Curr. Genet. 28, 374-379.
100 Niederpruem, D.J. (1980) Arch. Microbiol. 128, 162-171.
101 Niederpruem, D.J. (1980) Arch. Microbiol. 128, 172-178.
102 Trueheart, J., Boeke, J.D. and Fink, G.R. (1987) Mol. Cell. Biol. 7, 2316-2328.
103 Wessels, J.G.H. and Marchant, R. (1974) J. Gen. Microbiol. 83, 359-368.
104 Powers, S., Michaelis, S., Broek, D., Santa Anna, A.S., Field, J., Herskowitz, I. and Wigler, M. (1986) Cell 47, 413-422.
105 Maltese, W.A. (1990) FASEB J. 4, 3319-3328.
106 Clarke, S. (1992) Annu. Rev. Biochem. 61, 355-386.
107 Glomset, J.A., Gelb, M.H. and Farnsworth, C.C. (1990) TIBS 15, 139-142.
108 Shafer, W.R. and Rine, J. (1992) Annu. Rev. Genet. 30, 209-237.
109 He, B., Chen, P., Chen, S.Y., Vancura, K.L., Michaelis, S. and Powers, S. (1991) Proc. Nat. Acad. Sci. U.S.A. 88, 11373-11377.
110 Hrycyna, C.A., Sapperstein, S.A., Clarke, S. and Michaelis, S. (1991) EMBO J. 10, 1699-1709.
111 Duntze, W., Betz, R. and Popping, B. (1993) in Signal Transduction, Prokaryotic and Simple Eukaryotic Systems (Kurjan, J. and Taylor, B., eds.) pp. 239-254, Academic Press, San Diego, CA.
112 Fujino, M., Kitada, C., Sakagami, Y., Isogai, A., Tamura, S. and Suzuki, A. (1980) Naturwissenschaften 67, 406-408.
113 Kitada, C., Fujino, M., Sakagami, Y., Yoshida, M., Isogai, A. and Suzuki, A. (1981) Agric. Biol. Chem. 45, 1049-1051.
114 O'Dowd, B., Collins, S., Bouvier, M., Caron, M.G. and Lefkowitz, R.J. (1992) in Applications of Molecular Genetics to Pharmacology: Molecular Biology of G-Protein-Coupled Receptors (Brann, M.R., ed.) pp. 31-61, Birkhauser, Boston, MA.
115 Clarke, K.L., Davis, N.G., Wiest, D.K., Hwang-Shum, J.J. and Sprague, G.F. (1988) Cold Spring Harbor Symp. Quant. Biol. 53, 611-620.
116 Poyner, D.R. and Hanley, M.R. (1992) in Applications of Molecular Genetics to Pharmacology: Molecular Biology of G-Protein-Coupled Receptors (Brann, M.R., ed.) Birkhauser, Boston, MA.
117 Schwyzer, R. (1987) EMBO J. 6, 2255-2259.
118 Dohlman, H.G., Thorner, J., Caron, M.G. and Lefkowitz, R.J. (1991) Annu. Rev. Biochem. 60, 653-688.
119 Savarese, T.M. and Fraser, C.M. (1992) Biochem. J. 283, 1-19.
120 Konopka, J.B., Jenness, D.D. and Hartwell, L.H. (1988) Cell 54, 609-620.

CHEMICAL SYNTHESIS AND APPLICATIONS OF PHOSPHOPEPTIDES

Kazuyasu Sakaguchi[*], Peter P. Roller[#] and Ettore Appella[*]

[*]Laboratory of Cell Biology and [#]Laboratory of Medicinal Chemistry
National Cancer Institute
National Institutes of Health
Bethesda, MD 20892

INTRODUCTION

Many biologically important molecules including oncogenes, growth factor receptors, transcription factors, immune cell receptors, protein kinases and cytoskeletal proteins are specifically and reversibly phosphorylated in response to physiological stimuli as well as during differentiation and normal eukaryotic cell cycle progression (1-3). Phosphorylation occurs primarily on the hydroxyl group of serine, threonine or tyrosine residues in proteins. While other residues, including histidine, lysine and glutamic acid may be phosphorylated, these phosphoamino acids are chemically less stable and their roles in physiological processes are relatively poorly understood. Protein phosphorylation is accomplished by protein kinases, which catalyze the transfer of phosphate from a nucleotide triphosphate, usually ATP, to the substrate protein. Protein kinases traditionally are grouped by amino acid homology as protein tyrosine kinases or as protein serine/threonine kinases, although a few dual specificity tyrosine/serine/threonine kinases have been described (4,5), and recently, an evolutionarily separate group of serine/threonine kinases related to the phospholipid kinases was discovered (6). Many tyrosine-specific kinases are cell surface receptors or are associated with the plasma membrane as illustrated by the epidermal growth factor, insulin and platelet-derived growth factor receptors (7,8). Serine/threonine kinases more frequently are involved in intracellular signaling and may be linked in cascades with receptor tyrosine kinases to transmit signals from the extracellular environment to the cell nucleus.

For regulation, phosphorylation must be opposed by dephosphorylation, which, for serine, threonine, and tyrosine, requires the action of protein phosphatases to hydrolyze the phosphate group from amino acids in proteins. In general, phosphatases appear to have broader substrate specificities than protein kinases, but again, they can generally be classified on the basis of homology and substrate specificity as protein tyrosine phosphatases or serine/threonine phosphatases. The balance of kinase and phosphatase activities determine the steady state level of phosphorylation of many key cellular proteins implicated in a variety of biochemical pathways.

Phosphorylation may affect protein function in different ways. Phosphates add a bulk charge and hydrophilicity to a protein's surface. These changes may produce or destroy sites for interaction with other proteins or they induce local, or even global shifts in

Genetic Engineering, Vol. 18, Edited by J.K. Setlow
Plenum Press, New York, 1996

conformation, that affect enzymatic activity. For example, phosphorylation directly activates or inhibits several protein kinases (e.g., MAP and cdk kinases), other enzymes (e.g., glycogen synthase, phosphorylase) and transcription factors (e.g., Jun, Fos, SRF, CREB). Sequential, independent phosphate-mediated changes often are connected to form signaling cascades. Extracellular ligands initiate intracellular signaling by binding to specific receptors which are often associated with tyrosine kinases (9). The receptor-associated kinases are activated by ligand-induced receptor homo- or heterodimerization; they then phosphorylate tyrosines on receptor components, which function as docking sites that bind specific downstream target proteins. One outcome of this sequence of events is the activation of specific transcriptional regulators that bind their DNA targets to induce the synthesis of proteins and thus allow a distinct response to a particular extracellular signaling protein.

The formation of signaling complexes is a ubiquitous feature of many signal transduction pathways, of which transmembrane receptor tyrosine kinases provide just one example. Interactions involving Src homology 2 (SH2) domains, protein modules of approximately 100 amino acids that bind to phosphotyrosine sites on transmembrane and cytoplasmic signaling proteins, are an essential feature of several signaling pathways. These interactions are regulated specifically by tyrosine phosphorylation or dephosphorylation. Understanding the basis of specificity inherent in the SH2 protein-phosphoprotein interactions may provide valuable insights into discrimination among divergent signaling pathways. The binding of short phosphorylated peptides, which mimic a particular receptor kinase, to SH2 domains of signaling molecules can be used to study the energetics of SH2 domain specificity.

Here we have reviewed solid-phase chemical methods for the preparation of phosphopeptides and we have outlined basic protocols for the synthesis and purification of these reagents. Then, as alluded to above, we describe several applications of phosphopeptides to solving a wide range of important problems of biological interest.

CHEMICAL SYNTHESIS OF PHOSPHOPEPTIDES

The preparation of phosphopeptides which are subdomains of biologically occurring phosphoproteins can be useful for the study of the biological and structural properties of such proteins. While enzymatic methods can be used for obtaining phosphopeptides, they are limited to specific peptide sequences. Thus, enzymatic phosphorylation by a protein kinase is not suitable as a general approach. Chemical modification with solid-phase peptide synthesis would seem to be a more convenient method of obtaining various phosphopeptides. However, chemical synthesis of phosphopeptides has been very challenging, and, until recently, has not consistently resulted in good yields of phosphopeptides of high purity.

In the chemical synthesis of phosphopeptides, the major problems are the decomposition of free or protected phosphate groups by strong acids such as HF (10-12) and the ready loss of the phosphate groups in the presence of base by a β-elimination mechanism (13-15). In the synthesis of phosphoserine- and phosphothreonine-containing peptides which have alkyl phosphate groups, the problems of β-elimination are of special concern, while phosphotyrosine, which has a phenol phosphate group, is less susceptible to strong acid decomposition and is tolerant of treatment with bases (Figure 1A).

Synthesis in solution has been applied successfully to the preparation of phosphopeptides (16,17). However, this method is relatively time-consuming and appears to be practical only for small peptides. In this chapter, we will focus on phosphopeptide synthesis by the solid-phase methods. Solid-phase peptide synthesis is based on sequential addition of

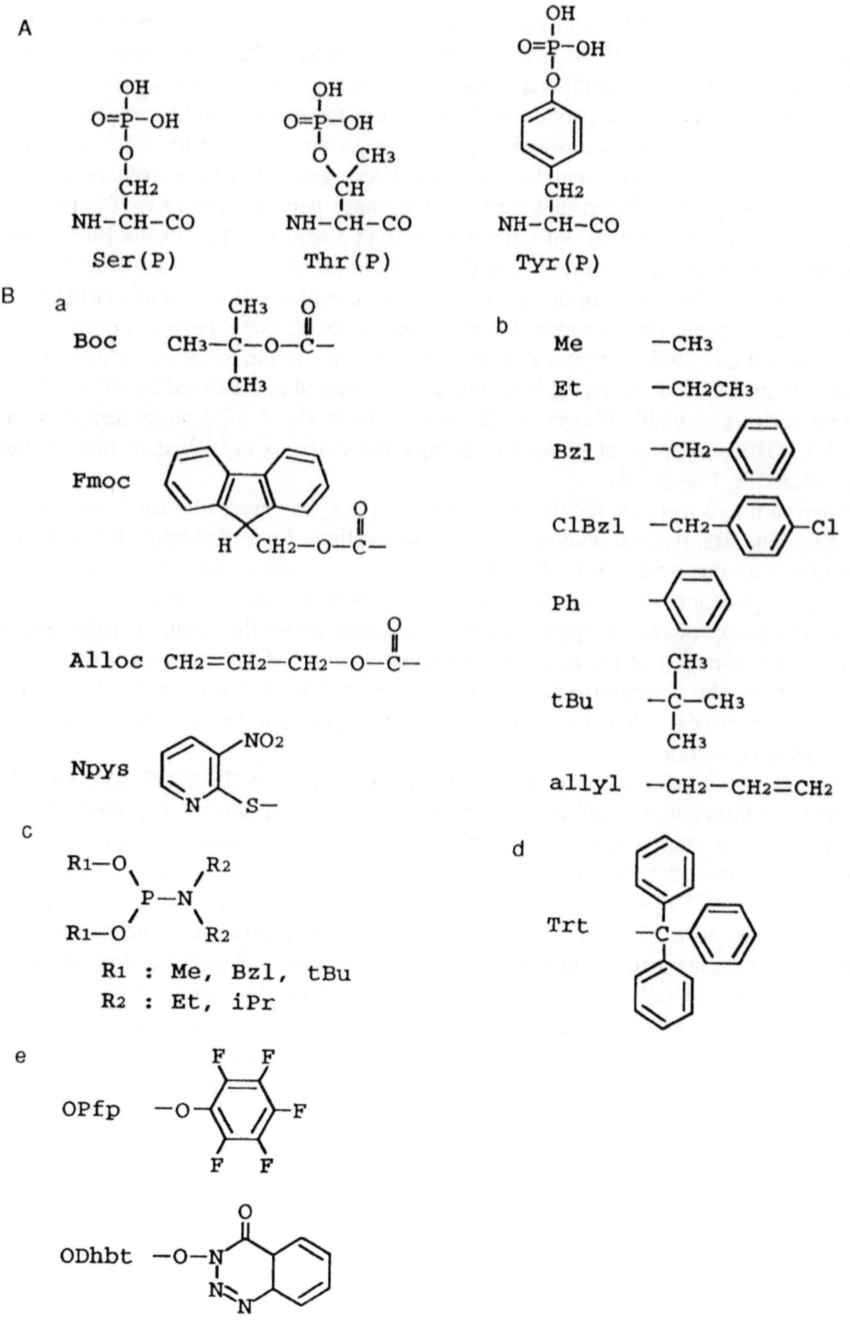

Figure 1. A. The structures of phosphoamino acid residues. Ser(P): Phosphoserine. Thr(P): Phosphothreonine. Tyr(P): Phosphotyrosine. B. The chemical structures of N-α-protecting groups (a); protecting groups for phosphate (b); phosphoramidites (c); hydroxyl-protecting group (d); active esters (e). Abbreviations: Boc=*t*-butyloxycarbonyl; Fmoc=9-fluorenylmethyloxycabonyl; Alloc=allyloxycarbonyl; Npys=3-nitro-2-pyridinesulfenyl; Me=methyl; Bzl=benzyl; ClBzl=4-chlorobenzyl; Ph=phenyl; tBu=*t*-butyl; Et=ethyl; iPr=isopropyl; Trt=trityl; OPfp=pentafluorophenyl ester; ODhbt=3-hydroxy-2,3-dihydro-4-oxo-benzotriazine ester.

amino acids with α-amino and side-chain protecting groups to an insoluble polymeric support. After removal of the α-amino protecting group, the next protected amino acid is coupled. The repeat of additional cycles (deprotection and coupling) yields an extended peptide chain. Finally, the peptide is cleaved from the support and the protecting groups are removed. Presently, two α-amino protecting groups are used for solid-phase peptide synthesis. The acid-labile *t*-butyloxycarbonyl (Boc) group has been used as an α-amino protecting group since Merrifield started solid-phase peptide synthesis. Removal of the Boc group is usually carried out with trifluoroacetic acid (TFA). In the Boc chemistry, cleavage of peptide from the resin and deprotection of side-chains is achieved by a strong acid, such as hydrogen fluoride (HF) or trifluoromethanesulfonic acid (TFMSA). The other α-amino protecting group is 9-fluorenylmethyloxycabonyl (Fmoc) group. The mild base piperidine is used for removal of the Fmoc group. In the Fmoc chemistry, side-chain protecting groups and anchoring linkages are deprotected and cleaved by TFA. There are several reviews available illustrating the general methods of solid-phase peptide synthesis (18-20). The structures of protecting groups and reagents described in this chapter are summarized in Figure 1B.

There are two methods for the preparation of phosphopeptides by solid-phase synthesis (21-23). The first method involves direct incorporation of phosphoamino acids (Figure 2). The target amino acids (Ser, Thr and/or Tyr) are coupled onto the peptide chain as previously phosphorylated amino acids during peptide chain assembly (14,24). The protecting group of the phosphate needs to be stable under the acidic or basic conditions used for deprotection of the α-amino protecting groups. After complete assembly of the peptide chain, the phosphopeptide is cleaved from the resin and the side-chain protecting groups are removed. This "incorporation" method appears to be preferred for synthesizing large phosphopeptides.

The other method involves post-assembly phosphorylation (25-27) (Figure 3A). In this method, the target amino acid with an unprotected, or with a selectively protected hydroxyl group, is incorporated into the peptide chain and the hydroxyl group is chemically phosphorylated on the resin after the peptide chain assembly is complete. In early studies, $POCl_3$ and $PO(OBzl_2)Cl$ were used for this type of phosphorylation. However, these reagents failed to achieve good yields of phosphorylated product. The phosphorylation reaction was dramatically improved by the introduction of a phosphitylation and oxidation method based on P(III) chemistry (26). In this method, the free hydroxyl group of the target amino acid is phosphitylated by phosphoramidite in the presence of 1*H*-tetrazole and subsequent oxidation to yield a corresponding phosphoamino acid (Figure 3B). With this method, both phosphorylated and unphosphorylated peptides can be obtained in a single peptide synthesis.

We will present separately the syntheses of phosphotyrosine-containing peptides and the syntheses of phosphoserine- and phosphothreonine-containing peptides, since these two peptide forms differ significantly in terms of chemical stabilities.

Synthesis of Phosphotyrosine-Containing Peptides

The synthesis of phosphotyrosine-containing peptides is relatively easy to accomplish due to the stability of phosphotyrosine to strong acid and base. Either Boc or Fmoc chemistries can be used for the direct incorporation of phosphoamino acids. In the Boc chemistry, Boc-Tyr(PO_3Bzl_2), Boc-Tyr(PO_3Ph_2) and Boc-Tyr(PO_3Me_2) are incorporated directly into the peptide chain during assembly (17,24,28,29). However, the benzyl (Bzl) group is not stable to the acidic deprotection cycle of the α-Boc group. The phenyl (Ph) groups can be deprotected by catalytic hydrogenolysis with the use of platinum oxide after

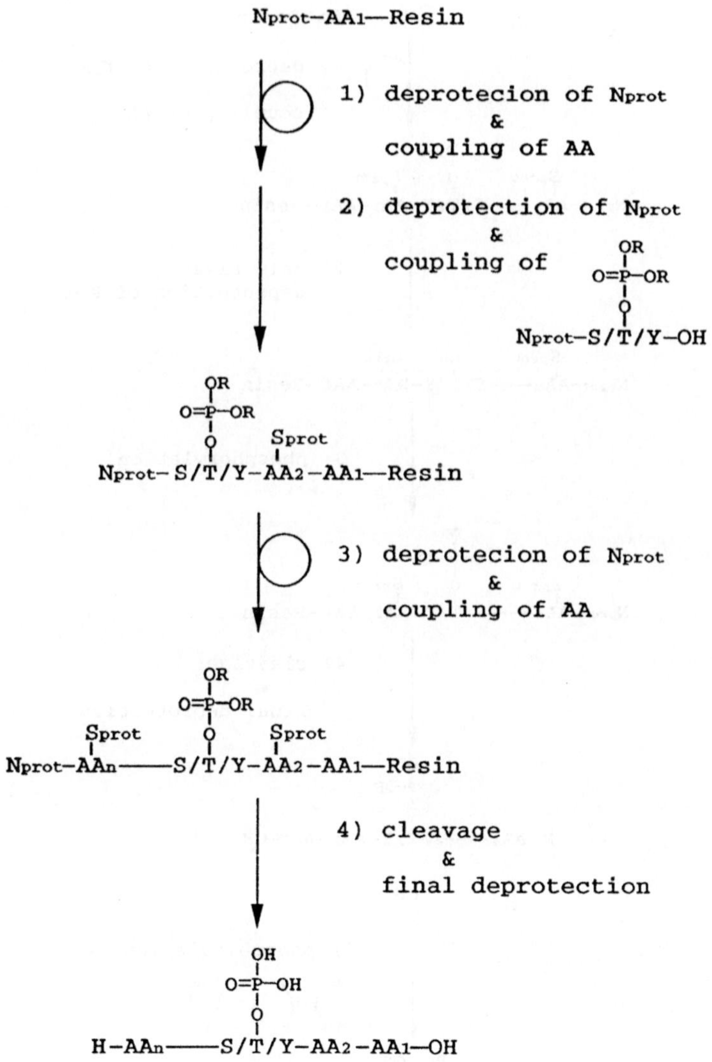

Figure 2. The synthetic scheme for the "incorporation" method. S/T/Y: target amino acid for phosphorylation (Ser, Thr and/or Tyr). N_{prot}: N-α-amino group protecting-group; S_{prot}: side-chain protecting group; R: phosphate-protecting group. 1) Repeat cycle for elongation of peptide chain; 2) phosphorylated amino acid is incorporated by a standard protocol; 3) elongation of peptide chain; 4) cleavage and final deprotection.

the peptide has been cleaved from the resin, but this group is hard to remove in large peptides. The methyl (Me) group has been shown to be stable to Boc chemistry procedures and can be deprotected by TFMSA/TFA (23). This group is also stable to piperidine treatment and is therefore suitable for Fmoc chemistry as well. In this case, deprotection is achieved with a milder acid such as trimethylsilyl bromide (TMSBr)/TFA (23,30), although the methyl (Me) deprotection generally requires a relatively long treatment (6 to 16 hr). The *t*-butyl (tBu) group is also utilized in Fmoc chemistry as well (23,31,32). This group is stable to base but is labile to TFA, and would appear to be an optimal protecting group for phosphotyrosine. Unfortunately, Fmoc-Tyr(PO_3tBu_2) cannot be stored due to its high acid lability. Recently, Ottinger, et al. have introduced a novel

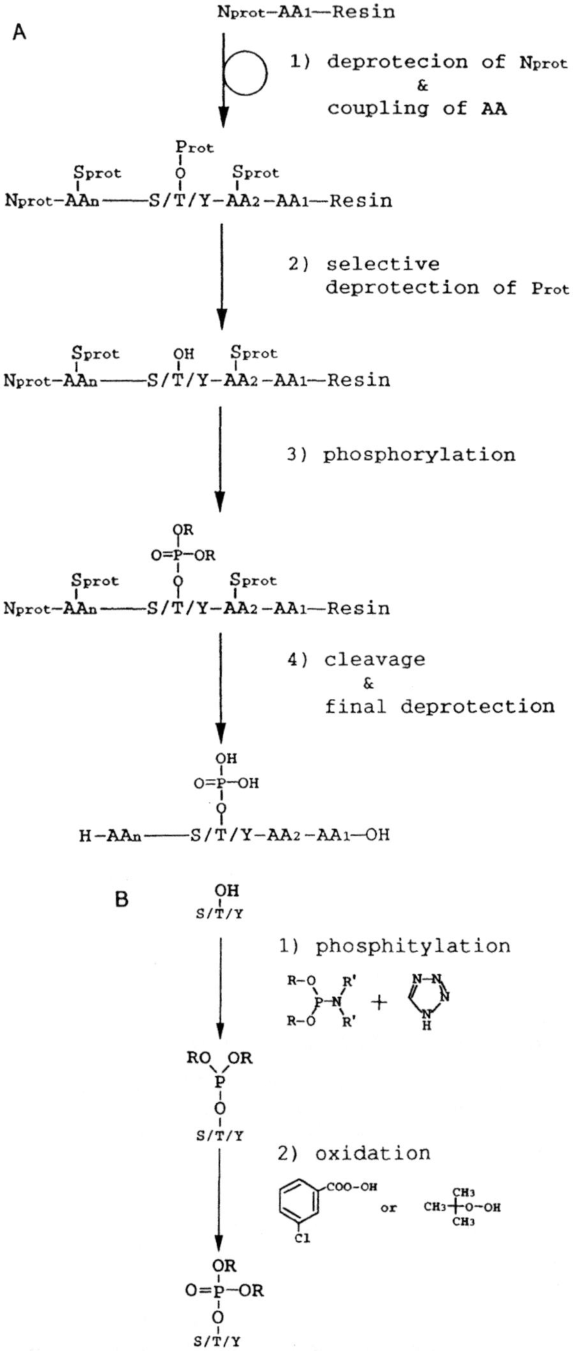

Figure 3. A. The synthetic scheme for the "post-assembly" method. S/T/Y: target amino acid for phosphorylation (Ser, Thr and/or Tyr). N_{prot}: N-α-amino group protecting-group. S_{prot}: side-chain protecting-group. X_{prot}: selective hydroxyl-protecting group for the target amino acid. R: phosphate-protecting group. 1) Repeat cycle for elongation of peptide chain. The target amino acid with selective hydroxyl-protecting group is also incorporated by a standard protocol. 2) Selective deprotection of the hydroxyl-protecting group of the target amino acid. 3) Phosphorylation of the hydroxyl group on resin. 4) Cleavage and final deprotection. B. The scheme for the phosphitylation-oxidation method. 1) The hydroxyl group is phosphitylated by phosphoramidite in the presence of 1*H*-tetrazole. 2) The P(III) atom is oxidized to P(V) by MCPBA or *t*-butylhydroperoxide.

phosphotyrosine derivative which has an unprotected phosphate side chain, Fmoc-Tyr(PO_3H_2) (33). This derivative is completely suitable for Fmoc chemistry and this approach minimizes the problems of deprotection and instability. In the latter case, the final deprotection and cleavage from the resin can be performed under mild conditions with the usual TFA reagents such as reagent K (34). We are using this method in our laboratory to synthesize relatively large phosphotyrosine-containing peptides (> 20 amino acid residues) with high quality and yield.

Post-assembly phosphorylation has also been used to prepare phosphotyrosine-containing peptides (35). The tyrosine residue, with an unprotected hydroxyl side chain, is incorporated into a peptide in the usual manner. After complete assembly of the peptide chain, the hydroxyl group is phosphitylated and oxidized on the resin (Figure 3B). A variety of phosphoramidites have been used to phosphitylate peptides with either Boc and Fmoc chemistry. These reagents include the dimethyl-, di-*t*-butyl-, dibenzyl- and di-4-chlorobenzyl-N,N-dialkyl phosphoramidites (21,23,36,37). All of these derivatives can be used for either chemistry. The *t*-butyl derivative is especially useful for Fmoc chemistry because of its acid lability. Two reagents, 3-chloroperoxybenzoic acid (MCPBA) and *t*-butylhydroperoxide, are often used for oxidation (21,26). Short treatment (< 5 min) with the mild oxidant *t*-butylhydroperoxide is preferred for use in the synthesis of Met- and/or Cys-containing peptides. The phosphitylation-oxidation method is quite successful for obtaining mono-phosphopeptides. However, it is usually difficult completely to phosphorylate peptides containing two or more phosphorylation sites (22). In addition, a side reaction has been reported in the phosphitylation-oxidation approach. A significant peptide by-product, differing from the unphosphorylated peptide, is sometimes found in crude preparations of phosphorylated peptides (33). This by-product has been described as an H-phosphonate peptide with a molecular mass of 16 fewer mass units than that of the correct phosphopeptide.

Synthesis of Phosphoserine- and Phosphothreonine-Containing Peptides

Serine and threonine residues with unprotected hydroxy side chains can be incorporated into the peptide chain for subsequent phosphorylation with the use of post-assembly phosphorylation methods. The active ester methods, such as Pfp (38) and Dhbt (39), are often used for coupling in order to avoid acylation of the unprotected hydroxyl group. However, complete prevention of acylation is very difficult to achieve when the amino acid is located at or near the carboxyterminus. The trityl group is utilized as a temporary protecting group for the hydroxyl side chain in Fmoc chemistry (40). After peptide chain assembly, this group can be selectively removed under very mild acidic conditions without removal of the other protecting groups. The phosphoramidites, dibenzyl and *t*-butyl derivatives, are commonly used for phosphorylation in Boc and Fmoc chemistry, respectively. MCPBA and *t*-butylhydroperoxide are used for oxidation.

Most of the previous work with the incorporation method was done with Boc rather than Fmoc chemistry because phosphodiesters undergo β-elimination upon exposure to piperidine (15). The phenyl (Ph) group is often used as a phosphate-protecting group to synthesize phosphoserine- and phosphothreonine-containing peptides (41-43). However, these phenyl groups are not adequately stable to HF treatment and, with tyrosine present in the peptide, they are hard to remove (44). The methyl (Me) and benzyl (Bzl) groups can be used with Boc chemistry, although the methyl group is not easy to remove and the benzyl group is not stable to TFA treatment. While the methyl and benzyl groups have also been used with Fmoc chemistry, the removal of the α-amino-protecting Fmoc group requires 1,8-diazabicyclo[5.4.0]undec-7-ene (DBU) instead of piperidine to prevent degradation of the phosphate-protecting groups (23). The allyl group also has been reported to be stable to piperidine treatment; however this group requires an additional deprotecting step by [$Pd^0(PPh_3)_4$] (23). Incorporation of the free Fmoc-Ser(PO_3H_2)-OH into peptides has also been reported; however major side-products were formed, one of these being the peptide dimer through a pyrophosphate linkage (45). Recently, Wakamiya and co-workers have found that the phosphomonobenzyl ester, Fmoc-Ser(PO_3HBzl), is stable to piperidine treatment and is suitable for phosphopeptide synthesis *via* Fmoc-chemistry (46,47). This new method appears to be the most promising for preparation of complex phosphoserine-containing peptides. More recently, the utility of the sodium salts of Fmoc-Ser(PO_3HBzl)-OH and Fmoc-Thr(PO_3HBzl)-OH has been demonstrated for the synthesis of a number of peptides, including the preparation of protected peptide segments suitable for fragment condensation (48).

Cleavage and Deprotection

After the phosphopeptide is cleaved from the resin and the other side-chain protecting groups are removed, the deprotection of the phenyl (Ph) group can be performed by catalytic hydrogenation with the use of an equimolar amount of platinum oxide (PtO_2) with respect to the phenyl group (13). The phosphopeptides containing the methyl phosphate-protecting groups are cleaved and deprotected by strong acid such as TFMSA or trimethylsilyl trifluoromethanesulfonate (TMSOTf)/TFA in the presence of dimethyl sulfide (DMS)-thioanisole when Boc chemistry is used, and by 1M TMSBr/thioanisole in TFA when Fmoc chemistry is used (23,30,49). However, complete removal of the phenyl or methyl groups becomes difficult as the phosphopeptides increase in size or in the number of phosphorylated residues. The *t*-butyl (tBu) and benzyl groups can be removed by mild acid TFA. The cleavage and deprotection of peptides containing Tyr(PO_3H_2) and Ser(PO_3HBzl) when Fmoc chemistry is employed are achieved easily with general reagents such as Reagent K without any further deprotection of phosphate groups (33,46). For preparation of mixed phosphoamino acid-containing peptides, an optimized procedure has been reported (49,50). Accordingly, the synthesis of 19-residue peptides containing multiple phosphorylated tyrosines, threonines and/or serines were achieved, with the incorporation of *O*-dimethyl-protected phosphorylated amino acids by standard Boc chemistry. Importantly, the resin cleavage and side-chain deprotection, including the removal of the phosphate methyl groups, were accomplished with a one pot two-step deprotection method consisting of high and low acidic treatments. This method is compatible with Met and Trp in the peptide sequence. The reagents used for peptide cleavage from the resin and final deprotection of side chains are summarized in Table 1.

Purification and Characterization of Synthesized Phosphopeptides

All of the purification methods used following standard peptide synthesis can be equally well applied to the purification of phosphopeptides. These methods include gel filtration, ion exchange and reverse-phase HPLC chromatography. Reverse-phase HPLC is the most powerful purification method for phosphopeptides, as it is for standard peptide purification. C-8 and C-18 columns are commonly used. Small phosphopeptides are easily purified with reverse-phase HPLC and a linear gradient of 0.05-0.01% aqueous TFA with acetonitrile. Under such conditions, phosphopeptides usually elute earlier than the corresponding unphosphorylated peptides.

In addition to the standard methods, iron(III)-chelating columns can be very useful for separating phosphorylated peptides from crude peptides, and particularly from unphosphorylated peptides and/or incompletely deprotected peptides (12). The phosphorylated peptides are retained on the iron-chelating column by the interaction between iron and phosphate, while unphosphorylated and phosphate-protected peptides flow through the column. After the column is washed, the phosphopeptides are eluted with high pH buffer. Ion exchange chromatography also can be very useful because phosphorylated peptides are more negatively charged than unphosphorylated or blocked-phosphate-containing peptides (17,41). These two methods are very useful for purifying large and/or complex phosphopeptides prior to the use of reverse-phase HPLC purification.

The purity of synthesized phosphopeptides can be monitored by analytical reverse-phase HPLC (C-8 and C-18). Capillary electrophoresis is also very useful (23,33,51).

Mass spectrometry is the most rapid and accurate method for the confirmation of the synthesis of phosphorylated peptides. Mass spectrometry also detects impurities such as unphosphorylated or incompletely deprotected peptides, and phosphonopeptides. FAB, ion-spray ionization and MALDI-TOF mass spectroscopy are all excellent for the analysis of phosphopeptides (52). Amino acid analysis and Edman peptide sequencing are sometimes used to confirm the synthesis of target peptides. However, chemical modification of the phosphoamino acid residue is required to identify the phosphoamino acid since phosphoamino acids are not stable to hydrolysis and PTZ-phosphoamino acids are unstable and difficult to extract (53,54). Characterization of phosphopeptides may also be performed by use of ^1H-, ^{13}C- ^{31}P-NMR spectroscopy (17,41). In the ^{13}C-NMR spectra of phosphopeptides, the Cα and the Cβ carbons of the phosphoserine and phosphothreonine residues are observed as phosphorous-coupled doublet signals with a coupling constant from 3 to 9 Hz.

Other Chemistries and Stable Analogs

Chemistries other than those based on Boc and Fmoc protection of α-amino groups would be interesting to explore for obtaining optimal phosphopeptide synthesis. Such chemistries should satisfy several criteria: the α-amino protecting group should be selectively removable by a very mild reagent, under weakly acidic to neutral conditions; the final removal of side-chain protecting groups and cleavage from the resin should be achievable by mild acid, such as TFA. The N-α-allyloxycarbonyl (Alloc) group has been used to prepare phosphopeptides (14). This group is easily removed by catalytic hydrostannolytic cleavage. The use of 3-nitro-2-pyridinesulfenyl (Npys) would be another good candidate for this purpose since this α-amino protecting group can be removed by a very mild neutral reagent, triphenyl phosphine (55). Also, this chemistry can be employed with the fragment condensation method to obtain large peptides (56-58). With the combination of the Npys chemistry and fragment condensation, very large phosphopeptides (~ 100 amino acid residues) might be prepared with high purity and yield.

Another possible approach is to use stable analogs of phosphoamino acids. One such analog is a phosphono-amino acid that has a stable P-C bond instead of a P-O bond (Figure 4). The phosphono-amino acid is free from base-mediated β-elimination and is not subject to dephosphorylation by phosphatases.

Phosphono-amino acid analogs of phosphotyrosine and phosphoserine have been synthesized and incorporated into peptides. Such suitable phosphotyrosine analogs, which contain $-CH_2PO_3H_2$ instead of $-OPO_3H_2$, include: the unprotected Boc-*p*($CH_2PO_3H_2$)Phe-OH (Boc-Pmp-OH) and its dimethyl or diethyl protected derivatives (59-61), as well as a di-*t*-butyl-protected Fmoc analog (62). The removal of the methyl or ethyl phosphonate

Table 1. Phosphate-protecting groups and corresponding deprotection methods.

Chemistry	Protecting group	Step 1	Step 2	Ref.
Boc	diphenyl (PO_3Ph_2)	HF, TFMSA/TFA	PtO_2	(12,13,42)
	dimethyl (PO_3Me_2)	TFMSA/TFA	not required	(23)
		TMSOTf/TFA	TMSOTf-DMS/TFA	(49)
	dibenzyl (PO_3Bzl_2)	TFMSA/TFA	not required	(16,25)
Fmoc	dimethyl (PO_3Me_2)	TMSBr/TFA	not required	(30)
	dibenzyl (PO_3Bzl_2)	TFA	not required	(21,38)
	di-*t*-butyl (PO_3tBu_2)	TFA	not required	(32,40)
	diallyl (PO_3allyl_2)	TFA	$[Pd^0(PPh_3)_4]$	(23)
	monobenzyl (PO_3HBzl)	TFA	not required	(46)
	free (PO_3H_2)	TFA	not required	(33)

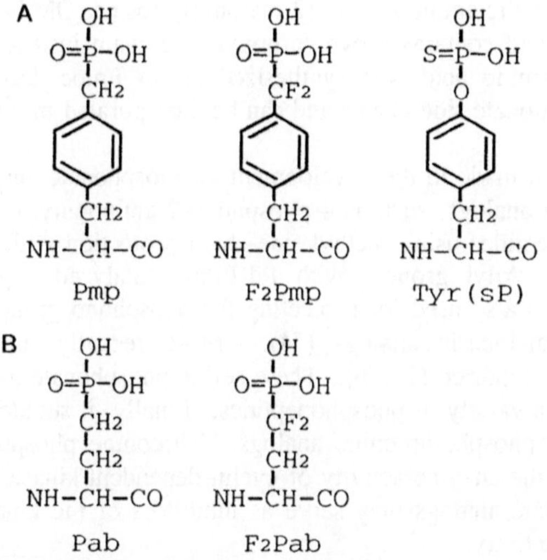

Figure 4. Structures of stable phosphoamino acid analogs. A. Phosphotyrosine analogs. Pmp: 4-phosphonomethyl-phenylalanine; F_2Pmp: 4-phosphono(difluoromethyl)-phenylalanine; OMT: *O*-(2-malonyl)tyrosine. B. Phosphoserine analogs. Pab: 4-phosphono-2-aminobutyric acid; F_2Pab: 4-phosphono-4,4-difluoro-2-aminobutylic acid.

protecting groups requires strong acid conditions such as TFMSA/TFA or TMSBr/TFA and long reaction times (63). For this reason, the use of *t*-butyl-protected phosphonate derivatives appears to be the most convenient. With Fmoc chemistry and mild acidolytic cleavage conditions, quantitative deprotection of phosphonate is achieved in the resulting peptides (64). The Pmp residues have been incorporated into specific peptides which contain tyrosine phosphorylation segments of growth factor receptors, and these peptides showed good affinity for the Src homology-2 (SH2) domains (65).

In a significant development, two fluorine atoms were introduced into the methylene carbon bridging the phenyl and the phosphonate groups, resulting in a derivative termed F_2Pmp. Both Boc- and Fmoc-derivatized F_2Pmp amino acids were prepared with diethyl phosphonate protection (64,66,67). For the efficient deprotection of ethyl groups, special conditions (1M TMSOTf-2M DMS/TFA) are required (68). Unfortunately, the chemical lability of the di-*t*-butyl derivative limits its usefulness (64). The side-chain unprotected Fmoc-F_2Pmp-OH also has been prepared, and its direct incorporation into short peptides was demonstrated (69). The workup for this method only requires mild acidolytic treatment for standard Fmoc chemistry. Thus, this approach has potential advantages in applications involving synthetic libraries, where the simplicity of deprotection methodology is an important consideration.

F_2Pmp-containing peptides have been demonstrated to bind to SH2 domains with affinities very similar to those observed with native phosphotyrosine-containing peptides, while Pmp-containing peptides showed 5- to 10-fold reduced affinity (65). It appears that, while the pKa2 of the Pmp phosphono group is higher than that of the natural tyrosine phosphate (70), the pKa2 of the F_2Pmp phosphonate is lower, closely approximating the phosphate. In addition, the fluorine atoms on the methylene carbon in F_2Pmp may participate in hydrogen bonding interactions in binding to proteins. The chemically and enzymatically stable phosphono-amino acid F_2Pmp is considered to be the best phosphotyrosine mimic currently available. Stable F_2Pmp-containing peptides could have significant potential for modulating cellular processes both *in vitro* and *in vivo*.

There are a number of potentially interesting phosphotyrosine mimetics under development and this review cannot do justice to all of them. A promising phosphotyrosine mimetic amino acid *O*-(2-malonyl)tyrosine (OMT) was reported recently (71). This amino acid contains a two carboxyl side chain instead of phosphate (Figure 4A). The OMT amino acid was synthesized as an Fmoc derivative with di-*t*-butyl protection of the malonate side-chains and can be incorporated into peptides with standard Fmoc chemistry.

Progress has been made in the development of phosphonate analogs of phosphoserine (Figure 4B). These analogs, such as 4-phosphono-2-aminobutyric acid (Pab), have been incorporated into peptides using methyl side-chain protection with Boc or Fmoc peptide synthesis (72,73). Allyl groups, with $Pd(PPh_3)_4$-catalyzed deprotection with Fmoc chemistry, have been also used for protecting the phosphono group in the preparation of neuromedin and calcineurin analogs (74). More recently, the synthesis of F_2Pab derivatives has been reported (75,76). These serine phosphonate mimetics are potentially useful for studying a variety of phosphopeptides. Finally, it should be noted that there is a lack of available phosphothreonine analogs. Threonine phosphorylation status is an important factor in the enzyme activity of cyclin-dependent kinases. Peptides containing threonine phosphonate analogs may serve as inhibitors of the kinases and phosphatases controlling kinase activity.

APPLICATIONS OF SYNTHETIC PHOSPHOPEPTIDES

Synthetic Phosphopeptides as Substrates

Several groups have devoted considerable effort to the identification and characterization of sequence-specific protein phosphatases. However, efforts have been severely hampered by the limited quantities of enzymes that can be isolated from cells. The use of synthetic phosphopeptides has been an alternative means of assessing phosphatase activity and achieving biochemical characterization of various phosphatases. Two enzymes, a rat brain phosphatase (PTPase) (77) and a human adipocyte acid phosphatase (78-80) have been studied in detail. A complete characterization of these enzymes required the use of pure synthetic phosphopeptides of 4 and 10 amino acid residues as substrates. The K_m and k_{cat}/K_m values for both tyrosine phosphatases were measured by a colorimetric assay (81), which quantifies the release of Pi by formation of a blue molybdate complex. Both peptides were dephosphorylated by the enzymes with different rates and affinities (33). These two enzymes also readily dephosphorylate *p*-nitrophenyl phosphate (*p*NP) and phospho-L-tyrosine, but they cannot be distinguished from each other by these assays. The rat brain PTPase hydrolyzed each of the phosphopeptides with simple Michaelis-Menten kinetics, with a k_{cat}/K_m value of approximately 10^5. In contrast, dephosphorylation of both phosphopeptides by the adipocyte acid phosphatase exhibited linear kinetics with no apparent saturation at concentrations of peptide up to 2 mM. These results demonstrate the utility of using synthetic tyrosine phosphopeptides to assess the catalytic efficiency of different phosphatases.

Recent work has drawn attention to phosphatases that act on cytoplasmic phosphotyrosine residues of growth factor receptors (82). These phosphatases possess non-catalytic domains of approximately 100 residues, termed Src homology 2 (SH2), that associate non-covalently with the cytoplasmic carboxyterminus of the growth factor receptors. Two related cytoplasmic protein-tyrosine phosphatases, SHP (83) and SH-PTP2 (84,85), which have two tandem aminoterminal SH2 domains linked to a single catalytic domain have been described. Use of well-defined, synthetic phosphopeptides has helped to clarify the role of SH2 *versus* catalytic domains in the specificity and efficiency of these phosphatases.

The effects of carboxyterminal truncation and SH2 domain removal on the catalytic activity of these two phosphatases were recently reported (86,87). Table 2 shows the peptide sequences and experimental results for phosphatase activity. It is apparent that while SH-PTP2 shows a 3-fold increase in activity following removal of the aminoterminal SH2 domains, SHP(ΔSH2/ΔCT), a construct lacking both the SH2 domains and part of the carboxyterminus, demonstrated an increased catalytic efficiency as compared to SHP(ΔCT), a construct lacking the carboxyterminus. A comparison of the kinetic constants for SHP(ΔCT) and SHP(ΔSH2/ΔCT) reveals that substrate specificity of this PTPase was unaltered by the linked SH2 domains. An analysis of the sequences flanking the phosphotyrosine residue in the peptides yields some clues regarding PTPase/substrate interactions. SH-PTP2 and SH-PTP2(ΔSH2) appear to favor acidic residues at the aminoterminal side of the phosphotyrosine, while basic residues at either side diminished substrate recognition and dephosphorylation rates. SHP also shared this preference for acidic residues, although it appeared to be less influenced by positively-charged residues flanking the phosphotyrosine. Thus, the substrate preferences of SHP(ΔSH2/ΔCT), SH(ΔCT) and SH-PTP2(ΔSH2) were similar overall and appear to differ considerably from those of other phosphatases, such as the widely-expressed PTB-1B phosphatase. This overall similarity demonstrated by the catalytic domains suggests that differences in the *in vivo* behavior of these molecules might stem not from differences in the substrate specificity, but instead from the specificity of the SH2 domains. The definition of the region in SH-PTP2(ΔSH2) molecules required for isolated SH2 binding is a goal for future studies.

SH2 Domain Specificity Determination with Phosphopeptides and Oriented Phosphopeptide Libraries

Molecular biological studies have furnished information concerning the binding of SH2-containing proteins to their targets. Table 3 lists, for example, the sites on the platelet-derived growth factor (PDGF) receptor that serve as binding sites for different signal transduction proteins (88). In general, individual SH2 domains can bind preferentially to one of a number of phosphotyrosines on a given target; however, the level of discrimination is not particularly specific since a given site can interact with more than one SH2 domain. The ability to synthesize phosphorylated peptides has contributed vastly to our understanding of SH2 binding specificity. For example, phosphopeptides corresponding to the 751 site of the PDGF receptor block the association of phosphatidylinositol (PI)-3 kinase, but do not block binding by ras GTPase activating protein (GAP) or phospholipase C-γ (PLC-γ) (89). Unphosphorylated peptides had no effect. A quantitative analysis showed that half-maximal competition was observed at about 10 μM. Eleven synthetic phosphopeptides containing Tyr(P)-Met-X-Met or Tyr(P)-Val-X-Met motifs were compared in competition assays with the 85 kD subunit (p85) of PI-3-kinase amino- and carboxyterminal SH2 domains. All Tyr(P)-Met-X-Met or Tyr(P)-Val-X-Met phosphopeptides bound within a 4-fold range of affinity (0.5 to 1.8 μM). Nonspecific tyrosine-phosphopeptides showed 30- to 1000-fold reduced affinity (90).

The binding affinities with sequence motifs predicted to interact with alternative SH2 domains were further analyzed to assess the sequence specificity of the p85-SH2 domain interactions, since several phosphorylation sites have been predicted for SH2 domain interactions. Peptides corresponding to the ras GTPase activating GAP or phospholipase PLC-γ SH2 domain binding motifs of the PDGF, FGF and EGF receptors bound to the p85 SH2 domain with 30- to 100-fold reduced affinity, indicating that this affinity range

Table 2. Kinetic constants for SHP and SH-PTP2.

Substrate[a]	Sequence[b]	Activity[c]											
		SHP(ΔSH2/ΔCT)			SHP(ΔCT)			SH-PTP2(ΔSH2)			SH-PTP2		
		k_m (μM)	k_{cat} (sec^{-1})	k_{cat}/K_m ($M^{-1}sec^{1}$)	K_m (μM)	k_{cat} (sec^{-1})	k_{cat}/K_m ($M^{-1}sec^{1}$)	K_m (μM)	k_{cat} (sec^{-1})	k_{cat}/K_m ($M^{-1}sec^{1}$)	K_m (μM)	k_{cat} (sec^{-1})	k_{cat}/K_m ($M^{-1}sec^{1}$)
PDGF-R	NEGDNDY(P)^{1021}IIPLPD	83	3.3	$39.8x10^3$	217	0.18	830	59	7.0	$110x10^3$	138	4	$29x10^3$
	DNDY(P)^{1021}IIP	190	4.9	$26x10^3$	844	0.25	291	306	9.8	$32x10^3$	1106	6.1	$5.5x10^3$
	DTSSVLY(P)^{1009}TAVQPN	135	3.2	$24x10^3$	515	0.23	452	171	7.2	$42x10^3$	705	3.3	$4.3x10^3$
	IMRDSNY(P)^{857}ISKGST	578	0.93	$1.6x10^3$	n.d.	n.d.	n.d.	n.d.	n.d.	n.d.	n.d.	n.d.	n.d.
Src	LIEDNEY(P)^{416}TARQGA	100	4.8	$49x10^3$	221	0.15	656	196	8.7	$44x10^3$	219	2.8	$13x10^3$
	TSTEPQY(P)^{527}QPGENL	252	3.3	$13x10^3$	1390	0.2	146	99	5.7	$57x10^3$	581	4.3	$7.4x10^3$
CFS-1R	PEGGVDY(P)^{699}KNIHLE	649	2.8	$4.3x10^3$	2870	0.25	85	n.d.	n.d.	n.d.	n.d.	n.d.	n.d.
*p*NP	-	14570	10.95	751	27730	1.24	45	14400	5.3	368	11820	36.2	$3x10^3$

[a] PDGF-R: platelet-derived growth factor receptor; CFS-1R: colony-stimulating factor-1 receptor; *p*NP: p-nitrophenyl phosphate. [b] Y(P): phosphotyrosine; numbers indicate the position of phosphotyrosine residues in substrate proteins. [c] n.d.: kinetic constants not determined due to lack of sufficient activity toward the substrate. For details, see Dechert et al., 1995 (86).

Table 3. Phosphorylation sites on the PDGF receptor and binding proteins for these phosphorylated peptides.

Site	Sequence[a]	Binding Protein[b]
Tyr579	SDGHEY(P)IYVDP	Src family, SHC
Tyr581	GHEYIY(P)VDPM	Src family
Tyr740	ESDGGY(P)MDMSK	PI-3-kinase, SHC
Tyr751	DESVDY(P)VPMLD	PI-3-kinase, NCK, SHC
Tyr771	IESSNY(P)MAPYD	GAP, SHC
Tyr1009	TSSVLY(P)TAVQP	PLCγ, SYP
Tyr1021	RGDNDY(P)IIPLP	PLCγ

[a] Y(P) indicates phosphotyrosine residue. [b] PI-3-kinase: phosphatidylinositol 3-kinase; GAP: RAS GTPase activating protein; PLC-γ: phospholipase C-γ. For details, see Shauffhausen (88).

confers specificity to SH2/phosphopeptide interactions in general. These findings also suggest that *in vivo* specificity may derive from differences in relative binding affinity and that an increase in *in vivo* concentrations of an SH2-domain counterpart could force an interaction that does not normally occur, with resulting pathological consequences.

A semi-quantitative analysis of the affinity of SH2 domains for phosphotyrosine-containing peptides was carried out by surface plasmon resonance. In the application of this technology, a phosphotyrosine-containing peptide is immobilized on a solvent-exposed dextran surface which coats a gold film. Purified SH2 domain-containing proteins are passed over this surface and monitored for interaction with the immobilized peptide by measurement of the change in refractive index at the liquid-solid surface which correlates with mass near the surface. As this change can be monitored as a function of time, the relative kinetics of SH2 domain binding can be evaluated. A significant advantage of this methodology is that for many binding interactions, the apparent rate constants K_a and K_d can be determined.

For phosphopeptide/SH2 domain binding experiments, different concentrations of GST-SH2 domain fusion proteins were passed over the sensor-chip surface containing covalently bound phosphopeptides. The apparent binding affinities and kinetics of p85 SH2 domains binding to phosphopeptides derived from a major insulin receptor substrate, IRS-1, were the first to be analyzed (91). Associations were specific and appeared to have high affinities with a dissociation constant of 0.3 to 3.0 nM. Both SH2 domains of p85 bound to IRS-1 tyrosine-phosphorylated peptide with high affinity. However, the apparent affinities of SH2 domains of other signaling molecules for the phosphopeptides were very low, with affinities reduced by 500- and 1,000-fold, relative to the binding of the aminoterminal p85 SH2 domain to the same phosphopeptide. Interestingly, the binding of the p85 SH2 domain to the tyrosine-phosphorylated peptides was of high apparent affinity as a result of a very high on-rate of 3×10^7 to 40×10^7 M^{-1} sec^{-1}. At the same time, the rate of dissociation of 0.11 to 0.19 sec^{-1} was relatively high, suggesting that the SH2 domains can rapidly displace each other from tyrosine-phosphorylated binding sites. Additionally, the affinity of unphosphorylated peptides is greatly reduced. Thus, the interactions with the SH2 domains could be controlled rapidly by phosphorylation and dephosphorylation.

Phosphotyrosine-containing peptides, corresponding to two autophosphorylation sites on the human PDGF β-receptor that are responsible for PI-3-kinase binding, were also synthesized and used for determination of apparent kinetic parameters by the biosensor approach (92). The results demonstrate fast association and dissociation rates as was seen for the interaction of SH2 domains with IRS-1 phosphopeptides (91). These rapid on/off interactions might allow other phosphorylation sites within the SH2 domains to interact with downstream molecules and mediate multiple assembly of signaling complexes.

Several autophosphorylation sites have been mapped on growth factor receptors. The PDGF receptor is the best-studied example. Tyr740 and Tyr751 are the sites of interaction with PI-3-kinase (89,93). Mutation of either tyrosine to a phenylalanine markedly reduces the amount of PI-3-kinase bound to the PDGF-β receptor; mutation of both abolishes binding. The association rate constants for the interaction of the amino- and carboxyterminal SH2 domains with the two Tyr(P)751 and Tyr(P)740 phosphopeptides were measured by surface plasmon resonance (Table 4). The data indicate that the carboxyterminal SH2 domain does not discriminate between the two peptides and interacts with them with high apparent association rates; the aminoterminal SH2 domain binds with approximately 100-fold higher affinity to the Tyr(P)751 than Tyr(P)740 phosphopeptide (92). These data suggest that the high affinity interaction of the intact p85 protein with the autophosphorylated PDGF-β receptor may be mediated by the binding of the aminoterminal SH2 domain to the Tyr751 and the carboxyterminal SH2 domain to the Tyr740 site. If this model is correct, one would expect p85 to have higher affinity towards a peptide containing both phosphorylated tyrosines. In order to address this question, the synthesis of the above peptide should be attempted.

Table 4. Association and dissociation rate constants of the interactions between SH2 domain-containing proteins and phosphopeptides Tyr(P)751 and Tyr(p)740 of the PDGF receptor.

SH2 Domain[a]	Tyr(P)740 peptide[b]			Tyr(P)751 peptide[c]		
	k_{ass} ($10^5 M^{-1} sec^{-1}$)	k_{diss} (s^{-1})	K_a (M^{-1})	k_{ass} ($10^5 M^{-1} sec^{-1}$)	k_{diss} (s^{-1})	K_a (M^{-1})
p85	19.3	0.100	1.93×10^7	92.4	0.127	7.28×10^7
p85-N	0.14	0.095	1.47×10^5	33.4	0.141	2.37×10^7
p85-C	15.9	0.102	1.56×10^6	16.9	0.098	1.72×10^7
PLC-γC	1.16	0.045	2.58×10^6	16.4	0.049	3.35×10^7
PLC-γN+C	1.51	0.034	4.44×10^6	12.0	0.045	2.67×10^7
GAP-N	2.06	0.039	5.28×10^6	0.40	0.054	7.41×10^5

[a] p85: 85 kD subunit of phosphatidylinositol 3-kinase; p85-N: aminoterminal SH2 domain of p85; p85-C: carboxyterminal SH2 domain of p85; PLC-γC: carboxyterminal SH2 domain of phospholipase C-γ; PLCγ-N+C: combined amino- and carboxyterminal SH2 domain of PLC-γ; GAP-N: ras GTPase activating protein. [b]The sequence is DMSKDESVDY(P)VPMLDMK. [c] The sequence is GESDGGY(P)MDMSK. For details, see Panayotou et al., 1993 (92).

The binding sites on the PDGF-β receptor for PLC-γ are distinct from those for PI-3-kinase. A test of the binding of PLC-γ SH2 domains to Tyr(P)751 and Tyr(P)740 was carried out. The aminoterminal SH2 domain of PLC-γ did not display any binding to the peptides tested. However, the phosphopeptide corresponding to the region surrounding Tyr1021 of the PDGF-β receptor was found to compete with PLC-γ binding, and bound with a K_d of 80 μM to the SH2 domains. In contrast, the phosphopeptide bound to the aminoterminal SH2 domain with a K_d of 1071 μM, and to the carboxyterminal SH2 domain with a K_d of 385 μM. The higher apparent affinity of the N+C SH2 domains arises from the combination of a fast association rate and a slow dissociation rate and suggests that the individual SH2 domains might act synergistically during PLC-γ binding.

In another complementary study, individual phosphopeptides corresponding to the hamster polyoma middle-sized tumor antigen sequence were independently assessed by ^{125}I-peptide competition assay or surface plasmon resonance assay for binding by both Src and Lck SH2 domains (94). The results show that there is a good agreement between the two methods. Both Src and Lck SH2 domains bound, with high affinity, to a phosphopeptide having a Tyr(P)-Glu-Glu-Ile motif which was observed in the hamster polyoma virus middle-sized tumor antigen. As isolated SH2 domains, Lck and Src bind phosphopeptides with near-identical specificity. These binding studies confirm that the specificity in the SH2/phosphoprotein interaction can be reconstituted with isolated SH2 domains and corresponding phosphopeptides.

The nature of the binding site has also been examined by comparing the behavior of different phosphopeptides. In a study addressing the interaction with the carboxyterminal SH2 of p85, surface plasmon resonance analyses have been made to compare the effects of substitutions in the phosphopeptides. Changing the +1 position residue from Met to Ile, Val, Glu, Thr, Ser or Ala, had little effect on binding, whereas a change to Gly or Arg had a greater effect (91). However, this method requires extensive synthetic effort.

To circumvent this problem, an "oriented phosphopeptide library" was developed (95). This library, in contrast to other synthetic peptide libraries in which peptides are degenerate at all positions, has a critical residue, phosphotyrosine, which remained invariant, with only three positions having random substitutions. The first peptide library was made with the following sequences: Gly-Asp-Gly-Tyr(P)-X-X-X-Ser-Pro-Leu-Leu, where X indicates all natural amino acids except Cys and Trp. The synthesis was carried out by standard Fmoc chemistry. For the degenerate positions, the resin was deprotected as usual and then divided into equivalent amounts by weight. Each aliquot was coupled with a different N-α-Fmoc-amino acid, recombined and the procedure repeated for each degenerate position. For affinity purification, the mixture of soluble degenerate peptides is incubated with a GST-SH2 domain fusion protein coupled to glutathione beads. A solution of sodium phenyl phosphate was passed through the column to elute the peptides. The eluates were collected, reduced to a small volume and sequenced on a protein sequencer. To determine selectivity for a particular sequence, a comparison was made of the purified mixture to the crude mixture or to the GST column eluate at the same sequencer cycle and a relative enrichment value was calculated. A large number indicates a stronger selection. In addition, a comparison of enrichment values between different degenerate positions indicates which position is more selective.

The results obtained for the aminoterminal SH2 domain of the p85 demonstrate the importance of the +1 position (the first residue after Tyr(P)) as well as that of the +3 position (the third residue after Tyr(P)). The aminoterminal SH2 of p85 has a strong preference for Met at the +3 position, no preference at the +2 position and preference for Met, Val, Ile, Leu or Glu at the +1 position. Interestingly, the optimal sequences predicted from this analysis are precisely the sequences that have been identified as the p85-binding sites on polyoma virus middle-sized tumor antigen or the PDGF receptor. The same phosphopeptide library was used to determine optimal sequences for binding to 11 additional SH2 domains. The amino acids that are preferentially selected at the positions +1, +2, and +3 carboxyterminal to the Tyr(P) are summarized in Table 5. Interestingly, members of the Src family all selected the sequence Tyr(P)-Glu-Glu-Ile; all other SH2 domains selected different and unique motifs. Nck and Crk SH2 domains are most selective whereas the SH-PTP2 aminoterminal SH2 domain is the least selective.

While this analysis was in progress, the crystal structures of the Src SH2 domain associated with two different low and high affinity phosphopeptides were determined (96,97). These structures provided an explanation for the ability of SH2 domains specifically to bind phosphotyrosine rather than free tyrosine or other phosphoamino acids. For the Tyr(P)-Glu-Glu-Ile motif, the crystal structure revealed an explanation for the selection of this motif (Figure 5). The Ile at the +3 position fits into a hydrophobic pocket that is induced during binding. The Glu residue at the +1 position forms hydrogen bonds with Lys200 (βD3, the third residue of the fourth β structure) and Tyr202 (βD5). The +2 position Glu residue makes close contact with Arg205 (βD'1, the first residue of the fourth β structure involved in a distinct sheet) (97,98). Consistent with the observation that members of the Src family select identical motifs, the side chains predicted to make contacts are conserved in this family. Conversely, all other SH2 domains have distinct amino acid replacements at one or more of the contact residues. Based on a comparison of optimal motifs for 13 SH2 domains and differences in their βD5 residues, the known SH2 domains have been divided into four distinct groups (Table 5). The group 1 SH2 domains, which include the Src family, have Tyr or Phe at the βD5 position and select phosphopeptides with the general motif Tyr(P)-hydrophilic-hydrophilic-hydrophobic residue. Group 3 SH2 domains have Cys, Ile, or Val at the βD5 position and select a Tyr(P)-hydrophobic-X-hydrophobic residue motif (99-101). The Vav SH2 domain is thus far the sole member of the group 2 family which has Thr at βD5 and a motif of the type Tyr(P)-Met-Glu-Pro. Finally, group 4 contains additional SH2 domains that have distinct amino acids at βD5. No phosphopeptide motif interacting with members of this family has been determined as yet. The major value of the data presented in Table 5 is that a search of protein sequence data bases has revealed sites in proteins that have been previously sequenced and a subgroup of proteins potentially involved in signaling has been uncovered.

Conformational Changes and Thermodynamic Studies of Tyrosyl-Phosphopeptides Binding to SH2 Domains

The best evidence for alteration of conformation of the SH2 domain upon peptide binding comes from a study of the aminoterminal SH2 domain of p85. Isolated aminoterminal PI-3-kinase p85 SH2 domain, isolated PDGF Tyr(P)740 phosphopeptide, and an equimolar complex of the two were examined for conformational status by circular dichroism (CD) (102). The 123 amino acid residue SH2 domain contains primarily β sheet with an estimated α-helix content of ~ 13%. No ordered structure is observed in the 11 amino acid phosphopeptide. On the other hand, the difference spectra of the complex showed a minimum at 230 nm, indicating a change in structure either in the SH2 domain or in the phosphopeptide. The same change was observed upon binding of the phosphopeptide IRS-1 Tyr(P)628. Addition of the non-phosphorylated peptide to the SH2 domain yielded no significant difference spectra. A truncated form of the same peptide, containing only 5 amino acids and exhibiting ~ 60-fold weaker binding induced qualitatively identical CD changes. These results suggest that the difference spectra reflect a change in SH2 structure rather than peptide folding.

Table 5. Phosphopeptide motifs for SH2 domains: residues predicted to interact with side chains of associated phosphopeptides.

Protein	Phosphopeptide[a]			Binding Pocket of SH2 Domain[b]							
	+1	+2	+3	I		II			III		
				200 βD3	202 βD5	205 βD'1	202 βD5	214 βE4	215 EF1	230 αB9	237 BG4
Group 1A											
SRC	E	E	**I**	K	Y	R	Y	I	T	Y	L
FYN	**E**	**E**	**I**	K	Y	R	Y	I	T	Y	L
LCK	**E**	E	**I**	K	Y	R	Y	I	S	Y	L
FGR	**E**	**E**	IV	K	Y	R	Y	I	T	Y	L
Group 1B											
SYKC	QIE	eqt	**L**	L	Y	D	Y	I	S	H	L
ABL	E	**N**	P	Y	Y	N	Y	V	S	H	L
CSK	**T**	n	mr	E	Y	M	Y	I	D	Y	?
CRK	D	H	**P**	S	Y	N	Y	A	G	Y	T?
NCK	**D**	**E**	**P**	K	F	Q	F	I	G	Y	T?
fes/ftp	**E**	-	vi	R	F	Q	F	R	L	L	G
SEM5	LV	**N**	vp	Q	F	L	F	L	W	H	R?
DGBR2	y	**N**	-	Q	F	L	F	L	W	H	R?
GRB2	qy	N	y	Q	F	L	F	L	W	H	R?
3BP2	**E**	**N**	-	R	Y	F	Y	E	G	Y	P?
Group 2											
VAV	**M**	E	P	K	T	I	T	I	T	Y	?
Group 3											
p85aN	MIVE	-	**M**	K	I	F	I	F	S	Y	A?
p85aC	mli	-	**M**	K	C	N	C	F	A	Y	V?
PLC-γ1C	VI	IL	**P**IV	K	C	N	C	L	G	Y	Y?
PLC-γ1N	L**I**V	Ed	LIV	Q	C	H	C	K	F	Y	L?
SHPTP1N	F	-	**F**	T	I	Q	I	D	L	Y	V
SHPTP2N	IV	-	VI	T	I	Q	I	D	L	Y	L
SHC	EI	-	**ILM**	K	L	V	L	T	K	H	P?
Group 4											
ShB				M	M	A	M	L	G	Y	?
SHPTP2C				T	V	R	V	D	V	Y	E
CSWC				T	V	R	V	D	V	Y	E

[a]Columns +1, +2 and +3 are the first, second and third residues carboxyterminal to phosphotyrosine residue of the optinal phosphopeptide selected by each SH2 domain. Bold, upper case and lower case letters indicate strong, medium and weak selection, respectively. [b]Binding pockets I, II and III are predicted to bind residues +1, +2 and +3, respectively. For details see Songyand et al., 1993 (95).

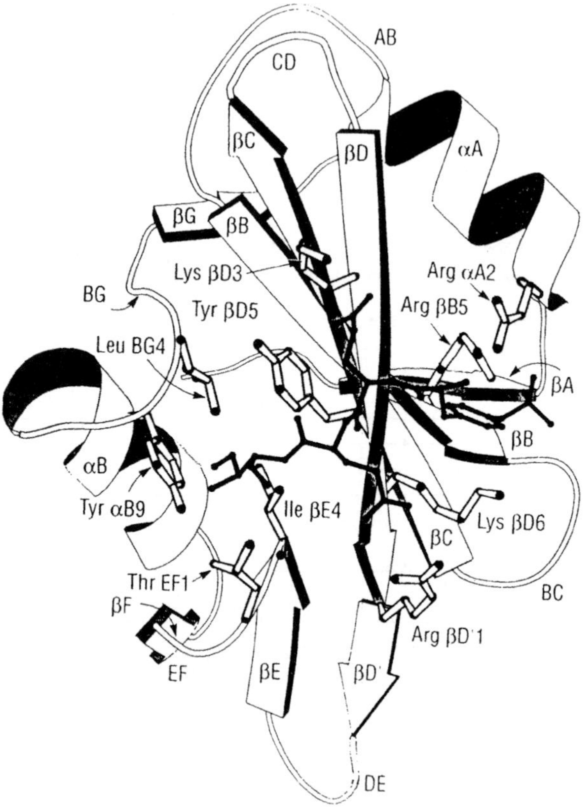

Figure 5. Schematic diagram of the Src SH2 domain. The phosphotyrosine (at the right), Glu(+1), Glu(+2) and Ile(+3) (at the left) of the peptide are shown with solid black bonds and are not labeled explicitly. α helix and β strands are shown as ribbons and arrows, respectively. Several of the side chains involved in peptide binding are shown as stick figures and are labeled according to the secondary structure notation. For details see Waksman et al., 1993 (97).

^{1}H-NMR analysis demonstrates significant perturbation in several aliphatic resonances upon binding of either IRS-1 Tyr(P)628 or PDGF receptor Tyr(P)740 phosphopeptides. In particular, Trp N^{ϵ}-H aromatic resonances were sensitive markers for environmental perturbations. A perturbation is observed in a Trp residue, which in the three-dimensional structure of SH2 domain, would be expected to be sensitive to the orientation of α-helix 1. No such perturbation is observed in the SH2 spectrum upon addition of phosphotyrosine or non-phosphorylated IRS-1 Tyr628 peptide. Rates of amide proton exchange with solvent, which correlate with the solvent accessibility of amides, have been used to interpret the local and global dynamics of proteins in solution. Amide proton exchange in the isolated SH2 domain in freshly-prepared D_2O solutions occurs within minutes at 45°C and pD 6.0. Addition of IRS-1 Tyr(P)628 peptide retards the exchange of ~ 30% of the SH2 domain amide resonances. This difference clearly demonstrates that complex formation is accompanied by reduced solvent accessibility of specific amides. The SH2 domain of p85 is cleaved by trypsin into two fragments with a $T_{1/2}$ for the first order loss of intact SH2 domain of 2.1 min. In the presence of the IRS-1 Tyr(P)628 peptide, the $T_{1/2}$ is 62 min, which is 30-fold slower than in the absence of the peptide. The site of cleavage is Lys31, which is located in a highly flexible loop between β-strand 1 and 2. Such a location strongly suggests that the loop affects protease sensitivity. These spectroscopic and biochemical studies were performed in the absence of a known SH2 domain three-dimensional structure. However, it is apparent that some conformational change is occurring. Moreover, the data strongly support an important role for the loop between β-strand 1 and 2 which is reflected by changes in protein structure and dynamics.

X-ray diffraction comparison of free Src SH2 with that complexed with a high affinity phosphopeptide, in part, supports the idea that structural changes can occur upon peptide binding (97). The changes observed in Src are relatively small and are localized to the immediate vicinity of the phosphotyrosine and Ile binding site. In contrast, the p85 SH2 domain undergoes a large conformational change upon binding of phosphopeptides, one source being the presence of a loop which is larger than in any other SH2 domain. However, the Src results demonstrate highly localized conformational change so that extensive conformational changes are not likely to be a general feature.

The technique of isothermal titration calorimetry was used to gain more information concerning the thermodynamics of phosphopeptide recognition by SH2 domains (103). This technique involves the direct measurement of the heat of reaction; it permits an accurate determination of the change of enthalpy (ΔH), the stochiometry of the interaction and the binding constant, K_B, which can be calculated from the free energy (ΔG°) and the entropy change (ΔS°) upon ligand binding. Tyrosyl-phosphopeptides that bound to the SH2 domain of Lck were used. Analysis by curve fitting demonstrated that a single phosphopeptide molecule was bound to each SH2 domain. The average binding constants were 0.14 x 10^5 M^{-1} and 2.2 x 10^5 M^{-1} for two phosphopeptides (carboxyterminal autophosphorylation sites of $p185^{neu}$ and EGF receptor). A third peptide corresponding to the carboxyterminal of Lck, Tyr(P)505, was also analyzed. The binding stochiometry was 1:1 and the binding constant (K_B) was 3.4 x 10^5 M^{-1}, corresponding to a K_D of 2.9 μM. This value is different from that reported by Payne et al. (94). With a competition-binding assay as well as surface plasmon resonance, the K_D for binding of the Lck Tyr(P)505 peptide to the Lck SH2 domain was estimated to be 19 to 34 nM. This binding is a full 2 orders of magnitude stronger than that suggested by the thermodynamic analysis. Recent work demonstrates that diffusion of the reactant in free solution in and out of the immobilized reactant-dextran matrix used in plasmon resonance studies has a profound effect on apparent on/off rates (104). Thus, these studies appear to be quantitatively distinct. Furthermore, the K_D for binding of a tyrosine-phosphorylated protein may be different than that for a small peptide.

Despite the similarity in the binding affinities of the Tyr(P)505 and EGF receptor phosphopeptides and their free energy change of binding ($\Delta G^\circ \sim 7.0$ kcal mol^{-1} at pH 6.8), significant differences in the change of the enthalpy and entropy were observed. These differences could be due to interactions other than those involving the phosphotyrosine side chain. The value of ΔS° is higher in the case of EGF receptor peptide binding than in the case of the Tyr(P)505 peptide. This difference could be due to a conformational change in the Tyr(P)505 peptide binding to the Lck SH2 domain, including a relative ordering of the SH2 domain structure. ^{15}N relaxation NMR spectroscopic studies appear to show a conformational change in the Lck SH2 domain upon phosphopeptide binding. In summary, isothermal titration calorimetry produces a detailed analysis of the binding reaction of SH2 domain/phosphopeptides, which is based upon reversible thermodynamics of reactants in free solution.

Synthetic Phosphopeptides as Immunogens

Protein-tyrosine kinases are important regulators of growth and differentiation. One of these kinase genes is *neu* which encodes a 185 kD phosphoprotein that is a receptor for a group of Neu differentiation factors. Neu is frequently amplified in human adenocarcinomas; however, this amplification is a poor predictor of the clinical course of the disease. Since the activated form of a tyrosine kinase influences cell growth and differentiation, the extent of tyrosine phosphorylation should be a good indicator of receptor activity. Patients with tumors expressing activated Neu could have a different prognosis than patients whose tumors express inactivated receptors.

Tyrosine phosphorylation of receptors has been routinely measured by anti-phosphotyrosine immunoblots (105). However, the antibodies against phosphotyrosine are not specific. The development of antibodies that distinguish the kinase active form of Neu from the inactive protein is very important, since these reagents could be used directly on tumor cells to assay for Neu activation. Therefore, a phosphopeptide corresponding to a major p185 autophosphorylation site was used to produce Neu-specific anti-phosphotyrosine antibodies (106,107). The synthetic phosphorylated peptide was coupled to bovine serum albumin or activated limpet hemocyanin and used to immunize rabbits. The anti-Neu phosphopeptide antibodies were purified by a series of adsorption and affinity steps, including a final elution from a Neu phosphopeptide Affi-gel 10 column. Immunoblots experiments showed that purified antibodies are highly specific for the tyrosine-phosphorylated forms of Neu and do not recognize five other receptor kinases or other phosphorylated substrates. Moreover, the purified antibodies detected tyrosine phosphorylation in immunofluorescence experiments on permeabilized cells and on lysates of various tumor cell lines.

Anti-phosphotyrosine peptide antibodies were also obtained from rabbits immunized with a phosphopeptide corresponding to amino acids 123 to 136 of a cyclic adenosine monophosphate response element binding protein (anti-pCREB) (108,109). These antibodies, after purification on an affinity chromatography column containing unphosphorylated peptide, specifically recognized CREB phosphorylated *in vitro* on Ser133. Anti-pCREB also recognized the phosphorylated form of CREB present in whole cell lysates of various cell types after stimulation with forskolin, which activates adenylate cyclase, or after membrane depolarization, which stimulates Ca^{2+} influx. In addition to recognizing CREB, anti-pCREG immunoprecipitated two other phosphoproteins from cell extracts, probably members of the CREB family. Thus, anti-pCREB antibodies are useful reagents in determining the phosphorylation state of CREB and other related proteins under conditions of cell stimulation. The Ser133 phosphorylated form of CREB was also detected in extracts of cells treated with nerve growth factor (NGF) or epidermal growth factor (EGF); thus CREB phosphorylation appears to be a general response to cells exposed to neutrophin as well to mitogenic growth factors.

These data represent the first systematic development of anti-phosphotyrosine peptide antibodies against differentially-activated tyrosine kinases and other proteins phosphorylated on serine residues. These reagents have already been invaluable in tumor biology and in studies of gene transcription. Further development of mixtures of monoclonal antibodies that recognize multiple autophosphorylation sites on tyrosine kinases and other substrates should have numerous applications in basic and clinical science. Sera that can discriminate between various phosphorylation sites in a single protein might allow separation of differentially-phosphorylated populations of cells with different functional properties. These sera will be valuable in analyzing the function of tyrosine kinases and perhaps will find diagnostic and therapeutic application in a variety of human diseases.

Applications of Phosphonopeptides

Protein phosphorylation is an important aspect of cellular signal transduction. Many receptors that mediate growth factor functions undergo phosphorylation, and also act as tyrosine kinases. Specific tyrosine-phosphorylated growth factor receptor peptide segments are involved in the assembly of protein complexes, as well as modulation of kinase and phosphatase activity of proteins involved in the cascade of events leading to activation of cellular processes, including cell proliferation.

As described in the previous sections, synthetic phosphopeptides serve as useful tools in mechanistic studies of intermolecular protein associations, conformational rearrangements and kinase activation, and can be used as phosphatase substrates. On the other hand, peptide analogs with incorporated phosphono-amino acids (Figure 4), can be useful as hydrolytically stable and phosphatase-resistant agents for *in vivo* or *in vitro* mechanistic studies, as phosphatase inhibitors, and as first-stage peptidomimetic models for development of therapeutic agents.

An initial application of 4-phosphonomethyl-phenylalanine (Pmp) peptides involved a study of the association between the $pp60^{c\text{-}src}$/polyoma middle-sized tumor antigen (mT) complex and PI-3-kinase as a prototype of phosphoprotein/SH2 domain interactions (Table 6) (110). In an *in vitro* system, an 11 amino-acid long Pmp-containing peptide, mT312-322(Pmp315), containing the Pmp-Met-X-Met motif, was able to inhibit protein association with an IC_{50} of ~ 0.8 μM. This peptide showed good affinity (IC_{50}=7.2 μM) for the aminoterminal SH2 domain of the p85. In comparison, the Tyr(P) analog exhibited a two-fold tighter binding (IC_{50}=3.6 μM). Studies by Fantl et al. showed that phosphotyrosine-containing peptides as short as 5 amino-acids long, such as Tyr(P)-Val-Pro-Met-Leu (PDGF receptor residues 751-755), were effective in inhibiting PDGF receptor/PI 3-kinase interactions (89). In comprehensive structure-activity studies, with the same protein association phenomenon, a number of phosphonopeptides were evaluated with Pmp and F_2Pmp incorporated to replace Tyr(P) (65,111). The phosphopeptide, Gly-Tyr(P)-Val-Pro-Met-Leu, binds with high affinity to a GST-fusion protein derived from the carboxyterminal SH2 domain of PI 3-kinase p85 subunit (IC_{50}=0.17 μM), and this serves as a reference point. Pmp substitution for Tyr(P) resulted in moderate decrease in potency (IC_{50}=1.0 μM). Further replacement of Met with Nle in the Pmp+4 position decreased potency by a negligible amount (IC_{50} = 1.6 μM) (112). Inversion of the Pmp configuration decreased potency by a factor of 29 (IC_{50} = 29 μM), but surprisingly, inversion of only the Met configuration in the Pmp peptide resulted in the abrogation of binding activity (113). In efforts to make a more conformationally rigid molecule, a backbone cyclized hexameric Pmp peptide was prepared, producing a moderately potent agent with an IC_{50} of 5.2 μM.

In a continuation of the above studies, the difluoro analog of Pmp, i.e., F_2Pmp, was incorporated also into the same PDGF receptor segment (65). The hexapeptide, or its N-acetyl, C-terminal amide analogs, showed binding affinities comparable to the native autophosphorylation segment with Tyr(P) at position 751. A similar effectiveness of F_2Pmp peptides was also demonstrated in other systems. For example, the hexapeptide incorporating F_2Pmp in a sequence surrounding Tyr324 of the polyoma middle-sized tumor antigen, Ac-Gln-F_2Pmp-Glu-Glu-Ile-Pro-amide, exhibited a 6-fold higher binding affinity to the Src SH2 domain (IC_{50}=1.0 μM) than the corresponding phosphotyrosine-containing peptide, and in other cases somewhat lower affinities, implying a differential selectivity for F_2Pmp peptides binding to SH2 domains. These and other results demonstrate that F_2Pmp is a superior agent when incorporated into relevant peptides. In comparison, all peptides incorporating the Pmp amino acid were less potent in SH2 binding interactions. This can be explained partly by recent work on the crystal structures of the SH2 domain of $p56^{lck}$ complexed with two Pmp phosphonopeptides (114). Overall, the conformation of the SH2 domain is similar to that of Lck complexed with the phosphotyrosine peptide, except in the loop that connects the second and the third β-strands. This loop, which was involved in hydrogen-bond interactions with the phosphotyrosine moiety, is displaced by about 7 Å in phosphonopeptide complexes, leaving the Pmp moiety accessible to solvent. There are no crystal structures available to date on F2Pmp peptide/SH2 domain complexes.

Table 6. Relative affinities of phosphopeptides and phosphonopeptides to various SH2 domains.

SH2 domain[a]	Phosphopeptide[b]	Sequence source[c]	IC_{50}[d] (μM)	References
p85-N	EEE**Y(P)**MPMEDLY	mT(Y315)	3.2	(110)
	EEE(**Pmp**)MPMEDLY		7.2	(110)
	EEE(**D-Pmp**)MPMEDLY		131	(110)
p85-C	G**Y(P)**VPML	PDGF receptor (Y751)	0.17	(65)
	G(**Pmp**)VPML		1.0	(65)
	G(**D-Pmp**)VPML		29	(65)
	cyclo(G(**Pmp**)VPML)		5.2	(111)
	G(**Pmp**)VP(Nle)L		1.6	(112)
	G(**Pmp**)VP(D-Met)L		> 500	(113)
	G(**F_2Pmp**)VPML		0.17	(65)
	cyclo(G(**F_2Pmp**)VPML)		2.2	(65)
SH-PTP2(Syp)	Ac-LN**Y(P)**IDLDLV-amide	IRS-1(Y1172)	1.5	(71)
	Ac-LN(**F_2Pmp**)IDLDLV-amide		~ 23	(71,119)
	Ac-LN(**OMT**)IDLDLV-amide		22	(71)

[a] p85-N: aminoterminal SH2 domain of 85 kD subunit of phosphatidylinositol 3-kinase; p85-C: carboxyterminal SH2 domain of p85. [b] Y(P): phosphotyrosine; (Pmp): 4-phosphono-phenylalanine; Nle: norleucine; (F_2Pmp): 4-phosphono(difluoromethyl)-phenylalanine; (OMT): *O*-(2-malonyl)tyrosine. [c] mT: polyoma middle-sized tumor antigen; PDGF receptor: platelet-derived growth factor receptor; IRS-1: insulin receptor substrate 1. [d] IC_{50}: half-maximal concentration of competitive inhibitor of ^{32}P-labeled ligand.

It appears that the high-affinity activation of PI 3-kinase requires simultaneous binding of two Tyr(P)-Met-X-Met motifs to its two SH2 domains (115). Insulin receptor substrate 1 (IRS-1) is a major substrate for insulin receptor tyrosine kinase and has nine potential tyrosine phosphorylation sites. With tandem peptides constructed with two phosphotyrosines, PI 3-kinase activity could be maximally stimulated at levels comparable to those obtained *in vivo* (cultured cells). The tandem phosphonopeptide was much less effective in binding and in stimulation of PI 3-kinase activity.

A salient characteristic of Pmp and F_2Pmp peptides is their phosphatase resistance. When Pmp is incorporated into peptide sequences that are good substrates for protein tyrosine kinases, several examples show that such peptides are competitive inhibitors with K_i values comparable to the K_m of the corresponding phosphotyrosine-containing peptide substrates (116,117). Moreover, when F_2Pmp was incorporated into such peptides very good inhibitors were obtained (117,118). For example, the short F_2Pmp-containing peptide derived from EGF receptor showed potent inhibition of PTPase PTP1 with a K_i of 180 nM. This inhibitory activity is observed at a concentration three orders of magnitude lower than the analogous Pmp peptide (*D*/*L* enantiomeric mixture, $K_i = 181\ \mu M$). Since the surrounding amino acid sequence is also expected to contribute to some degree to the binding interactions, such peptide-based inhibitors offer the possibility of developing highly selective agents acting on specific PTPases.

The use of hydrolytically stable and phosphatase-resistant phosphonate analogs in peptides provides an opportunity for evaluating the physiological role of phosphorylation in cellular or *in vivo* studies. In one study an IRS-1-derived phosphorylation segment, F_2Pmp analog at Tyr1172, Ac-Leu-Asn-F_2Pmp-Ile-Asp-Leu-Asp-Leu-Val-amide, was microinjected into mammalian fibroblasts (119). This peptide, *in vitro*, inhibits the binding of the cytosolic tyrosine phosphatase, SH-PTP2(Syp), to IRS-1 with an IC_{50} of 23 μM. In cells Syp binds to activated growth factor receptors, such as PDGF receptor, EGF receptor and insulin receptor, through one of its SH2 domains, and at the same time it becomes phosphorylated on Tyr1172. The phosphorylated site provides a binding region for IRS-1 through its SH2 domain, resulting in an upregulated phosphatase activity for Syp. Thus, the activation of Syp correlates with the phosphorylation status of activated growth factor receptors, and it is believed that Syp plays a direct role in growth factor-induced signal transduction. In this study, microinjection of F_2Pmp 9-residue peptide inhibited the growth factor (insulin, ILGF-1, EGF)-stimulated DNA synthesis through modulation of Syp function.

In another study, the mechanism of T cell receptor signaling was studied with a tyrosine-based activation motif 17-residue peptide, in which the two tyrosines were replaced with F_2Pmp (120). Normally this region of the T cell receptor becomes Tyr phosphorylated upon TCR stimulation, creating a high-affinity binding site for the tandem SH2 domains of ZAP-70, which is also a kinase. Results indicated that the phosphatase-resistant, tandem repeat F_2Pmp peptide inhibited this protein interaction in permeabilized T cells. Inhibition of this association prevented TCR-stimulated tyrosine phosphorylation of ZAP-70 and reduced ZAP-70 kinase activity to basal levels. The use of phosphatase-resistant phosphotyrosine-mimicking peptides that disrupt key SH2 domain-mediated cellular signaling should be useful for both mechanistic studies and for developing antiproliferative drugs.

Transport and internalization of peptides and other agents into cells is an important factor for consideration. Highly-charged molecules do not cross the cellular membrane appreciably unless there is an active transport mechanism for that particular agent. In the cellular studies cited here, the peptides were internalized with harsh and artificial conditions. One approach would be to protect the polar amino acid side chains with functional groups that could be removed once the peptide is internalized (121). One non-phosphorus-containing Tyr(P) mimic was reported recently, *O*-(2-malonyl)tyrosine (OMT)(Figure 4A), where the two carboxyl side chains of OMT could be derivatized with enzymatically labile functional groups. When OMT was replaced for Tyr(P)1172 of the 9-residue peptide in the IRS-1, it showed good binding affinity for the Syp SH2 domain, with an IC_{50} of 22 μM (71).

In a separate study, OMT was inserted in place of Tyr(P) in the EGF receptor 988-993(Tyr992) 6-residue peptide; the resulting agent was a good inhibitor of PTPase(PTP-1B) with an IC_{50} of 10 μM (122). This level of inhibitory activity is obtained at a concentration at least an order of magnitude lower than that of the corresponding Pmp-containing analog. Development of phosphoamino acid-containing peptidomimetics that are amenable to in-cell studies is an exciting prospect for future work and efforts are being devoted to the development and application of these new reagents.

REFERENCES

1 Hunter, T. and Karin, M. (1992) Cell 70, 375-387.
2 Hunter, T. (1995) Cell 80, 225-236.
3 Meek, D.W. and Street, A.J. (1992) Biochem. J. 287, 1-15.
4 Lindberg, R.A., Quinn, A.M. and Hunter, T. (1992) Trends Biochem. Sci. 17, 114-119.
5 Hunter, T. (1991) Methods Enzymol. 200, 3-37.
6 Hunter, T. (1995) Cell 83, 1-4.
7 Heldin, C.H. (1995) cell 80, 213-223.
8 Cantley, L.C. (1995) Trends Biochem. Sci. 20, 470-475.
9 Karin, M. and Hunter, T. (1995) Curr. Biol. 5, 747-757.
10 Perich, J.W. and Johns, R.B. (1989) J. Org. Chem. 54, 1750-1752.
11 Perich, J.W., Valerio, R.M. and Johns, R.B. (1986) Tetrahedron Lett. 27, 1377-1380.
12 Arendt, A., Palczewski, K., Moore, W.T., Caprioli, R.M., McDowell, J.H. and Hargrave, P.A. (1989) Int. J. Pept. Protein Res. 33, 468-476.
13 Paquet, A. and Johns, M. (1990) Int. J. Pept. Protein Res. 36, 97-103.
14 Lacombe, J.M., Andriamanampisoa, F. and Pavia, A.A. (1990) Int. J. Pept. Protein Res. 36, 275-280.
15 Perich, J.W. (1990) in Peptides and Protein Phosphorylation (Kemp, B.E., ed.) pp. 289-314., CRC Press, Boca Raton, Florida.
16 de Bont, H.B.A., Liskamp, R.M., O'Brian, C.A., Erkelens, C., Veeneman, G.H. and van Boom, J.H. (1989) Int. J. Pept. Protein Res. 33, 115-123.
17 Perich, J.W. (1991) Methods Enzymol. 201, 234-245.
18 Atherton, E. and Sheppard, R.C., eds. (1989) Solid Phase Peptide Synthesis: A Practical Approach, IRL Press at Oxford University Press, Oxford, England.
19 Bodansky, M. and Trost, B., eds. (1993) Principle of Peptide Synthesis, Springer-Verlag Inc., New York, NY.
20 Grant, G.A., eds. (1992) Synthetic Peptides: A User's Guide, W. H. Freeman and Company, New York, NY.
21 Andrews, D.M., Kitchin, J. and Seale, P.W. (1991) Int. J. Pept. Protein Res. 38, 469-475.
22 Bannwarth, W. and Kitas, E.A. (1992) Helv. Chim. Acta 75, 707-714.
23 Kitas, E.A., Knorr, R., Trzeciak, A. and Bannwarth, W. (1991) Helv. Chem. Acta 74, 1314-1328.

24 Valerio, R.M., Alewood, P.F., Johns, R.B. and Kemp, B.E. (1989) Int. J. Pept. Protein Res. 33, 428-438.
25 de Bont, H.B.A., van Boom, J.H. and Liskamp, R.M.J. (1990) Tetrahedron Lett. 31, 2497-2500.
26 Bannwarth, W. and Trezeciak, A. (1987) Helv. Chem. Acta 70, 175-186.
27 Perich, J.W. and Johns, R.B. (1988) Tetrahedron Lett. 29, 2369-2372.
28 Gibson, B.W., Falick, A.M., Burlingame, A.L., Nadasdi, L., Nguyen, A.C. and Kenyon, G.L. (1987) J. Amer. Chem. Soc. 109, 5343-5348.
29 Tian, Z., Gu, C., Roeske, R.W., Zhou, M. and van Etten, R.L. (1993) Int. J. Pept. Protein Res. 42, 155-158.
30 Kitas, E.A., Perich, J.W., Tregear, G.W. and Johns, R.B. (1990) J. Org. Chem. 55, 4181-4187.
31 Perich, J.W. and Reynolds, E.C. (1991) Int. J. Pept. Protein Res. 37, 572-575.
32 Perich, J.W., Ruzzene, M., Pinna, L.A. and Reynolds, E.C. (1994) Int. J. Pept. Protein Res. 43, 39-46.
33 Ottinger, E.A., Shekels, L.L., Bernlohr, D.A. and Barany, G. (1993) Biochemistry 32, 4354-4361.
34 King, D.S., Fields, C.G. and Fields, G.B. (1990) Int. J. Pept. Protein Res. 36, 255-266.
35 Pennington, M.W. (1994) in Methods in Molecular Biology, vol. 35: Peptide Synthesis Protocols (Pennington, M.W. and Dunn, B.M., eds.) pp. 195-200, Humana Press Inc., Totowa, NJ.
36 Perich, J.W., Nguyen, D.L. and Reynolds, E.C. (1991) Tetrahedron Lett. 32, 4033-4034.
37 Perich, J.W. (1992) Int. J. Pept. Protein Res. 40, 134-140.
38 Otvos, L.J., Elekes, I. and Lee, V.M.-Y. (1989) Int. J. Pept. Protein Res. 34, 129-133.
39 Stærkær, G., Jakoben, M.H., Olsen, C.E. and Holm, A. (1991) Tetrahedron Lett. 32, 5389-5392.
40 Pullen, N., Brown, N.G., Sharma, R.P. and Akhtar, M. (1993) Biochemistry 32, 3958-3964.
41 Perich, J.W. (1991) Methods Enzymol. 201, 225-233.
42 Perich, J.W., Terzi, E., Carnazzi, E., Seyer, R. and Trifilieff, E. (1994) Int. J. Pept. Protein Res. 44, 305-312.
43 Perich, J.W., Kelly, D.P. and Reynolds, E.C. (1992) Int. J. Pept. Protein Res. 40, 81-88.
44 Arendt, A. and Hargrave, P.A. (1994) in Methods in Molecular Biology, vol. 35: Peptide Synthesis Protocols (Pennington, M.W. and Dunn, B.M., eds.) pp. 187-193, Humana Press Inc., Totowa, NJ.
45 Wakamiya, T., Saruta, K., Yasuoka, J. and Kusumoto, S. (1995) in Peptide Chemistry 1994 (Ohno, M., ed.) pp. 5-8, Protein Research Foundation, Osaka.
46 Wakamiya, T., Saruta, K., Yasuoka, J. and Kusumoto, S. (1994) Chem. Lett., 1099-1102.
47 Wakamiya, T., Saruta, K., Kusumoto, S., Nakajima, K., Yoshizawa-Kumagaye, K., Imajoh-Ohmi, S. and Kanegasaki, S. (1993) Chem. Lett., 1401-1404.
48 Vorherr, T. and Bannwarth, W. (1995) Bioorg. Medicinal Chem. Lett. 5, 2661-2664.
49 Otaka, A., Miyoshi, K., Kaneko, M., Tamamura, H., Fujii, N., Nomizu, M., Burke, Jr.,T.R. and Roller, P.P. (1995) J. Org. Chem. 60, 3967-3974.
50 Otaka, A., Miyoshi, K., Roller, P.P., Burke, Jr.,T.R., Tamamura, H. and Fujii,

N. (1995) J. Chem. Soc., Chem. Commun. , 387-389.
51 Adamson, N., Riley, P.F. and Reynolds, E.C. (1993) J. Chromatogr. 646, 391-396.
52 Amankwa, L.N., Harder, K., jirik, F. and Aebersold, R. (1995) Protein Sci. 4, 113-125.
53 Meyer, H.E., Hoffmann, P.E. and Heilmeyer, L.J. (1991) Methods Enzymol. 201, 169-185.
54 Meyer, H.E., Hoffmann, P.E., Donella, D.A. and Korte, H. (1991) Methods Enzymol. 201, 206-224.
55 Matsueda, R. and Walter, R. (1980) Int. J. Pept. Protein Res. 16, 392-401.
56 Dawson, P.E., Muir, T.W., Clark, L.I. and Kent, S.B. (1994) Science 266, 776-779.
57 Hojo, H. and Aimoto, S. (1993) Bull. Chem. Soc. Jpn. 66, 3004-3008.
58 Liu, C.F. and Tam, J.P. (1994) Proc. Nat. Acad. Sci. U.S.A. 91, 6584-6588.
59 Marseigne, I. and Roques, B.P. (1988) J. Org. Chem. 53, 3621-3624.
60 Garbay-Jaurequiberry, C., McCort-Tranchepain, I., Barbe, B., Ficheux, D. and Roques, B.P. (1992) Tetrahedron: Asymmetry 3, 637-650.
61 Cushman, M. and Lee, E.-S. (1992) Tetrahedron Lett. 33, 1993-1996.
62 Burke, Jr.,T.R., Russ, P. and Lim, B. (1991) Synthesis , 1019-1020.
63 Garbay-Jaureguiberry, C., Ficheux, D. and Roques, B.P. (1992) Int. J. Pept. Protein Res. 39, 523-527.
64 Burke, Jr.,T.R., Smith, M.S., Nomizu, M., Otaka, A. and Roller, P.P. (1993) J. Org. Chem. 58, 1336-1340.
65 Burke, Jr.,T.R., Smyth, M.S., Otaka, A., Nomizu, M., Roller, P.P., Wolf, G., Case, R. and Shoelson, S.E. (1994) Biochemistry 33, 6490-6494.
66 Wrobel, I. and Dietrich, A. (1993) Tetrahedron Lett. 34, 3543-3546.
67 Smyth, M.S. and Burke, Jr., T.R. (1994) Tetrahedron Lett. 35, 551-554.
68 Otaka, A., Burke, Jr.,T.R., Smyth, M.S., Nomizu, M. and Roller, P.P. (1993) Tetrahedron Lett. 34, 7039-7042.
69 Gordeev, M.F., Patel, D.V., Barker, P.L. and Gordon, E.M. (1994) Tetrahedron Lett. 35, 7585-7588.
70 Smyth, M.S., Ford, J.H. and Burke, Jr.,T.R. (1992) Tetrahedron Lett. 33, 4137-4140.
71 Ye, B., Akamatsu, M., Shoelson, S.E., Wolf, G., Giorgetti-Peraldi, S., Yan, X., Roller, P.P. and Burke, Jr., T.R. (1995) J. Med. Chem. 38, 4270-4275.
72 Tong, G., Perich, J.W. and Johns, R.B. (1990) Tetrahedron Lett. 31, 3759-3762.
73 Perich, J.W. (1994) Int. J. Pept. Protein Res. 44, 288-294.
74 Shapiro, G., Buecheler, D., Enz, A. and Pombo-Villar, E. (1994) Tetrahedron Lett. 35, 1173-1176.
75 Otaka, A., Miyoshi, K., Burke, Jr.,T.R., Roller, P.P., Kubota, H., Tamamura, H. and Fujii, N. (1995) Tetrahedron Lett. 36, 927-930.
76 Berkowtiz, D.B., Shen, Q. and Maeng, J.-H. (1994) Tetrahedron Lett. 35, 6445-6448.
77 Guan, K.L., Haun, R.S., Watson, S.J., Geahlen, R.L. and Dixon, J.E. (1990) Proc. Nat. Acad. Sci. U.S.A. 87, 1501-1505.
78 Boivin, P., Galand, C. and Bertrand, O. (1987) Int. J. Biochem. 19, 613-618.
79 Shekels, L.L., Smith, A.J., Van, E.R. and Bernlohr, D.A. (1992) Protein Sci. 1, 710-721.
80 Zhang, Z.Y. and Van, E.R. (1990) Arch. Biochem. Biophys. 282, 39-49.
81 Black, M.J. and Jones, M.E. (1983) Anal. Biochem. 135, 233-238.

82 Mayer, B.J., Jackson, P.K. and Baltimore, D. (1991) Proc. Nat. Acad. Sci. U.S.A. 88, 627-631.

83 Matthews, R.J., Bowne, D.B., Flores, E. and Thomas, M.L. (1992) Mol. Cell. Biol. 12, 2396-2405.

84 Freeman, R.M., Plutzky, J. and Neel, B.G. (1992) Proc. Nat. Acad. Sci. U.S.A. 89, 11239-11243.

85 Bastin, L., Ramanchandran, C., Liu, S. and Adam, M. (1993) Biochem. Biophys. Res. Comm. 196, 124-133.

86 Dechert, U., Affolter, M., Harder, K.W., Metthews, J., Owen, P., Clarks-Lewis, I., Thomas, M.L., Aebersold, R. and Jirik, F.R. (1995) Eur. J. Biochem. 231, 673-681.

87 Dechert, U., Adam, M., Harder, K.W., Clark, L.I. and Jirik, F. (1994) J. Biol. Chem. 269, 5602-5611.

88 Schauffhausen, B. (1995) Biochim. Biophys. Acta 1242, 61-75.

89 Fantl, W.J., Escobedo, J.A., Martin, G.A., Turck, C.W., del Rosario, M., McCormick, F. and Williams, L.T. (1992) Cell 69, 413-423.

90 Piccione, E., Case, R.D., Domchek, S.M., Hu, P., Chaudhuri, M., Backer, J.M., Schlessinger, J. and Shoelson, S.E. (1993) Biochemistry 32, 3197-3202.

91 Felder, S., Zhou, M., Hu, P., Urena, J., Ullrich, A., Chaudhuri, M., White, M., Shoelson, S.E. and Schlessinger, J. (1993) Mol. Cell. Biol. 13, 1449-1455.

92 Panayotou, G., Gish, G., End, P., Truong, O., Gout, I., Dhand, R., Fry, M.J., Hiles, I., Pawson, T. and Waterfield, M.D. (1993) Mol. Cell. Biol. 13, 3567-3576.

93 Kashishian, A., Kazlauskas, A. and Cooper, J.A. (1992) EMBO J. 11, 1373-1382.

94 Payne, G., Shoelson, S.E., Gish, G.D., Pawson, T. and Walsh, C.T. (1993) Proc. Nat. Acad. Sci. U.S.A. 90, 4902-4906.

95 Songyang, Z., Shoelson, S.E., Chaudhuri, M., Gish, G., Pawson, T., Haser, W.G., King, F., Roberts, T., Ratnofsky, S., Lechleider, R.J., Neel, B.G., Birge, R.B., Hanafusa, H., Schauffhausen, B. and Cantley, L.C. (1993) Cell 72, 767-778.

96 Waksman, G., Kominos, D., Robertson, S.C., Pant, N., Baltimore, D., Birge, R.B., Cowburn, D., Hanafusa, H., Mayer, B.J., Overduin, M., Resh, M.D., Rios, C.B., Silverman, L. and Kuriyan, J. (1992) Nature 358, 646-653.

97 Waksman, G., Shoelson, S.E., Pant, N., Cowburn, D. and Kuriyan, J. (1993) Cell 72, 779-790.

98 Eck, M.J., Shoelson, S.E. and Harrison, S.C. (1993) Nature 362, 87-91.

99 Lee, C.H., Kominos, D., Jacques, S., Margolis, B., Schlessinger, J., Shoelson, S.E. and Kuriyan, J. (1994) Structure 2, 423-438.

100 Pascal, S.M., Singer, A.U., Gish, G., Yamazaki, T., Shoelson, S.E., Pawson, T., Kay, L.E. and Forman, K.J. (1994) Cell 77, 461-472.

101 Songyang, Z., Shoelson, S.E., McGlade, J., Olivier, P., Pawson, T., Bustelo, X.R., Barbacid, M., Sabe, H., Hanafusa, H., Yi, T., Ren, R., Baltimore, D., Ratnofsky, S., Feldman, R.A. and Cantley, L.C. (1994) Mol. Cell. Biol. 14, 2777-2785.

102 Shoelson, S.E., Sivaraja, M., Williams, K.P., Hu, P., Schlessinger, J. and Weiss, M.A. (1993) EMBO J. 12, 795-802.

103 Lemmon, M.A. and Ladbury, J.E. (1994) Biochemistry 33, 5070-5076.

104 Schuck, P. (1996) Biophys. J. (in press).

105 Stern, D.F., Kamps, M.P. and Cao, H. (1988) Mol. Cell. Biol. 8, 3969-3973.

106 Bangalore, L., Tanner, A.J., Laudano, A.P. and Stern, D.F. (1992) Proc. Nat. Acad. Sci. U.S.A. 89, 11637-11641.

107 Epstein, R.J., Druker, B.J., Roberts, T.M. and Stiles, C.D. (1992) Proc. Nat. Acad. Sci. U.S.A. 89, 10435-10439.

108 Ginty, D.D., Bonni, A. and Greenberg, M.E. (1994) Cell 77, 713-725.

109 Ginty, D.D., Kornhauser, J.M., Thompson, M.A., Bading, H., Mayo, K.E., Takahashi, J.S. and E., G.M. (1993) Science 260, 238-241.

110 Domchek, S.M., Auger, K.R., Chatterjee, S., Burke, Jr.,T.R. and Shoelson, S.E. (1992) Biochemistry 31, 9865-9870.

111 Burke, Jr.,T.R., Nomizu, M., Otaka, A., Smyth, M.S., Roller, P.P., Case, R.D., Wolf, G. and Shoelson, S.E. (1994) Biochem. Biophys. Res. Commun. 201, 1148-1153.

112 Roller, P.P., Otaka, A., Nomizu, M., Smyth, M.S., Barchi, Jr., J.J. and Burke, Jr., T.R. (1994) Bioorg. Med. Chem. Lett. 15, 1879-1882.

113 Roller, P.P., Otaka, A., Nomizu, M., Yan, X.-J., Smyth, M.S., Barchi, Jr., J.J., Shoelson, S.E. and Burke, Jr., T.R. (1995) in Peptides 1994 (Maia, H.L.S., ed.) pp. 355-356, ESCOM publishers, Leiden, The Netherlands.

114 Mikol, V., Bauman, G., Keller, T.H., Manning, U. and Zurini, M.G.M. (1995) J. Mol. Biol. 246, 344-355.

115 Herbst, J.J., Andrews, G., Contillo, L., Lamphere, L., Gardner, J., Lienhard, G.E. and Gibbs, E.M. (1994) Biochemistry 33, 9376-9381.

116 Zhang, Z.-Y., Maclean, D., MacNamara, D.J., Sawyer, T.M. and Dixon, J.E. (1994) Biochemistry 33, 2285-2290.

117 Chen, L., Wu, L., Otaka, A., Smyth, M.S., Roller, P.P., Burke, Jr., T.R., den Hertog, J. and Zhang, Z.-Y. (1995) Biochem. Biophys. Res. Commun. 216, 976-984.

118 Burke, Jr.,T.R., Kole, H.K. and Roller, P.P. (1994) Biochem. Biophys. Res. Commun. 204, 129-134.

119 Xiao, S., Rose, D.W., Sasaoka, T., Maegawa, H., Burke, Jr., T.R., Roller, P.P., Shoelson, S.E. and Olefsky, J.M. (1994) J. Biol. Chem. 269, 21244-21248.

120 Wange, R.L., Isakov, N., Burke, Jr., T.R., Otaka, A., Roller, P.P., Watts, J.D., Aebersold, R. and Samelson, L.E. (1995) J. Biol. Chem. 270, 944-948.

121 Aungst, B.J., Blake, J.A., Rogers, N.J., Saitoh, H., Hussain, M.A., Ensinger, C.L. and Pruitt, J.R. (1995) Pharmaceutical Res. 12, 763-767.

122 Kole, H.K., Akamatsu, M., Ye, B., Yan, X., Barford, D., Roller, P.P. and Burke, Jr.,T.R. (1995) Biochem. Biophys. Res. Commun. 209, 817-822.

INDEX

Δ9 *see* stearoyl-ACPΔ9 desaturase
ø29, 194
σ^{32}, 207
σ^{s}, 205

Ace, 138
Acetyl CoA, 112
Achlya mating, 223
ACP *see* acyl carrier protein
Actin, 205
Actinomycin D, 92,93
Acyl-ACP thioesterases, 111-133
Acyl carrier protein, 111-113,115-117,121-123,125-129
Adenovirus replication, 194
Adipocyte acid phosphatase, 260
Adrenocorticotropic hormones, 59
Aequoria victoria, 146
AGG arginine codon, 158
Agglutinin, 69
    in mating, 223,225
Agrobacterium transformation, 83,84
Alfalfa mosaic virus, 86
Allium porrum see leek
Alloc, 251,257
Allyloxycarbonyl *see* alloc
Altermorph, 229,231,232
Alveolar buds, 58,59
Amber-imbedded DNA, 15,16,22
AMV *see* alfalfa mosaic virus
Anarhichas lupus see wolffish
Ancient DNA, 13-32
    avoiding contamination, 21-23
isolation, 16-21
Angiosperms, 120,122,123,127
Anti-hemophilic Factor IX, 64,69
Antisense RNA, 186,195-197
α_1-antitrypsin 64,69,71,72
Apocarotenoids in mating, 223
Arabidopsis thaliana, 114, 115,117,119,120,127
    and *B.t.* toxin, 88
    release factor, 170
    rubisco, 90
Artichoke plume moth, 136
Asci, 219
Ascomycetes, 219-221,223,226, 228,232,233,235,243
α_1AT *see* α_1-antitrypsin
ats1A, 90
AUG, 157
Avocado, 111,112
Axon guidance, 33-36,43
    tip, 33,38

BAC clones, 51,53
Bacillus and palindromes, 3
Bacillus cereus plasmid, 185
    coagulans, 185
    liquefaciens, 185
    subtilis, codon preference, 162
        plasmids, 183,185,186,191
        release factors, 170
    thuringiensis, 135,151
        plasmid, 183,185,186
        toxin genes, 83-99
Bacteriophage λ, 203
    λgtll, 102
    λO, 204,205,210
    λP, 205,207
    palindromes, 4,6-8
    red gam, 6,7,10
bap1(1), *bap1*(2), *bap1*(3), 234,236
bar see bialaphos resistance
bar1, *bar2*, 236,241,242
BDNF *see* brain-derived neurotrophic factor
Basidia, 219

Basidiomycetes pheromones and
  receptors, 219-247
Basidiospores, 219,226
bbp1(1), *bbp1*(2), *bbp1*(3), 234,236
bbr1, 236,242
Beet armyworm, 85,87-89
Bioreactor, 57-59,64,68,72,75
BL21(DE3), 102,105
Blackheaded fireworm, 88
Black vine weevil, 136
BLAST, 55
Boc, 251,252,255-257,259
Brachydanio rerio see zebra fish
Brain-derived neurotrophic factor,
  34,38,41-43
Brassica napus, 114,116-118,123
  rapa, 114
B.t. see Bacillus thuringiensis

CaaX box, 240
Cabbage looper, 88,89
Cacao, 128
Cadherin, 35,37
Caenorhabditis elegans, 137,138,146
  genes in nematodes, 138,139,
  141,149,150
  heat-shock genes, 146,147,149
  microinjection, 141
  palindromes, 9
  transposons, 144,145
CAI *see* Codon Adaptation Index
California bay, 113-115,117,119,122,
  126,127
CAM *see* cell adhesion molecule
Camphor, 114,122
Canola, 116,126,128
γ-carboxylation, 68,69
Carrot and *B.t.* toxin, 88
Carthamus tinctorius see safflower
Casein, 61-65,70,71,73,74
Casein kinase II, 101-110
Casiella burnetii release factor, 170
CAT, 140
Cathepsin D, 71
Cauliflower mosaic virus, 85,90
CbpA, 203-205
cdk kinases and phosphorylation, 250
Cell adhesion molecule, 37
Central nervous system, 36,37,39
c-*fos*, 92
CFS-1R, 262
Chaperones of *E. coli*, 203-207
Chaperonins, 211
chi site, 6
Chloramphenicol acetyl transferase *see*
  CAT
3-Chloroperoxybenzoic acid *see*
  MCPBA
Chloroplast transformation, 95
CHX *see* cycloheximide
Ciliary neurotrophic factor, 42
Cinnamonum camphorum see camphor
CKII *see* casein kinase II
Clostridium butyricum plasmid, 183,185
ClpA, 203-205,207-211
ClpB, 204,207,209,211
ClpP, 204,207-210
ClpX, 204,207,209-211
ClpY, 204,207
cmp, 192,193
CMV *see* cauliflower mosaic virus
CMV *see* cytomegalovirus
c-*neu*, 62
CNS *see* central nervous system
CNTF *see* ciliary neurotrophic factor
Coconut, 102,122
Cocos nucifera see coconut
Codon Adaptation Index, 160,161
ColE1, 183,195
Coleoptera, 87
Colony-stimulating factor-1 receptor *see*
  CFS-1R
Colorado potato beetle, 88
CopF, 196,197
CopG, 197
CopR, 196,197
CopS, 197
Coprinus cinereus, 220
  mating, 222,228,236,237,239
Copy number control of plasmids, 183-
  201
Coriander, 114,116
Coriandrum sativum see coriander
Corn borer, 87
Corn earworm, 85,89
Corynebacterium xerosis plasmid, 185
Cos7, 50,51,55
Cosmid clones, 49,51,53
Cosmids and palindromes, 9
Cotton and *B.t.* toxin, 88

bollworm, 85,88
Cranberry and *B.t.* toxin, 88
CREB, 270
Cruciforms, 2,5,9
in plasmids, 184,188
CryI, 83,84,86,89-95
CryII, 83
CryIII, 83,89
CryIV, 83
CryV, 83,84
CryVI, 84
Cryptococcus neoformans, 220
mating, 222,233,234
Crystal protein, 83
c-*src*, 67
Cuphea hookeriana, 114,117-120,122, 126,129
lanceolata, 114,116-119,122, 123,125,126,128,129
palustris, 114,122
wrightii, 114,128
Curved DNA binding protein A *see* CbpA
Cyclic adenosine monophosphate response element binding protein *see* CREB
Cyclin D1, 62,67
Cycloheximide, 92,93
Cytokines, 60,61
Cytomegalovirus, 74
Cytosolic tyrosine phosphatase *see* SH-PTP2

DAG *see* 1,2-diacylglycerol
DAG acyltransferase, 112
DAGAT *see* DAG acyltransferase
Dam methylase and palindromes, 5
DBU, 255
deoA, 8
Dephosphorylation, 249,260
des(1-3)hIGF-1, 63,64,67,70-72
1,2-diacylglycerol, 112
1,8-Diazabicyclo[5.4.0] undec-7-ene *see* DBU
Dichelobacter nodosus release factor, 170
Di-deoxy chain termination, 7,9
Dikaryon, 220,227,230,232,237-239
Dimethyl sulfide *see* DMS
Dinosaur DNA, 14,24,25
Diptera, 87
DMS, 256,258
DNA A, 206
DNA B, 205
DNA J, 203-210
DNA K, 203-210
DNA hybridization and palindromes, 2,7
DNA polymerase I, 194
III, 184,187,194
DNA replication origins and palindromes, 1
DNA sequencing and palindromes, 2,7
Dopamine-β-hydroxylase, 40
Dorsal root ganglion, 37,39
DOTAP, 35
Double-strand origin of plasmids, 184, 186-188, 192,197
DRG *see* dorsal root ganglion
Drosophila CKII, 101-109
codon usage, 159,164
melanogaster,
genes in nematodes, 138,139
heat-shock genes, 146-148,150
microprojectile, 140
transposons, 145
dso *see* double-strand origin
Dysplasia, 63

ECM *see* extracellular matrix
EC-SOD *see* extracellular superoxide dismutase
EF-1α, 171
EF-G, 169,176-178
EF-Tu, 169,171,176-178
EGF *see* epidermal growth factor
Electroporation, 140
Elm, 114,117,122
Embryonic stem cell, 74
Endoplasmic reticulum, 112
δ-endotoxins, 83
Enterococcus faecalis, 193,194
Eph receptors, 42
Epidermal growth factor, 66,261,269, 270,273,274
ER *see* endoplasmic reticulum
eRF, 168,170,171,176
Erythropoietin transgene, 73
ES *see* embryonic stem cell
Escherichia coli chaperones, 203-217

codon preference, 158,162,163
palindromes, 2,3,8
release factors, 170,172
TransTerm data, 161
EST *see* expressed sequence tag
Euascomycetes, 219-222,233,235
European corn borer, 87,88
Exon amplification *see* exon trapping
Exon trapping, 49-56
Expressed sequence tags, 55,56,127
Extracellular matrix, 33,37-39
Extracellular superoxide dismutase, 69,70

Farnesyl group, 223-225,227,240,241
Fast protein liquid chromatography, 104,106
FatA, 112,113,115-122,125,127-129
FatB, 112,113,115,117-120,122-129
Fatty acid synthase, 111-133
Fatty acyl-ACP thioesterase *see* Fat
Federal Insecticide Fungicide Agency, 136
FGF *see* fibroblast growth factor
FGFR *see* fibroblast growth factor receptor
Fiber-mediated DNA delivery, 141
Fibroblast growth factor, 34-42,67,261
receptor, 34,37
FIFRA *see* Federal Insecticide Fungicide Agency
FISH *see* fluorescence *in situ* hybridization
FIX *see* Anti-hemophilic Factor-IX
Fleas, pests of turfgrass, 136
9-Fluorenylmethyloxycabonyl *see* Fmoc
Fluorescence *in situ* hybridization, 73
Fmoc, 251-253,255-257,259,260,265
Follicle-stimulating hormone, 60,64
Forest tent caterpillar, 88
F_2Pab, 259,260
FPLC *see* fast protein liquid chromatography
F_2Pmp, 259,271,273
Frog neuromuscular junction, 43-45
FSH *see* follicle-stimulating hormone
FUZ, 239
fuz7, 232
fuz$^+$ reaction, 230-232

β-galactosidase, 145,147
fusion protein, 208
Galleria mellonella, 136,149
GAP, 261,263,264
Gene gun *see* microprojectile identification, 49-56
gfp, 146,150
GH *see* growth hormone
Glucanase, 239
Glucocorticoids, 61
β-glucuronidase reporter, 90
Glycerolipid synthesis, 111
Glycerol-3-phosphate, 111,112
Glycine max see soybean
Glycogen synthase and phosphorylation, 250
Glycosylation, 68-70
G3P *see* glycerol-3-phosphate
G3P acyltransferase, 112
gpa1$^+$, 225
GPAT *see* G3P acyltransferase
G-protein, 169,171,176-178,224,225
Gramineae, 87
Gram-negative bacteria, plasmids, 183,198
Gram-positive bacteria, plasmids, 183-201
Green fluorescent protein *see gfp*
GroEL, 203,204,206,207-209,211,212
GroES, 203,204,206,209,211,212
Growth cone, 33,34,36,37,39-41,43
factors, 33-47,63,68
hormone, 60,62-64,73
Growth factor receptor
phosphatases, 260
phosphorylation, 249
GrpE, 203-210
GSSG *see* oxidized glutathione
GUS *see* β-galactosidase
Gypsy moth, 88

Haemophilus influenzae release factors, 170
Hairpins, 2,4,5,9
snapbacks, 7
HA-*ras*, 62
Helicase, 184,187,194,196,206
Helicobacter pylori plasmid, 183
Helicoverpa zea see corn earworm

Heliothis virescens see tobacco budworm
Heliothis zea see tomato fruitworm
Hemiascomycetes, 219-222,225,226, 235,237,239
Hemibasidiomycetes, 221,222,226,228, 237,239
Heparin, 101-104,106-108
Heparan sulfate proteoglycans, 34,41
Heterorhabditidae, 135-137,145
Heterorhabditis bacteriophora, 136,138, 150,151
  electroporation, 140,141
  heat-shock genes, 146,148-150
  microinjection, 142
  microprobe transformation, 143
  rol6, 146
Heterothallic fungi, 221
Hexokinase, 205
HF *see* hydrogen fluoride
hGH *see* human growth hormone
hLF *see* lactoferrin, human
HMGCoA-reductase *see* hydroxymethylglutaryl CoA-reductase
Holliday junctions and cruciforms, 5
Homobasidiomycetes, 220,221,235-237,239,243
Homo sapiens release factors, 170
Homothallic fungi, 221
Hook cell, 238,239
hPC *see* human protein C
HSA *see* human serum albumin
Hsc66, 203,204
HscA, 203,204
HscB, 203,204
hsp16, 139,141,146
hsp70, 139,148,149
Hsp60, 211
Hsp70, 203-206,211
Hsp78, 207
Hsp100, 204
Hsp104, 204,207,209,211
HslU, 207
HSPG *see* heparan sulfate proteoglycan
1*H*-tetrazole, 252,254
Human CKII, 101-109
Human growth hormone, 60,63,64,73
Human protein C, 64,70,71
Human serum albumin, 72
Hydrogen fluoride, 250,251,258
3-Hydroxy-2,3-dihydro-4-oxo-benzotriazine ester *see* ODhbt
β-hydroxylation, 69
Hydroxymethylglutaryl CoA-reductase, 63,65
Hyperplasia, 63,67
Hyphae, 220,230,233,238

"Ice-man", 24
ICP *see* insecticidal crystal protein
Ideomorph, 228,229,235
IF-2, 178
IF-3 *see* initiation factor
IGF *see* insulin-like growth factor
Immune cell receptor phosphorylation, 249
incI8, 194
incFII, 183
infC, 158
Initiation codon, 157
  factor, 158,177
Insecticidal crystal protein, 83
Insect parasite genetic engineering, 135-155
Insulin-like growth factor, 64,66,70-72
Insulin receptor substrate *see* IRS-1
int-2, 62,66,67
int-3, 62
Integrin, 35
γ-interferon, 64,72
Interleukin, 36
Interrupted palindromes, 1,2
Introns in transgene expression, 71
Inverted repeat, 184,187,191,196
IPTG, 102,104
IR *see* inverted repeat
IRS-1, 263,264,268,272-274
Isoprenoids in mating, 223,240,241
Isopropyl-β-D-thiogalactopyranoside *see* IPTG
Iterons, 183,188

Jelly fungi, 226
Jojoba, 112
Jumbling test, 174
Jumping genes *see* transposons

β-ketoacyl-ACP synthase, 127
Kinase phosphorylation, 249
Kinases, 249,250,261,273

Kissing complex, 195,196
Kluyveromyces lactis release factor, 170
Knock-out mice, 35,43

α-*lac see* α-lactalbumin
α-lactalbumin, 59,61,64,65,69,74
Lactation, 59-63,68
Lactobacillus hilgardii plasmid, 185
    pentosus plasmid, 185,196
    plantarum plasmids, 183,185
Lactococcus lactis plasmids, 183,185
Lactoferrin, 64,69,74,75
β-lactoglobulin, 61,65,71
Lagging strand origin *see* single-strand origin
Lambda phage *see* bacteriophage λ
Lateral geniculate nucleus, 42
Lauraceae, 116,117,122,128
Laurate, 125
Lauric acid, 113
Leading strand origin *see* double-strand origin
Leaffolder, 88
Lectin, 69
Leek, 114,120,122
Leguminosae, 87
Leishmania, 140
Leptomonas, 140
Leukemia inhibitory factor, 39
LGN *see* lateral geniculate nucleus
LIF *see* leukemia inhibitory factor
Lipid biosynthesis, 111,112
Listeria monocytogenes release factor, 170
Loach, 140
LPAAT *see* lysophosphatidic acid acyltranferase
L2 protein, 172
luc see luciferase, firefly
Luciferase, bacterial, 146
    firefly, 146,204,206,207,209
    reporter gene, 86
lux see luciferase, bacterial
Lysophosphatidic acid acyltranferase, 112,125,127
Lythraceae, 122,129

M13, 102
    and palindromes, 9,10
Maize and *B.t.* toxin, 88-90
mam2$^+$, 225
Mammary gland development, 57, 59,62,63
    transgenes, 57-81
Manduca sexta see tobacco hornworm
Mangifera indica see mango
Mango, 114,120
Mannopine synthase promoter, 86
MAP and phosphorylation, 250
map3$^+$, 225,242
MAR *see* matrix attachment region
Mastotermes in amber, 15
MAT-1 and 2, 232,233
MATa and Matα, 233
Matrix attachment region, 71,72
    lysozyme, 73
Maxam-Gilbert sequencing, 9
MCPBA, 254
MCS *see* multiple cloning site
mdr49, 139
mdr65, 139
mec4, 146
Metallothionein 1, 65
    genes, 138,139
MFα1 and *2*, 223,224,234
MFa1 and *2*, 223,224,229,231
mfm1$^+$, *mfm2*$^+$, *mfm3*$^+$, 224,225
Micrococcus luteus codons, 159
Microprobes, 142-144
Microprojectile, 140,141
milchlos, 60
Milk protein genes, regulation, 57,59, 68
mini-F, 203,206
mini-P1 *see* P1
Misgurnus fossilis see loach
MMTV *see* mouse mammary tumor virus
Mole cricket, 136
Mouse mammary tumor virus, 62,65-67,74
mRF-1, 173
mRNA, 158,159,165,169,171,174-178
mT *see* polyoma middle-sized tumor antigen
MT1 *see* metallothionein 1
mtn see metallothionein genes
MuA, 204,210
Mu DNA replication, 204
Multiple cloning sites, 51

Mushroom, 220
  pests, 136
myc, 62,67
Mycobacterium leprae release factor, 170
Mycoplasma capricolum codons, 159
  genitalium release factor, 170
  mucoides plasmid, 184

Nematode genetic engineering, 135-155
Nerve growth factor, 34,35,37-42,270
Netrin-1, 33,34,37
Neu, 269,270
neu-NT, 62
Neural connectivity, 33-47
Neurospora crassa, 220,235
Neurotrophins, 34-43
NGF *see* nerve growth factor
3-Nitro-2-pyridinesulfenyl *see* Npys
NMJ *see* frog neuromuscular junction
Npys, 251,257
nptII, 85,86
N-*ras*, 62
NT *see* neurotrophins

OCS *see* octopine synthase
Octopine synthase, 92,93
ODhbt, 251
Oleate, 112
Oleic acid, 116
Oleoyl thioesterase, 115,116,123,129
O-(2-malonyl)tyrosine *see* OMT
OMT, 259,260,272,274
Oncogene phosphorylation, 249
Oncogenes, 61
opd see organophosphate-based insecticides
OPfp, 251
O-phosphorylation, 68
Organophosphate-based insecticides, 138
Oryctolagus cuniculus release factor, 170
Ostrinia nubilalis see European corn borer
Oxidized glutathione, 105,106

P1, 183,203-205,209-211
p53, 63,67
p85, 261,263-269,272

PA *see* phosphatidic acid
Pab, 259
pADB201, 197
palA see single-strand origin
Palindromes, 1-11
Palm, 123
Palmitate, 116,124,125
Palmitoyl thioesterase, 123
pAMβ1, 193-197
Paromomycin, 176
PC *see* phosphatidyl choline
pC194, 184-186,189,191,197
pC221, 187-190
pC223, 187,190
P1 clones, 49,51,53
PCR, 50,53,54,73,113,119
  and ancient DNA, 13,14
  and palindromes, 2,7
  inhibition in isolation, 17,21-23
  jumping, 23
PcrA helicase, 194,196
PDGF receptor (PDGF-R), 261-266, 268,271-273
pE194, 184-192,197
P element, 145
Pentafluorophenyl ester *see* OPfp
PEPCK, 60,63,65
Peripheral nervous system, 36,38,39
Perithecium, 235
pET3a, 101,102,104,105
Petroselinic acid, 116
  -ACP thioesterase, 128
PHA *see* phytohemagglutinin
Phage clones, 49,51,53
Phagemids, 7,10
Phenylmethylsulfonyl fluoride, 102,103
Pheromones in Basidiomycetes, 219-247
Phosphatase, 260,261,273
Phosphatidic acid, 112
Phosphatidyl choline, 112
Phosphatidylinositol kinase *see* PI-3 kinase
Phosphitylation, 254,255
Phospholipase C-γ *see* PLC-γ
Phospholipid kinases, 249
Phospho-L-tyrosine, 260
4-Phosphono-2-aminobutyric acid *see* Pab
4-Phosphono-4,4-difluoro-2-aminobutylic acid *see* F_2Pab

4-Phosphono(difluoromethyl)-
phenylalanine *see* F_2Pmp
4-Phosphonomethyl-phenylalanine *see* Pmp
Phosphopeptides, synthesis and applications, 249-278
Phosphoramidites, 251,252,254,255
Phosphorylase and phosphorylation, 250,273
Phosphorylation, 68,249,250,261,263, 264,270
Photorhabdus spp, 135
luminescens, 143,148
Physarum polycephalum and palindromes, 3
Phytohemagglutinin, 95
Phytoseiid mites, 137
Pichia pinus release factor, 170
PI-3-kinase, 261,263-265,271,273
pIP501, 193-197
Piperidine to remove Fmoc, 252,253, 255
Plant fatty acid biosynthesis, 111-133
Plant thioesterases, 111-133
Plasmid clones, 49,51,53
copy number control, 183-201
Plastid, 111-113,123,127,128
Platelet-derived growth factor *see* PDGF
receptor *see* PDGF-R
pLB4, 197
PLC-γ, 261,263-265,267
pLS1, 194,195,197
Pmp, 259,271-273
PMSF *see* phenylmethylsulfonyl fluoride
pMV158, 184-186,188,191,197
p-Nitrophenyl phosphate *see* *p*NP
*p*NP, 260,262
PNS *see* peripheral nervous system
Poly(A), 50,51,53,55
Polymerase chain reaction *see* PCR
Polyoma middle-sized tumor antigen, 62,265,266,271,272
Popillia japonica, 135
Poplar and *B.t.* toxin, 88
Potato and *B.t.* toxin, 88
pPCZ1, 143
pral and *2*, 229,231,241,242
Primers for ancient DNA, 23
Progesterone, 58
Prolactin, 58,61
Proplebeia dominicana in amber, 15
Protein B, 71,72
Protein C, 64,70,71
Protein folding and degradation, 203-217
Protein kinases, 249
Protein remodeling in *E. coli*, 203-217
PS2, 62
pS194, 187,189,190
Pseudomonas aeruginosa release factor, 170
diminuta, 138
psi$^+$, 171,204,211
P site, 171
pSL2, 194
pSLA2, 194
pSLC, 194
pSM19035, 193,194,196,197
pSN2, 184,186,191,192
pT7-7, 101,102,104
pT181, 184-192,195,196
pTAG4, 50-52
PTB-1B, 261
PTPase *see* tyrosine phosphatase
pTX14-3, 184-186
pUB110, 184-186,188,191-193,195,197
pUB112, 186,190
pUC18, 148
and palindromes, 6,7,10
PV-mt *see* polyoma middle-sized tumor antigen
pWV01, 185,191

Rabbit CKII, 106
3′ RACE, 50,53
Rainbow trout, 140
RAM1, *RAM2*, 240
Rapeseed *see Brassica napus*
RARβ4-like, 62
Ras, 240
ras GTPase activating protein *see* GAP
Rattus norvegicus release factor, 170
RC *see* rolling circle
Reagent K, 256
recA and palindromes, 6-8
recBC and palindromes, 3
recD and palindromes, 6-8
Receptor tyrosine kinase, 34

Recombinases and palindromes, 1
Release factor in translational termination, 157-180
Rep, 184,186-190,193,195-197
RepA, 204-211
RepE, 206
Replication terminator protein, 193
Restriction enzymes and palindromes, 1
ret, 62
Retinal ganglion cell, 35,39,40,42
Reverse transcription, 50,53,55
RF-1, 169,172-174,178
RF-2, 167,169,172-175,178
RF-3, 169,176-178
RGC *see* retinal ganglion cell
RHA1, *RHA2*, *RHA3*, 224,227,240
Rhodanese, 206,207
Rhodosporidium toruloides, 220
mating, 222,224,226,227,230, 232,240
Rhodotorucine A, 226,227,230,240
Rice and *B.t.* toxin, 88,89
RK2, 183
R6K, 183
RNA processing and palindromes, 1
rol6, 139,141,142,146
Rolling circle, 183-188,191,192,194, 195,197
rRNA, 158,174-178
RT *see* reverse transcription
RTK *see* receptor tyrosine kinase
RTP *see* replication terminator protein
Rubisco, 86

Saccharomyces cerevisiae
as Hemiascomycetes, 219
codon usage, 159,164,168
fiber-mediated DNA delivery, 141
mating, 221-227,230,235,239, 240,242
palindromes, 3
release factors, 170
sup35p, 171
Safflower, 113-116
Salmo gairdneri see rainbow trout
Salmonella typhimurium release factors, 170,171
Sambucus nigra, 69
sbcB mutants and palindromes, 3,6,7
sbcC201, 6,8
sbcC mutants and palindromes, 3-7,10
sbcD and cruciforms, 5
and palindromes, 4,6
Scarabs, 137
αsl-casein, 63,64,70,74
Schizophyllum commune, 220
mating, 222,228,234-239,241-243
Schizosaccharomyces pombe as Hemiascomycetes, 219
mating, 221,224-226,235,240,242
Sciariid flies, 136
Sec *see* selenocysteine insertion
Secernentea, 145
SelB, 177,178
Selenocysteine insertion, 167,168,175
Serine/threonine kinases, 249
Serine/threonine phosphatases, 249
Sesquiterpenes in mating, 223
SH2 *see* Src homology 2
Shigella sonnei plasmid, 183
Shine-Dalgarno sequence, 158,162
Shmoos, 223
SHP, 260-262
SH-PTP2, 260-262,273,274
Signaling pathways, 250
Signal transduction and phosphorylation, 270
Single-stranded DNA binding protein, 187
Single-strand origin of plasmids, 184, 186,187,191,192
Site-directed mutagenesis, 102
Smuts, 220,226,232,233
Soybean, 86,114,116,117,123-125
SP-B *see* protein B
Spermine, 102,104,108
Spodoptera exigua see beet armyworm
Src homology 2, 250,259-261,263-269,271-274
phosphatase *see* SHP
SSB *see* single-stranded DNA binding protein
sso see single-strand origin
Staphylococcus aureus plasmids, 183-186
epidermidis plasmid, 186
STE, 223,230,240-243
Stearate, 116
Stearoyl-ACPΔ9 desaturase, 112,128
Steinernema carpocapsae, 136,137,145

glaseri, 135,139
Steinernematidae, 135-137,145
Sterile mutants *see* STE
Stop signal in translational termination, 157-182
Strand-slippage and palindromes, 4
Streptococcus and palindromes, 3
Streptococcus agalactiae plasmid, 183, 185,193
faecalis plasmid, 185
ferus plasmid, 185
gordonii release factor, 170
mutans release factor, 170
pneumoniae plasmid, 194
pyogenes plasmids, 193,194
Streptomyces and palindromes, 3
Streptomyces lividans plasmid, 183,185
Striped stemborer, 88
Stromelysin, 62,63,66,67
sul106 see rol6
Sunflower, 117
sup7 of *C. elegans*, 142
supE, 8,159
Superoxide dismutase, 64
Sup35p, 170
Sup45p, 170
SV40 early promoter, 50-52
SV40t, 63
Syp *see* SH-PTP2

TAG *see* triacylglycerol
TAG, 160,161,167
Taq polymerase, 53
t-butylhydroperoxide, 254,255
t-butyloxycarbonyl *see* Boc
Tcl, 144,145
Tc3, 145
tetR-VP16, 74
TFA, 253,256-259
TFMSA, 252,253,256,258,259
TGF-α *see* transforming growth factor-α
TGF-β *see* transforming growth factor-β
Thermoplasma acidophilum, 209
Theta-type plasmids, 193-196
Thioanisole, 256
Thymidine phosphorylase, 8
Tissue plasminogen activator, 64,69
TMSBr, 253,256,258,259
TMSOTf, 256,258

Tobacco, 116
and *B.t.* toxin, 84,88,90
budworm, 85
hornworm, 85
transgene, 86,90,93,94
Tomato, 117
and *B.t.* toxin, 84,88
fruitworm, 85
transgene, 86
tPA *see* tissue plasminogen activator
tpp, 8
Transcription factor phosphorylation, 249,250
Transcription regulators and palindromes, 1
Transforming growth factor, 36,39
Transforming growth factor α, 62,63,67
Transforming growth factor β, 58,66,67
Transgenes of mammary gland, 57-81
Transgenic mice, 57,60,64-67,70
Transgenic nematodes, risk, 149,150
Transgenic plants, 83-99
Translational termination, 157-182
Translational Termination database, 160,161
Transposons, 144,145
tagging, 145
TransTerm *see* Translational Termination database
Tremella spp., 220,228
brasiliensis mating, 222,224,227
mesenterica mating, 222,224,227
Tremerogens, 227,230,241
Triacylglycerol, 112,122,125
Trichogynes, 235,239
Trichoplusia ni see cabbage looper
Trifluoromethanesulfonic acid *see* TFMSA
Triglycerides, 122,125
Trimethylsilyl bromide *see* TMSBr
Trimethylsilyl trifluoromethanesulfonate *see* TMSOTf
Triphenyl phosphine, 257
Triponema denticola plasmid, 184
Trks, 34,38,41-43
tRNA, 158,165,166,171,172,175-178
Trophoblast interferon, 64
Trypanosoma brucei, 140
Tumorigenesis of mammary gland, 62
Turfgrass pests, 136

Tyrolean "ice-man", 24
Tyrosine kinases, 249,250
Tyrosine phosphatases, 249,260,261, 273,274

UAA, 159-169,172,174,175
UAG, 159-164,166,167,169,172,174, 175
UGA, 159-169,174,175,178
Ulmaceae, 122
Ulmus americana see elm
Umbellularia californica see California bay
UmuC, 206
unc22, 139,142
Urokinase in milk, 64,70
Ustilago spp., 220
  hordei mating, 222,232,233,235, 236
  maydis mating, 221,222,224,228-233,236,241-243
  violacea mating, 222,233

VP16, 74
Vsr, 161

WAP *see* whey acidic proteins
Whey acidic proteins, 57,60-63,67,70
  promoter, 61,64,65,72
wnt-1, 62,67
Wobble nucleotide, 159
Wolffish, 138,139

øX174, 184,189,191
Xenopus laevis, 35,40,42,43
  release factors, 170
  sup35p, 171
Xenorhabdus, 135

YAC clones, 49,51-53
YACs and palindromes, 9
Yeast *see Saccharomyces cerevisiae*
YfhE *see* HscB

ZAP-70, 273
Zebra fish, 138

GPSR Compliance

The European Union's (EU) General Product Safety Regulation (GPSR) is a set of rules that requires consumer products to be safe and our obligations to ensure this.

If you have any concerns about our products, you can contact us on ProductSafety@springernature.com

In case Publisher is established outside the EU, the EU authorized representative is:

Springer Nature Customer Service Center GmbH
Europaplatz 3
69115 Heidelberg, Germany

Batch number: 10755284

Printed by Printforce, the Netherlands